Revise
**AS**

# Physics

## Graham Booth

# Contents

Specification lists                                              04

AS/A2 Level Physics courses                                      12

Different types of questions in AS examinations                 14

Exam technique                                                   16

What grade do you want?                                         18

Essential mathematics                                           19

Four steps to successful revision                              24

## Chapter 1 Force and motion

1.1 Forces and vectors                                          25

1.2 Some effects of forces                                      29

1.3 Under stress                                                31

1.4 Solid materials                                             33

1.5 Motion in one and two dimensions                            39

     Sample question and model answer                           43

     Practice examination questions                             44

1.6 Force and acceleration                                      47

1.7 Newton I and III                                            52

1.8 Momentum and the second law                                 54

1.9 Energy to work                                              59

1.10 Drag and terminal velocity                                 64

     Sample question and model answer                           67

     Practice examination questions                             68

## Chapter 2 Electricity

2.1 Ohm's Law, resistance and resistivity                       71

2.2 Energy transfer in circuits                                 74

2.3 Resistance in circuits                                      77

2.4 Alternating current and the oscilloscope                    81

2.5 Electromagnetism                                            84

     Sample question and model answer                           86

     Practice examination questions                             87

## Chapter 3  Particle physics

3.1  A model gas                                    90

3.2  Internal energy                                94

3.3  Radioactive decay                              99

3.4  Forces and particles                          103

3.5  Stability and the nucleus                     109

Sample question and model answer                   113

Practice examination questions                     114

## Chapter 4  Electromagnetic radiation

4.1  What is a wave?                               117

4.2  Simple wave properties                        120

4.3  More complex wave behaviour                   125

4.4  Wave properties across the spectrum           130

Sample question and model answer                   135

Practice examination questions                     136

Practice examination answers                       139

Index                                              144

# Specification lists

## AQA A Physics

| MODULE | SPECIFICATION TOPIC | CHAPTER REFERENCE | STUDIED IN CLASS | REVISED | PRACTICE QUESTIONS |
|---|---|---|---|---|---|
| **Module 1 (M1)** *Particles, radiation and quantum phenomena* | Particles | 3.3, 3.4 | | | |
| | Electromagnetic radiation and quantum phenomena | 4.2, 4.4 | | | |
| **Module 2 (M2)** *Mechanics and molecular kinetic theory* | Mechanics | 1.1, 1.5, 1.6, 1.7, 1.8, 1.9, 1.10, 3.2 | | | |
| | Molecular kinetic theory model | 3.1 | | | |
| **Module 3 (M3)** *Current electricity and elastic properties of solids* | Current electricity | 2.1, 2.2, 2.3, 2.4 | | | |
| | Elastic properties of solids | 1.2, 1.3 | | | |

## Examination analysis

The specification comprises three compulsory modules. The module tests consist of short structured questions, all of which are compulsory.

| | | | |
|---|---|---|---|
| **Module 1** | 1 hr test | 50 marks | 30% |
| **Module 2** | 1 hr test | 50 marks | 30% |
| **Module 3** | 1 hr test | 50 marks | 25% |
| | 1 hr 45 min practical examination or coursework | | 15% |

# AQA B Physics

| MODULE | SPECIFICATION TOPIC | CHAPTER REFERENCE | STUDIED IN CLASS | REVISED | PRACTICE QUESTIONS |
|---|---|---|---|---|---|
| **Module 1 (M1)** *Foundation physics* | Scalars and vectors | 1.1 | | | |
| | Kinematics | 1.5, 1.6, 1.7, 1.10, 4.1 | | | |
| | Energy concepts | 1.3, 1.9 | | | |
| | Electricity | 2.1, 2.2, 2.3 | | | |
| | d.c. circuits | 2.2, 2.3 | | | |
| | Information and communication | | | | |
| **Module 2 (M2)** *Waves and nuclear physics* | Waves | 4.1, 4.2 | | | |
| | Diffraction and interference | 4.3 | | | |
| | Spectra | 4.4 | | | |
| | Radioactivity | 3.3 | | | |
| | Physics of particles | 3.3, 3.4 | | | |
| | Information and communication | | | | |

## Examination analysis

The specification comprises three compulsory modules. The module tests consist of short answer questions and structured questions, all of which are compulsory.

| **Module 1** | 1 hr 30 min test | 75 marks | 35% |
|---|---|---|---|
| **Module 2** | 1 hr 30 min test | 75 marks | 35% |
| **Module 3** | 2 hr practical examination | 78 marks | 30% |

# Edexcel A Physics

| MODULE | SPECIFICATION TOPIC | CHAPTER REFERENCE | STUDIED IN CLASS | REVISED | PRACTICE QUESTIONS |
|---|---|---|---|---|---|
| **Module 1 (M1)** Mechanics and radioactivity | Rectilinear motion | 1.5 | | | |
| | Forces and moments | 1.1, 1.2, 1.7, 1.10 | | | |
| | Dynamics | 1.6, 1.8 | | | |
| | Mechanical energy | 1.9 | | | |
| | Radioactive decay and the nuclear atom | 3.3, 3.4, 3.5 | | | |
| **Module 2 (M2)** Electricity and thermal physics | Electric current and potential difference | 2.1, 2.2 | | | |
| | Electrical circuits | 2.3 | | | |
| | Heating matter | 1.2, 3.1, 3.2 | | | |
| | Kinetic model of matter | 3.1 | | | |
| | Conservation of energy | 1.9, 3.2 | | | |
| **Module 3 (M3)** Topics | **A** Astrophysics | | | | |
| | **B** Solid materials | 1.3, 1.4 | | | |
| | **C** Nuclear and particle physics | 3.4, 3.5 | | | |
| | **D** Medical physics | | | | |

## Examination analysis

The specification comprises three compulsory modules. The module tests consist of structured questions, all of which are compulsory.

| | | |
|---|---|---|
| **Module 1** | 1 hr 15 min test | 30% |
| **Module 2** | 1 hr 15 min test | 30% |
| **Module 3** | 30 minute test, consisting of one question on each of the four topics. Candidates must answer the question on one topic only. | 20% |
| and | 1hr 30 min practical examination | 20% |

# Edexcel B Physics

| MODULE | SPECIFICATION TOPIC | CHAPTER REFERENCE | STUDIED IN CLASS | REVISED | PRACTICE QUESTIONS |
|---|---|---|---|---|---|
| **Module 1 (M1)** *Physics at work, rest and play* | The sound of music | 4.1, 4.2, 4.3, 4.4 | | | |
| | Technology in space | 1.9, 2.1, 2.2, 2.3, 3.2 | | | |
| | Higher, faster, stronger | 1.1, 1.5, 1.6, 1.9 | | | |
| **Module 2 (M2)** *Physics for life* | Good enough to eat | 1.2, 1.3, 1.10, 4.2 | | | |
| | Digging up the past | 2.1, 3.3, 4.3, 4.4 | | | |
| | Spare part surgery | 1.3, 1.4, 4.2, 4.4 | | | |

# Examination analysis

*The specification comprises three compulsory modules. The module tests consist of short-answer questions, all of which are compulsory.*

| **Module 1** | 1 hr 30 min test | 33.3% |
|---|---|---|
| **Module 2** | 1 hr 30 min test | 33.3% |
| **Module 3** | coursework | 33.3% |

# OCR A Physics

| MODULE | SPECIFICATION TOPIC | CHAPTER REFERENCE | STUDIED IN CLASS | REVISED | PRACTICE QUESTIONS |
|---|---|---|---|---|---|
| **Module 1 (M1)** *Forces and motion* | Scalars and vectors | 1.1 | | | |
| | Kinematics | 1.5 | | | |
| | Dynamics | 1.1, 1.2, 1.6 | | | |
| | Force, work and power | 1.2, 1.4, 1.9 | | | |
| | Deformation of solids | 1.3 | | | |
| | Forces on vehicles | 1.6, 1.7, 1.9, 1.10 | | | |
| | Car safety | 1.6 | | | |
| **Module 2 (M2)** *Electrons and photons* | Electric current | 2.1, 2.2, 2.3 | | | |
| | d.c. circuits | 2.1, 2.2, 2.3 | | | |
| | Magnetic effects of current | 2.5 | | | |
| | Quantum physics | 4.4 | | | |
| | Electromagnetic waves | 4.4 | | | |
| **Module 3 (M3)** *Wave Properties* | Reflection and refraction | 4.2 | | | |
| | Waves | 4.1, 4.2 | | | |
| | Superposition | 4.3 | | | |

# Examination analysis

The specification comprises three compulsory modules. In module tests all questions are compulsory; they consist of structured questions and questions requiring extended answers.

| **Module 1** | 1 hr test | 30% |
|---|---|---|
| **Module 2** | 1 hr test | 30% |
| **Module 3** | 45 min test | 20% |
| and | 1 hr 30 mins practical examination or coursework | 20% |

## OCR B Physics

| MODULE | SPECIFICATION TOPIC | CHAPTER REFERENCE | STUDIED IN CLASS | REVISED | PRACTICE QUESTIONS |
|---|---|---|---|---|---|
| **Module 1 (M1)** Physics in action | Imaging and signalling | 4.1, 4.2, 4.4 | | | |
| | Sensing | 2.1, 2.2, 2.3 | | | |
| | Designer materials | 1.2, 1.3, 1.4, 2.1, 4.2 | | | |
| **Module 2 (M2)** Understanding processes | Waves and quantum behaviour | 4.2, 4.3, 4.4 | | | |
| | Space, time and motion | 1.1, 1.5, 1.6, 1.9, 1.10 | | | |

## Examination analysis

The specification comprises three compulsory modules. In module tests all questions are compulsory; they consist of short focused questions and longer, structured questions with some questions being open-ended.

| **Module 1** | 1 hr 30 min test | 33% |
|---|---|---|
| **Module 2** | 1 hr 30 min test | 37% |
| **Module 3** | coursework | 30% |

# WJEC Physics

| MODULE | SPECIFICATION TOPIC | CHAPTER REFERENCE | STUDIED IN CLASS | REVISED | PRACTICE QUESTIONS |
|---|---|---|---|---|---|
| **Module 1** (M1) *Waves, light and Basics* | Basic physics | 1.1, 1.2, 1.6, 1.7 | | | |
| | Kinematics | 1.5, 1.6, 1.10 | | | |
| | Solids under stress | 1.3, 1.4 | | | |
| | Waves | 4.1, 4.2, 4.3 | | | |
| | Light | 4.2, 4.3 | | | |
| **Module 2** (M2) *Quanta and electricity* | Conduction of electricity | 2.1, 2.2, 2.3 | | | |
| | Nuclear | 3.3 | | | |
| | Quantum physics | 4.4 | | | |

## Examination analysis

*The specification comprises three compulsory modules.* In module tests all questions are compulsory and will be 5 short answer and 2 longer questions.

| | | | |
|---|---|---|---|
| **Module 1** | 1 hr 30 min test | 90 marks | 35% |
| **Module 2** | 1 hr 30 min test | 90 marks | 35% |
| **Module 3** | 2 hr  practical test | 60 marks | 30% |

# NICCEA Physics

| MODULE | SPECIFICATION TOPIC | CHAPTER REFERENCE | STUDIED IN CLASS | REVISED | PRACTICE QUESTIONS |
|--------|--------------------|--------------------|------------------|---------|---------------------|
| **Module 1 (M1)** Forces and electricity | Physical quantities and units | 1.1 | | | |
| | Static forces | 1.2 | | | |
| | Kinematics | 1.5 | | | |
| | Dynamics | 1.6, 1.7, 1.8, 1.9 | | | |
| | Electricity | 2.1, 2.2, 2.3 | | | |
| **Module 2 (M2)** Waves and photons | Waves | 4.1, 4.2, 4.4 | | | |
| | Superposition of waves | 4.3 | | | |
| | Photons and energy levels | 4.4 | | | |
| **Module 3 (M3)** Medical physics | Medical physics | | | | |

## Examination analysis

*The specification comprises three compulsory modules. The module tests consist of several compulsory short-answer questions.*

| | | | |
|--------|--------------------------|-----------|-------|
| **Module 1** | 1 hr test | 60 marks | 33.3% |
| **Module 2** | 1 hr test | 60 marks | 33.3% |
| **Module 3** | 45 mins test | 45 marks | 18.4% |
| and | 1 hr 15 mins practical examination | 45 marks | 15% |

# AS/A2 Level Physics courses

## AS and A2

All Physics A Level courses are in two parts, with three separate units or modules in each part. Most students will start by studying the AS (Advanced Subsidiary) course. Some will then go on to study the second part of the A Level course, called the A2. It is also possible to study the full A Level course, both AS and A2, in any order.

## How will you be tested?

### Assessment units

For AS Physics, you will be tested by three assessment units. For the full A Level in Physics, you will take a further three units. AS Physics forms 50% of the assessment weighting for the full A Level.

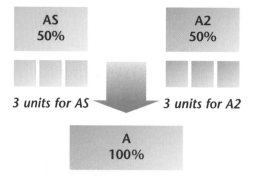

Each unit can normally be taken in either January or June. Alternatively, you can study the whole course before taking any of the unit tests. There is a lot of flexibility about when exams can be taken and the diagram below shows just some of the ways that the assessment units may be taken for AS and A Level Physics.

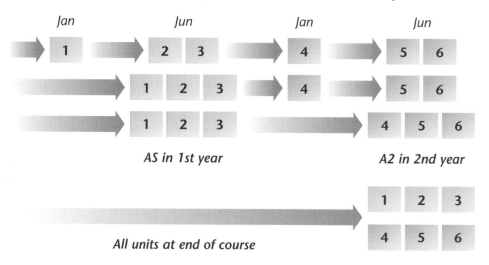

If you are disappointed with a module result, you can resit each module once. You will need to be very careful about when you take up a resit opportunity because you will have only one chance to improve your mark. The higher mark counts.

### A2 and Synoptic assessment

For those students who, having studied AS, decide to go on to study A2, there are three further modules to be studied. Similar assessment arrangements apply except some module tests, those that draw together various elements of the course in a 'synoptic' assessment, have to be taken at the end of the course.

### Coursework

Coursework may form part of your AS Level Physics course, depending on which specification you study. Where students have to undertake coursework, it may be a written project or for the assessment of practical skills or both. If in doubt, check with your teacher or the specification that you are studying.

### Key Skills

To gain the key skills qualification, which is equivalent to an AS Level, you will need to demonstrate that you have attained level 3 in the areas of Communication, Application of number and Information technology. Part of the assessment can be done as normal class activity and part is by formal test.

## What skills will I need?

For AS Physics, you will be tested by assessment objectives: these are the skills and abilities that you should have acquired by studying the course. The assessment objectives for AS Physics are shown below.

### Knowledge with understanding

- recall of facts, terminology and relationships
- understanding of principles and concepts
- drawing on existing knowledge to show understanding of the responsible use of Physics in society
- selecting, organising and presenting information clearly and logically

### Application of knowledge and understanding, analysis and evaluation

- explaining and interpreting principles and concepts
- interpreting and translating, from one form into another, data presented as continuous prose or in tables, diagrams and graphs
- carrying out relevant calculations
- applying knowledge and understanding to familiar and unfamiliar situations
- assessing the validity of physical information, experiments, inferences and statements

You must also present arguments and ideas clearly and logically, using specialist vocabulary where appropriate.

### Experimental and investigative skills

Physics is a practical subject and part of the assessment of AS Physics will test your practical skills. This may be done during your lessons or you may be tested in a more formal practical examination. You will be assessed on four main skills:

- planning
- implementing
- analysing evidence and drawing conclusions
- evaluating evidence and procedures

# Different types of questions in AS examinations

In order to assess these abilities and skills, a number of different types of question are used.

In AS Level Physics unit tests, these include short-answer questions, structured questions requiring both short answers and more extended answers, together with free-response and open-ended questions. Multiple choice question papers are not used, although it is possible that some short-answer questions will use a multiple choice format, where you have to choose the correct response from a number of given alternatives.

## Short-answer questions

A short-answer question may test recall or it may test understanding by requiring you to undertake a short, one–stage calculation. Short-answer questions normally have space for the answers printed on the question paper. Here are some examples (the answers are shown in blue):

What is the relationship between electric current and charge flow?

Current = rate of flow of charge.

The current passing in a heater is 6 A when it operates from 240 V mains. Calculate the power of the heating element.

$P = I \times V = 6\,A \times 240\,V = 1440\,W$

Which of the following is the correct unit of acceleration?

  A  m/s
  B  s/m
  C  $m/s^2$
  D  $m^2/s$

  C  $m/s^2$

## Structured questions

Structured questions are in several parts. The parts are usually about a common context and they often become progressively more difficult and more demanding as you work your way through the question. They may start with simple recall, then test understanding of a familiar or an unfamiliar situation. The most difficult part of a structured question is usually at the end, where the candidate is sometimes asked to suggest a reason for a particular phenomenon or social implication. Most of the practice questions in this book are structured questions, as this is the main type of question used in the assessment of AS Level Physics.

When answering structured questions, do not feel that you have to complete one question before starting the next. The further you are into a question, the more difficult the marks are to obtain. If you run out of ideas, go on to the next question. Five minutes spent on the beginning of that question are likely to be much more fruitful than the same time spent wracking your brains trying to think of an explanation for an unfamiliar phenomenon.

Here is an example of a structured question that becomes progressively more demanding.

(a)  A car speeds up from 20 m/s to 50 m/s in 15 s.

Calculate the acceleration of the car.

acceleration = increase in velocity ÷ time taken
= 30 m/s ÷ 15 s = 2 m/s²

(b) The total mass of the car and contents is 950 kg.

Calculate the size of the unbalanced force required to cause this acceleration.

force = mass × acceleration
= 950 kg × 2 m/s² = 1900 N

(c) Suggest why the size of the driving force acting on the car needs to be greater than the answer to (b)

The driving force also has to do work to overcome the resistive forces, e.g. air resistance and rolling resistance.

## Extended answers

In AS Level Physics, questions requiring more extended answers will usually form part of structured questions. They will normally appear at the end of structured questions and be characterised by having at least three marks (and often more, typically five) allocated to the answers as well as several lines (up to ten) of answer space. These questions are also used to assess your abilities to communicate ideas and put together a logical argument.

The correct answers to extended questions are less well-defined than to those requiring short answers. Examiners may have a list of points for which credit is awarded up to the maximum for the question, or they may first of all judge the quality of your response as poor, satisfactory or good before allocating it a mark within a range that corresponds to that quality.

As an example of a question that requires an extended answer, a structured question on the use of solar energy could end with the following:

Suggest why very few buildings make use of solar energy in this country compared to countries in southern Europe. [5]

Points that the examiners might look for include:

• the energy from the Sun is unreliable due to cloud cover
• the intensity of the Sun's radiation is less in this country than in southern Europe due to the Earth's curvature
• more energy is absorbed by the atmosphere as the radiation has a greater depth of atmosphere to travel through
• fossil fuels are in abundant supply and relatively cheap
• the capital cost is high, giving a long payback time
• photo-voltaic cells have a low efficiency
• the energy is difficult to store for the times when it is needed the most

Full marks would be awarded for an argument that put forward three or four of these points in a clear and logical way.

## Free-response questions

Little use is made of free-response and open-ended questions in AS Level Physics. These types of question allow you to choose the context and to develop your own ideas. Examples could include 'Describe a laboratory method of determining *g*, the value of free-fall acceleration' and 'Outline the evidence that suggests that light has a wave-like behaviour'. When answering this type of question it is important to plan your response and present your answer in a logical order.

# Exam technique

Advanced Subsidiary Physics builds from grade CC in GCSE Science: Double Award, or equivalent in Science: Physics. This Study Guide has been written so that you will be able to tackle AS Physics from a GCSE Science background.

You should not need to search for important Physics from GCSE Science because this has been included where needed in each chapter. If you have not studied Science for some time, you should still be able to learn AS Physics using this text alone.

## What are examiners looking for?

Examiners use instructions to help you to decide the length and depth of your answer.

If a question does not seem to make sense, you may have misread it – read it again!

### State, define or list

This requires a short, concise answer, often recall of material that can be learnt by rote.

### Explain, describe or discuss

Some reasoning or some reference to theory is required, depending on the context.

### Outline

This implies a short response, almost a list of sentences or bullet points.

### Predict or deduce

You are not expected to answer by recall but by making a connection between pieces of information.

### Suggest

You are expected to apply your general knowledge to a 'novel' situation, one which you have not directly studied during the AS Physics course.

### Calculate

This is used when a numerical answer is required. You should always use units in quantities and significant figures should be used with care.
Look to see how many significant figures have been used for quantities in the question and give your answer to this degree of accuracy.
If the question uses 3 (sig figs), then give your answer to 3 ( sig figs) also.

## Some dos and don'ts

### Dos

*Do answer the question*

- No credit can be given for good Physics that is irrelevant to the question.

*Do use the mark allocation to guide how much you write*

- Two marks are awarded for two valid points - writing more will rarely gain more credit and could mean wasted time or even contradicting earlier valid points.

*Do use diagrams, equations and tables in your responses*

- Even in 'essay-type' questions, these offer an excellent way of communicating physics.

*Do write legibly*

- An examiner cannot give marks if the answer cannot be read.

*Do write using correct spelling and grammar. Structure longer essays carefully*

- Marks are now awarded for the quality of your language in exams.

### Don'ts

*Don't fill up any blank space on a paper*

- In structured questions, the number of dotted lines should guide the length of your answer.
- If you write too much, you waste time and may not finish the exam paper. You also risk contradicting yourself.

*Don't write out the question again*

- This wastes time. The marks are for the answer!

*Don't contradict yourself*

- The examiner cannot be expected to choose which answer is intended. You could lose a hard-earned mark.

*Don't spend too much time on a part that you find difficult*

- You may not have enough time to complete the exam. You can always return to a difficult calculation if you have time at the end of the exam.

# What grade do you want?

Everyone would like to improve their grades but you will only manage this with a lot of hard work and determination. You should have a fair idea of your natural ability and likely grade in Physics and the hints below offer advice on improving that grade.

## For a Grade A

You will need to be a very good all-rounder.

- You must go into every exam knowing the work extremely well.
- You must be able to apply your knowledge to new, unfamiliar situations.
- You need to have practised many, many exam questions so that you are ready for the type of question that will appear.

The exams test all areas of the syllabus and any weaknesses in your Physics will be found out. There must be no holes in your knowledge and understanding. For a Grade A, you must be competent in all areas.

## For a Grade C

You must have a reasonable grasp of Physics but you may have weaknesses in several areas and you will be unsure of some of the reasons for the Physics.

- Many Grade C candidates are just as good at answering questions as the Grade A students but holes and weaknesses often show up in just some topics.
- To improve, you will need to master your weaknesses and you must prepare thoroughly for the exam. You must become a better all-rounder.

## For a Grade E

You cannot afford to miss the easy marks. Even if you find Physics difficult to understand and would be happy with a Grade E, there are plenty of questions in which you can gain marks.

- You must memorise all definitions.
- You must practise exam questions to give yourself confidence that you do know some Physics. In exams, answer the parts of questions that you know first. You must not waste time on the difficult parts. You can always go back to these later.
- The areas of Physics that you find most difficult are going to be hard to score on in exams. Even in the difficult questions, there are still marks to be gained. Show your working in calculations because credit is given for a sound method. You can always gain some marks if you get part of the way towards the solution.

## What marks do you need?

As a rough guide, you will need to score an average of 40% for a Grade E, 60% for a Grade C and 80% for a Grade A:

| average | 80% | 70% | 60% | 50% | 40% |
|---------|-----|-----|-----|-----|-----|
| grade   | A   | B   | C   | D   | E   |

# *Essential mathematics*

This section describes some of the mathematical techniques that are needed in studying AS Level Physics.

## Quantities and units

Physical quantities are described by the appropriate words or symbols, for example the symbol *R* is used as shorthand for the value of a *resistance*. The quantity that the word or symbol represents has both a numerical value and a unit e.g. $10.5\Omega$. When writing data in a table or plotting a graph, only the numerical values are entered or plotted. For this reason headings used in tables and labels on graph axes are always written as (physical quantity)/(unit), where the slash represents division. When a physical quantity is divided by its unit, the result is the numerical value of the quantity.

Resistance is an example of a **derived** quantity and the ohm is a derived unit. This means that they are defined in terms of other quantities and units. All derived quantities and units can be expressed in terms of the seven **base** quantities and units that make up the SI, or International System of Units.

The quantities, their units and symbols are shown in the table. The candela is not used in AS or A Level Physics.

| quantity | unit | symbol |
|---|---|---|
| length | metre | m |
| mass | kilogram | kg |
| time | second | s |
| electric current | ampere | A |
| temperature difference | kelvin | K |
| amount of substance | mole | mol |
| luminous intensity | candela | cd |

## Equations

The equations that you use in Physics are relationships between physical quantities. Since it is not possible for one quantity to equate to a different quantity, an equation must be homogenous, i.e. the units on each side of the equation must be the same. This is useful for:

- finding the units of a constant such as resistivity
- checking the possible correctness of an equation; if the units on each side are the same the equation may be correct but if they are different it is definitely wrong.

It is important to distinguish between a physical quantity and its unit. The value of a physical quantity includes both the numerical value and the unit it is measured in.

When working out the units of an equation, make it clear that you are dealing with units only by writing units of (quantity) = (unit).

Often equations need to be used in a different form to that in which they are given or memorised. The equation needs to be **rearranged**. When rearranging an equation it is important to remember that:

- the same mathematical operation must be applied to each side of the equation
- addition or subtraction should be done first, followed by multiplication or division and finally roots and powers.

For example, to rearrange the equation $v^2 = u^2 + 2as$ to enable $u$ to be calculated:

1 subtract $2as$ from each side of the equation to give $v^2 - 2as = u^2$

2 square root each side of the equation to give $\sqrt{(v^2 - 2as)} = u$

An equation that relates the quantity to the product or quotient of two others can be rearranged using the 'magic triangle', a triangle divided into three.

To use the triangle, start with the quantities that are multiplied or divided in the relationship. For example, in $V = IR$, you would write $I$ and $R$ into the triangle first. Two quantities that are multiplied together are written side-by-side at the bottom of the triangle. Where one quantity is divided by the other these are written into the triangle as they would be in a division, with one quantity 'over the other'.

The remaining quantity is now written into the vacant slot. This is illustrated below.

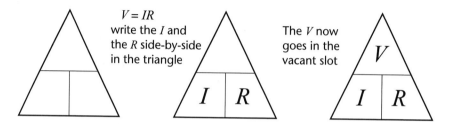

To use the triangle to make $I$ the subject of the equation, cover up the $I$ and you are left with '$V$ over $R$', so $I = V/R$.

## Drawing graphs

Graphs have a number of uses in Physics:

- they give an immediate, visual display of the relationship between physical quantities
- they enable the values of quantities to be determined
- they can be used to confirm or disprove a hypothesis about the relationship between variables.

When plotting a graph, it is important to remember that values determined by experiment are not exact. They are subject to both the accuracy and precision of any measuring instrument and the person using it, as well as any changes that the measuring instrument itself might cause. Consider measuring the temperature of a liquid with a simple thermometer.

Accuracy is the property of a measuring instrument. Precision refers to the reading taken; the precision depends on the user as well as the instrument being used.

- The manufacturer specifies the accuracy, the extent to which the thermometer is reliable. Typically, on a standard laboratory thermometer this is ±1°C.
- When taking a reading, the user may read to the nearest half degree or nearest degree. This describes the precision.
- Inserting the thermometer into a substance may change the temperature of the substance as the two reach thermal equilibrium.

For these reasons, having plotted experimental values on a grid, the graph line is drawn as the best straight line or smooth curve that represents the points. Where there are 'anomalous' results, i.e. points that do not fit the straight line or curve, these should always be checked. If in doubt, ignore them but do add a note in your experimental work to explain why you have ignored them and suggest how any anomalous results could have arisen.

## Measurements and graphs

Quantities such as velocity, acceleration and power are defined in terms of a **rate of change** of another quantity with time. This **rate of change** can be determined by calculating the gradient of an appropriate graph. For example, *velocity* is the *rate of change of displacement with time.* Its value is represented by the gradient of a displacement-time graph. Different techniques are used to determine the gradient of a straight line and a smooth curve. For a straight line:

- determine the value of $\Delta y$, the change in the value of the quantity plotted on the $y$-axis, using the whole of the straight line part of the graph

- determine the corresponding value of $\Delta x$

- calculate the gradient as $\Delta y \div \Delta x$

A common error when determining the gradient of a graph is to work it out using the gridlines only, without reference to the scales on each axis.

For a smooth curve, the gradient is calculated by first drawing a tangent to the curve and then using the above method to determine the gradient of the tangent. To draw a tangent to a curve:

- mark the point on the curve where the gradient is to be determined

- use a compass to mark in two points on the curve, close to and equidistant from the point where the gradient is to be determined

- join these points with a ruler and extend the line beyond each point.

These techniques are illustrated in the diagrams below.

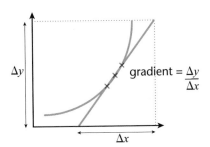

The area between a graph line and the horizontal axis, often referred to as the 'area under the graph' can also yield useful information. This is the case when the product of the quantities plotted on the axes represents another physical quantity, for example, on a *speed-time* graph this area represents the distance travelled. In the case of a straight line, the area can be calculated as that of the appropriate geometric figure. Where the graph line is curved, then the method of 'counting squares' is used.

- Count the number of complete squares between the graph line and the horizontal axis.

- Fractions of squares are counted as '1' if half the square or more is under the line, otherwise '0'.

- To work out the physical significance of each square, multiply together the quantities represented by one grid division on each axis.

These techniques are illustrated in the diagrams below

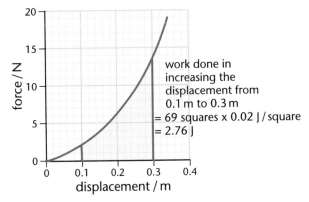

## Equations from graphs

By plotting the values of two variable quantities on a suitable graph, it may be possible to determine the relationship between the variables. This is straightforward when the graph is a straight line, since all straight line graphs have an equation of the form $y = mx + c$, where $m$ is the gradient of the graph and $c$ is the value of $y$ when $x$ is zero, i.e. the intercept on the $y$-axis. The relationship between the variables is determined by finding the values of $m$ and $c$.

The straight line graph below shows how the displacement, $s$, of an object varies with time, $t$.

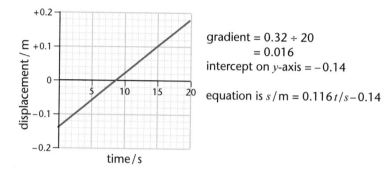

gradient = 0.32 ÷ 20
          = 0.016
intercept on $y$-axis = $-0.14$

equation is $s/m = 0.116 t/s - 0.14$

## Trigonometry and Pythagoras

In the right-angled triangle shown below, the sides are labelled o (opposite), a (adjacent) and h (hypotenuse). The relationships between the size of the angle θ and the lengths of these sides are:

- $\sin \theta = o/h$
- $\cos \theta = a/h$
- $\tan \theta = o/a$

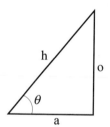

Pythagoras' theorem gives the relationship between the sides of a right-angled triangle: $h^2 = a^2 + o^2$

## Multiples and submultiples

For AS Physics, you are expected to be familiar with the following multiples of units:

| name | multiple | symbol |
|------|----------|--------|
| micro- | $10^{-6}$ | μ |
| milli- | $10^{-3}$ | m |
| kilo- | $10^3$ | k |
| mega- | $10^6$ | M |
| giga- | $10^9$ | G |

For example, the symbol MHz means $1 \times 10^6$ Hz and mN means $1 \times 10^{-3}$ N.

## Equations that you need to know

This is a list of equations that may be needed but are not provided in topic tests or end-of-course examinations. Not all of these equations are needed for every specification as some are in the A Level and not the AS specification. The table shows the equations needed for each AS specification.

All other equations that are needed will be provided on a data sheet, along with any data such as the values of physical constants.

| equation | symbol form | AQA A | AQA B | ED A | ED B | OCR A | OCR B | WJEC | NICCEA |
|---|---|---|---|---|---|---|---|---|---|
| speed = $\dfrac{\text{distance}}{\text{time taken}}$ | $v = \dfrac{\Delta d}{t}$ | • | • | • | • | • | • | • | • |
| force = mass × acceleration | $F = ma$ | • | • | • | • | • | • | • | • |
| acceleration = $\dfrac{\text{change in velocity}}{\text{time taken}}$ | $a = \dfrac{\Delta v}{t}$ | • | • | • | • | • | • | • | • |
| density = $\dfrac{\text{mass}}{\text{volume}}$ | $\rho = \dfrac{m}{V}$ | • | | • | • | • | | | |
| momentum = mass × velocity | $p = mv$ | • | A2 | • | A2 | A2 | A2 | A2 | • |
| work done = force × distance moved in direction of force | $W = Fs$ | • | A2 | • | • | • | • | A2 | A2 |
| power = $\dfrac{\text{energy transferred}}{\text{time taken}}$ = $\dfrac{\text{work done}}{\text{time taken}}$ | $P = \dfrac{\Delta W}{t}$ | • | A2 | • | • | • | • | A2 | A2 |
| weight = mass × gravitational field strength | $W = mg$ | • | • | • | • | • | • | • | • |
| kinetic energy = $\frac{1}{2}$ × mass × speed² | $E_k = \frac{1}{2}mv^2$ | • | • | • | • | • | • | A2 | • |
| change in gravitational potential energy = mass × gravitational field strength × change in height | $E_p = mgh$ | • | • | • | • | • | • | A2 | • |
| pressure = $\dfrac{\text{force}}{\text{area}}$ | $P = \dfrac{F}{A}$ | | | • | | • | • | | |
| pressure × volume = number of moles × molar gas constant × absolute temperature | $pV = nRT$ | • | A2 | • | A2 | A2 | A2 | A2 | A2 |
| charge = current × time | $q = It$ | • | • | • | • | • | • | • | • |
| potential difference = current × resistance | $V = IR$ | • | • | • | • | • | • | • | • |
| electrical power = potential difference × current | $P = IV$ | • | • | • | • | • | • | • | • |
| potential difference = $\dfrac{\text{energy transferred}}{\text{charge}}$ | $V = \dfrac{W}{q}$ | • | • | • | • | • | • | • | • |
| resistance = $\dfrac{\text{resistivity} \times \text{length}}{\text{cross-sectional area}}$ | $R = \dfrac{\rho l}{A}$ | • | • | • | • | • | • | • | • |
| energy = potential difference × current × time | $E = VIt$ | • | • | • | • | • | • | • | • |
| wave speed = frequency × wavelength | $v = f\lambda$ | A2 | • | A2 | • | • | • | • | • |

# Four steps to successful revision

## Step 1: Understand

- Study the topic to be learned slowly. Make sure you understand the logic or important concepts.
- Mark up the text if necessary – underline, highlight and make notes.
- Re-read each paragraph slowly.

**GO TO STEP 2**

## Step 2: Summarise

- Now make your own revision note summary:
  What is the main idea, theme or concept to be learned?
  What are the main points? How does the logic develop?
  Ask questions: Why? How? What next?
- Use bullet points, mind maps, patterned notes.
- Link ideas with mnemonics, mind maps, crazy stories.
- Note the title and date of the revision notes
  (e.g. Physics: Electricity, 3rd March).
- Organise your notes carefully and keep them in a file.

This is now in **short term memory**. You will forget 80% of it if you do not go to Step 3.
**GO TO STEP 3**, but first take a 10 minute break.

## Step 3: Memorise

- Take 25 minute learning 'bites' with 5 minute breaks.
- After each 5 minute break test yourself:
  Cover the original revision note summary
  Write down the main points
  Speak out loud (record on tape)
  Tell someone else
  Repeat many times.

The material is well on its way to **long term memory**.
You will forget 40% if you do not do step 4. **GO TO STEP 4**

## Step 4: Track/Review

- Create a Revision Diary (one A4 page per day)
- Make a revision plan for the topic, e.g. 1 day later, 1 week later, 1 month later.
- Record your revision in your Revision Diary, e.g.
  Physics: Electricity, 3rd March 25 minutes
  Physics: Electricity, 5th March 15 minutes
  Physics: Electricity, 3rd April 15 minutes
  ... and then at monthly intervals.

# Chapter 1
# Force and motion

**The following topics are covered in this chapter:**

- Forces and vectors
- Some effects of forces
- Under stress
- Solid materials
- Motion in one and two dimensions
- Force and acceleration
- Newton I and III
- Momentum and the second law
- Energy to work
- Drag and terminal velocity

## 1.1 Forces and vectors

**After studying this section you should be able to:**

- recognise and describe forces
- distinguish between a vector and a scalar quantity
- add together two vectors and subtract one vector from another
- split a vector into two parts at right angles to each other

**LEARNING SUMMARY**

### Describing forces

| | | | |
|---|---|---|---|
| AQA A | M2 | OCR A | M1 |
| EDEXCEL A | M1 | OCR B | M2 |
| EDEXCEL B | M1 | WJEC | M1 |
| NICCEA | M1 | | |

In this context 'normal' means 'at right angles to the surface'.

The first bullet point here is emphasising that all forces are caused by objects and they act on other objects.

Forces can have different effects; they can cause objects to start moving, stop moving, stay in the same place or rotate. They can also be of different types; **gravitational** forces are caused by and affect objects with mass, **friction** forces oppose relative motion and the force pushing up on you at the moment, the **normal contact** force, is due to compression of the material that you are sat on.

Forces also have things in common:

- all forces can be described as **object A pulls/pushes object B**
- all forces can be represented in both size and direction by an arrow on a diagram.

Here are some examples.

**A gravitational force**

the Earth pulls
the moon

**A friction force**

the ground pushes
the shoe

**A normal contact force**

the chair pushes
the person

### Drawing the arrows

| | | | |
|---|---|---|---|
| AQA A | M2 | OCR A | M1 |
| EDEXCEL A | M1 | OCR B | M2 |
| EDEXCEL B | M1 | WJEC | M1 |
| NICCEA | M1 | | |

Of the three examples given above, two of the forces are due to contact between the objects. In each case the arrow representing the force has been drawn in the centre of the area of contact. Gravitational forces act at a distance with no contact being necessary. These forces are always drawn as if they act at the centre of gravity, or centre of mass, of the object.

25

the centre of mass of a traffic cone, a rubber ring and a person

The gravitational force acting on an object due to the planet or moon whose surface it is on is known as its weight.

> **KEY POINT**
>
> Weight is related to mass by the formula:
>   weight = mass × gravitational field strength
>   $W = m \times g$

Gravitational field strength and free fall acceleration are the same physical quantity. Close to the Earth's surface, $g$ has the value 9.8 N kg$^{-1}$ or 9.8 m s$^{-2}$. In calculations, the value of 10 m s$^{-2}$ is often used as a good approximation.

Physical quantities that have direction as well as size are called **vectors**. Quantities with size only are **scalars**. Some examples are given in the table.

| vectors | scalars |
| --- | --- |
| force | mass |
| velocity | speed |
| acceleration | length |
| displacement | distance |
| field strength | energy |

When representing a vector quantity on a diagram, an arrow is always used to show its direction.

To add together two scalar quantities the normal rules of arithmetic apply, for example, 2 kg + 3 kg = 5 kg and no other answers are possible. When adding vector quantities, both the size and direction have to be taken into account.

What is the sum of a 2 N force and a 3 N force acting on the same object? The answer could be any value between 1 N and 5 N, depending on the directions involved.

## The sum of two vectors

| AQA A | M2 | NICCEA | M1 |
| AQA B | M1 | OCR A | M1 |
| EDEXCEL A | M1 | OCR B | M2 |
| EDEXCEL B | M1 | WJEC | M1 |

The sum, or resultant, of two vectors such as two forces acting on a single object is the single vector that could replace the two and have the same effect.

There are three steps to finding the sum of two vectors. These are illustrated by working out the sum of a 2 N force acting up the page and a 3 N force acting from left to right, both forces acting on the same object.

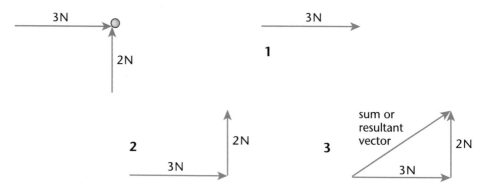

The order in which you draw the vectors does not matter, you should arrive at the same answer whichever one you start with.

- Draw an arrow that represents one of the vectors in both size and direction.
- Starting where this arrow finishes, draw an arrow that represents the second vector in size and direction.

- The sum, or resultant of the two vectors is represented (in both size and direction) by the single arrow drawn from the **start** of the first arrow to the **finish** of the second arrow.

In the example given, the size of the resultant force is 3.6 N and the direction is at an angle of 34° to the 3 N force. These figures were obtained by scale drawing.

## Variations on a theme

| AQA A | M2 | NICCEA | M1 |
|-------|----|--------|----|
| AQA B | M1 | OCR A | M1 |
| EDEXCEL A | M1 | OCR B | M2 |
| EDEXCEL B | M1 | WJEC | M1 |

The method described above can be used to find the sum of any number of vectors by drawing an arrow for each vector, starting each new arrow where the previous one finished. The resultant of the vectors is then represented by the single arrow that starts at the beginning of the first vector and ends where the last one finishes. If the vectors being added together form a closed figure, i.e. the last one finishes where the first one starts, it follows that the sum is zero. This is what you would expect to find when working out the resultant force at a point in a stable structure for example.

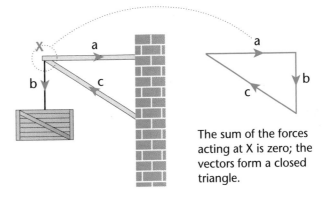

> For equilibrium at point X, the sum of the forces must be zero.

The sum of the forces acting at X is zero; the vectors form a closed triangle.

Although a scale drawing is often the quickest way of working out the resultant of two vectors, the size and direction can also be calculated. This is straightforward when the vectors act at right angles, but needs more complex mathematics in other cases.

In the worked example:

- the size of the resultant force can be calculated using Pythagoras' theorem as $\sqrt{(2^2 + 3^2)}$

> The notation 'tan⁻¹' means 'the angle whose tangent is'.

- the angle between the resultant force and the 3 N force can be calculated using the definition of tangent as $\tan^{-1}(2 \div 3)$.

> Bold letters are used here to represent both the size and direction of a vector. When normal type is used, it represents the size of the vector only.

The method used to add two vectors is also used to subtract one vector from another. For example, an aircraft has velocity **a** and changes its velocity by an amount **b** so that its new velocity is **c**. The relationship between these vectors is **a** + **b** = **c**. To work out the change in velocity **b**, the relationship can be written as **b** = **c** − **a**, the new velocity − the old velocity. Vector **b** is found by adding (−**a**) to **c**. Vector −**a** is the vector equal in size to **a** but opposite in direction. Vector subtraction is illustrated in the diagram below.

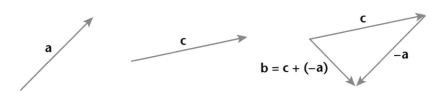

## Splitting a vector into two

| AQA A | M2 | NICCEA | M1 |
| AQA B | M1 | OCR A | M1 |
| EDEXCEL A | M1 | OCR B | M2 |
| EDEXCEL B | M1 | WJEC | M1 |

A vector has an effect in any direction except the one at right angles to it. Sometimes a vector has two independent effects which need to be isolated. Just as the combined effect of two vectors acting on a single object can be calculated, two separate effects of a single vector can be found by splitting the vector into two **components**. Provided that the directions of the two components are chosen to be at right angles, each one has no effect in the direction of the other so they are considered to act independently.

The process of splitting a vector into two components is known as **resolving** or **resolution of** the vector.

The diagram shows the tension (T) in a cable holding a radio mast in place. The force is pulling the mast both vertically downwards and horizontally to the left.

> For a system in equilibrium the sum of all the vertical components is zero and the sum of all the horizontal components is zero.

  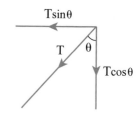

To find the effect of the tension in these directions, $T$ can be split into two components, shown as $T_H$ (the horizontal component) and $T_V$ (the vertical component) in the diagram. You should check that according to the rules of vector addition, $T_H + T_V = T$.

> **KEY POINT**
>
> To find the magnitude of $T_H$ and $T_V$ the following rules apply:
>
> the component of a vector **a** at an angle $\theta$ to its own angle is $a\cos\theta$
>
> the component of a vector **a** at an angle $(90° - \theta)$ to its own angle is $a\sin\theta$

The application of these rules is shown in the right-hand diagram above.

## Progress check

1   Work out the resultant of the two forces shown in the diagram.

2   A car is being towed with a rope inclined at 20° to the horizontal. The tension in the rope is 350 N.
    Work out the horizontal and vertical components of the tension in the rope.

3   A van is travelling north at a speed of 28 m s⁻¹.
    After turning a corner it is heading 40° east of north at 25 m s⁻¹.
    Work out the change in velocity of the van.

3   18 m s⁻¹ in a direction 29° S of E
2   Horizontal component = 329 N   Vertical component = 120 N
1   10.8 N at an angle of 22° to the 10 N force

# *1.2 Some effects of forces*

*After studying this section you should be able to:*

- *calculate the moment of a force and apply the principle of moments to a stable object*
- *calculate the density of a solid, liquid or gas*
- *calculate the pressure due to a force*

**LEARNING SUMMARY**

## *The turning effect*

| AQA A | M2 | NICCEA | M1 |
|---|---|---|---|
| AQA B | M1 | OCR A | M1 |
| EDEXCEL A | M1 | WJEC | M1 |

When forces are used to open doors, steer and pedal bicycles or turn a tap they are causing turning. The effect that a force has in turning an object round depends on:

- the size of the force
- the perpendicular (shortest) distance between the force line and the pivot (the axis of rotation).

Both of these factors are taken into account when measuring the turning effect, or moment, of a force.

> The 'line of action' of a force is the line drawn along the direction in which the force acts.

**KEY POINT**

Moment of a force = force × perpendicular distance from line of action of force to pivot.
The moment of a force, also known as torque, is measured in N m.

The diagram shows how the same force used to open a door can produce very different effects, according to the direction in which it is applied.

> The moment of a force is a vector that can only have one of two directions; either clockwise or anticlockwise.

moment = 50 N x 0.8 m = 40 N m    moment = 50 N x 0.6 m = 30 N m    moment = 50 N x 0 m = 0 N m

> If the forces that make the couple are equal in size, the moment of the couple = size of one force × shortest (perpendicular) distance between the force lines.

When turning on a tap or steering a bicycle, two forces are normally used. The forces act in opposite directions, but they each produce a moment in the same direction. A pair of forces acting like this is called a couple. The combined moment is equal to the sum of the moments of the individual forces.

## *A question of balance*

| AQA A | M2 | NICCEA | M1 |
|---|---|---|---|
| AQA B | M1 | OCR A | M1 |
| EDEXCEL A | M1 | WJEC | M1 |

For an object or a structure to be stable, two conditions apply:

- the resultant force in any direction is zero
- the resultant moment in either direction (clockwise or anticlockwise) is zero.

This second condition is known as the principle of moments.

**KEY POINT**

The principle of moments states that:
For an object to be balanced, the sum of the clockwise moments about any pivot is equal to the sum of the anticlockwise moments about the same pivot.

The principle of moments can be used to determine the state of balance of an object and also to work out the size of the forces acting on objects that are known to be balanced.

## Pressure

EDEXCEL A   M2
OCR A   M1
OCR B   M1

When a force is applied to a solid object such as a drawing pin or a nail, the force is transmitted through the solid. Factors that determine the effect of the force in cutting or piercing include:

- the size of the force
- the area of contact.

These two factors together are used to calculate the pressure.

> If the force that causes the pressure is not acting in a direction normal to the surface, then the normal component of the force is used to calculate the pressure.

> **KEY POINT**
>
> Pressure = normal force per unit area
> $$p = F/A$$
> Pressure is measured in pascal (Pa) where $1\ \text{Pa} = 1\ \text{N m}^{-2}$

small force applied here    large force produced here

Whereas a solid transmits a force in the direction in which it is applied, fluids (liquids and gases) transmit pressure. The pressure at any point in a fluid has the same value in all directions. The transmission of pressure by a fluid is used in hydraulic machinery, where the force exerted by the fluid is determined by its pressure and the area over which it acts. This enables a large force to be exerted by the application of a small force that causes the pressure in the fluid. The principle of a hydraulic lifting device is shown in the diagram.

## Density

AQA A   M3
EDEXCEL A   M1
EDEXCEL B   M2
OCR A   M1

The density of a material is a measure of how close-packed the particles are. Gases at atmospheric pressure are much less dense than solids and liquids because the particles are more widespread.

> Typical densities of some everyday materials in $\text{kg m}^{-3}$ are:
>
> air 1.2
> water 1000
> concrete 5000
> iron 9000

> **KEY POINT**
>
> Density of a material = mass per unit volume.
> $$\rho = m/V$$
> Density is measured in $\text{kg m}^{-3}$ or $\text{g cm}^{-3}$.

It is measured by dividing the mass of a specimen of the material by its volume.

## Progress check

1. Use the principle of moments to calculate the tension in the rope, T, in the diagram below.

0.6 m    0.3 m    0.3 m

150 N    100 N

2. Explain how the hydraulic system shown in the diagram above multiplies the input force.

3. Describe how you would measure the density of air.

# 1.3 Under stress

**After studying this section you should be able to:**

● *describe the difference between elastic and plastic behaviour*
● *state Hooke's law and appreciate its limitations*
● *contrast the behaviour of a ductile, a brittle and a polymeric solid when stretched*
● *explain how stress and strain are used in measurements of the Young modulus*

**LEARNING SUMMARY**

## Elastic or plastic

| | | | |
|---|---|---|---|
| AQA A | M3 | OCR A | M1 |
| EDEXCEL A | M3B | OCR B | M1 |
| EDEXCEL B | M2 | WJEC | M1 |
| NICCEA | A2 | | |

No material is absolutely rigid. Even a concrete floor changes shape as you walk across it. The behaviour of a material subjected to a tensile (pulling) or compressive (pushing) force can be described as either elastic or plastic.

> a material is **elastic** if it returns to its original shape and size when the force is removed
> a material is **plastic** if it does not return to its original shape and size when the force is removed

Most materials are elastic for a certain range of forces, up to the elastic limit, beyond which they are plastic. Plasticine and playdough are plastic for all forces.

Hooke attempted to write a simple rule that describes the behaviour of all materials subjected to a tensile force.

> *If the extension is proportional to the stretching force, then doubling the force causes the extension to double.*

**KEY POINT**

Hooke's law states that:

the extension of a sample of material is proportional to the stretching force

$e \propto F$

This can be written as $F = ke$

where $k$ represents the stiffness of the sample and has units of N m$^{-1}$

> *A polymeric solid is one made up of long chain molecules.*

Metals and springs 'obey' Hooke's law up to a certain limit, called the limit of proportionality. For small extensions, the extension is proportional to the stretching force. Rubber and other polymeric solids do not show this pattern of behaviour.

The graphs below contrast the behaviour of different materials subjected to an increasing stretching force.

> *Because the limits of proportionality and elasticity lie close together on the force-extension curve, the term 'elastic limit' is often used to refer to both points.*

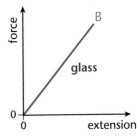

P = limit of proportionality     E = elastic limit     B = breaking point

● **Copper** is a ductile material, which means that it can be drawn into wires. It is also malleable, which means that it can be reshaped by hammering and bending without breaking. When stretched beyond the point E on the graph it retains its new shape.
● **Rubber** does not follow Hooke's law and it remains elastic until it breaks.
● **Glass** is brittle; it follows Hooke's law until it snaps.
● **Kevlar** is tough; it can withstand shock and impact.
● **Mild steel** is durable; it can withstand repeated loading and unloading.
● **Diamond** is hard; it cannot be easily scratched.

## Storing energy

| AQA A | M3 | EDEXCEL B | M2 |
|-------|----|-----------|----|
| AQA B | M1 | OCR A | M1 |
| EDEXCEL A | M3B | WJEC | A2 |
| NICCEA | A2 | | |

If seat belts and climbing ropes did not stretch, the forces exerted in stopping a person in an emergency would be too great for the body to withstand.

Some elastic materials are intended to absorb energy. The more a material is extended, the greater the force required.

- The energy stored as a material is deformed is represented by the area between the curve and the extension axis.
- If a force $F$ produces an extension $e$ below the limit of proportionality, then the energy stored $= \frac{1}{2}Fe$, as shown in the diagram opposite.
- since $F = ke$, energy stored $= \frac{1}{2}ke^2$.

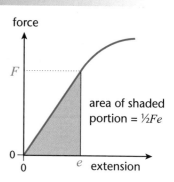

area of shaded portion $= \frac{1}{2}Fe$

## Stress, strain and the Young modulus

| AQA A | M3 | NICCEA | A2 |
|-------|----|--------|----|
| AQA B | A2 | OCR A | M1 |
| EDEXCEL A | M3B | OCR B | M1 |
| EDEXCEL B | M2 | WJEC | M1 |

100 N        200 N

The stress on these wires is the same, even though the forces are different.

For some materials, e.g. copper and mild steel, the stress when the material breaks is less than the ultimate tensile stress.

The Young modulus for mild steel is $2.1 \times 10^{11}$ Pa and that for copper is $0.7 \times 10^{11}$ Pa. This means that for the same stretching force, a sample of copper stretches three times as much as one of mild steel with the same dimensions.

The extension that a force produces depends on the dimensions of the sample as well as the material that it is made from. To compare the elastic properties of materials in a way that does not depend on the sample tested, measurements of **stress** and **strain** are used instead of force and extension.

> **KEY POINT**
>
> stress = normal force per unit area
> $$\sigma = F/A$$
> Like pressure, stress is measured in Pa, where 1 Pa = 1 N m$^{-2}$
>
> strain = extension per unit length
> $$\varepsilon = e/l$$
> Strain has no units.

The strength of a material is measured by its **breaking stress** or **ultimate tensile stress**. This is the maximum stress the material can withstand before it fractures or breaks and may not be the actual stress when breaking occurs. Its value is independent of the dimensions of the sample used for the test.

Using measurements of stress and strain, the stiffness of different materials can be compared by the **Young modulus**, $E$.

> **KEY POINT**
>
> Young modulus = stress ÷ strain
> $$E = \sigma/\varepsilon = F/A \div e/l = Fl/eA$$
> The Young modulus has the same units as stress, Pa.

The greater the value of the Young modulus the **stiffer** the material, i.e. the less it stretches for a given force.

## Progress check

1  Describe the difference between a **ductile** material and one that is **brittle**.

2  A force of 150 N applied to a spring causes an extension of 25 cm.
   a   Calculate the energy stored in the spring.
   b   What assumption is necessary to answer a?

3  The Young modulus for copper is $1.3 \times 10^{11}$ Pa.

   A tensile force of 150 N is applied to a sample of length 1.50 m and cross-sectional area 0.40 mm². (1.0 mm² $= 1.0 \times 10^{-6}$ m²).

   Assuming that the limit of proportionality is not exceeded, calculate the extension of the copper.

3  4.3 mm
   b   the limit of proportionality has not been exceeded
2  a  18.75 J
1  A ductile material can be drawn into wires; a brittle material snaps easily.

# 1.4 Solid materials

**After studying this section you should be able to:**

- *describe the structure of metals and polymers and relate the structure to the properties*
- *explain how the properties of metals and polymers can be changed*
- *explain the advantages of using composite materials in appropriate applications*

**LEARNING SUMMARY**

## Rubber

EDEXCEL A ▶ M3B    WJEC ▶ M1
EDEXCEL B ▶ M2

The difference between a force–extension graph and a stress–strain graph for the same material is that they have different values and units on the axes.

Graphs that show the variation of stress with strain have the same shape as the force-extension graphs on page 31. The gradient of these graphs represents the stiffness of the material, i.e. how difficult the material is to stretch. The structure of **natural rubber** is like a can of worms, consisting of long chain molecules with occasional cross-linking between the chains. This is shown in the diagram below

The long chain molecules in unstretched and stretched rubber

When rubber is stretched the stiffness gradually decreases and it becomes relatively easy to stretch as the chains uncoil; once this has happened the rubber is much stiffer.

If you repeatedly stretch and release a rubber band, you can feel the effect of heating caused by hysteresis.

The stress–strain graph for rubber (opposite) shows that the behaviour as a load is removed is not the same as that when the load is being increased. This is called **hysteresis** and the curves are said to make a **hysteresis loop**. In unit 1.3. Under stress, it was shown that the energy absorbed or released by a material is represented by the area between the curve and the extension axis of a force–extension graph. The equivalent area on a stress–strain graph represents energy per unit volume.

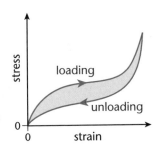

> **KEY POINT**
>
> On a graph of stress against strain:
> the area between the curve and the strain axis represents the energy per unit volume.

This is the energy absorbed when a material is being stretched and the energy that is released when the force is removed. Rubber absorbs more energy during loading than it releases in unloading. The difference is represented by the area of the hysteresis loop, shown shaded in the stress–strain graph.

Most tyres use the same rubber to do two different jobs; provide rigidity for the tyre wall and elasticity for the tread.

The effect of hysteresis in rubber is to transfer energy to its molecules, resulting in heating. A car tyre goes through hysteresis for each revolution of the wheel. If the tyre is under-inflated this can lead to a 'blow-out' when the air inside the tyre becomes hot and the tyre explodes due to the increase in pressure. Modern, energy-efficient tyres reduce the problem of hysteresis by using a rigid rubber for the tyre walls and a more flexible rubber for the tread.

### Steel

EDEXCEL A  M3B    OCR B  M1
EDEXCEL B  M2

The behaviour of steel under stress depends on the carbon content of the metal. High carbon steels such as cast iron are brittle; they snap just beyond the elastic limit. Mild steel is tougher, much more energy is needed to cause failure. The diagram below compares stress–strain graphs for high carbon steel and mild steel.

> Mild steel can be deformed plastically, so it is used in place of wrought iron to make structures such as iron gates.

The points marked on the graphs are:

- P, the limit of proportionality; this is the end of the Hooke's law region
- Y, the elastic limit or yield point; beyond this point the material is permanently deformed
- UTS, the ultimate tensile stress or breaking stress; the maximum stress that the material can withstand.

> The term 'poly' means 'many' and 'crystalline' refers to atoms 'arranged in a regular array' to form crystals. The crystalline nature of metals can be seen clearly on the zinc coating on the inside of a can of pineapple.

Metals have a polycrystalline structure, which means that they consist of many small crystals. Within each crystal the atoms are arranged in a regular pattern. This structure is shown in diagram (a), which also shows the slip planes within the crystals. The ductile behaviour of mild steel beyond its elastic limit is due to whole planes of atoms slipping against each other. Slip occurs readily by the movement of an edge dislocation; a defect in the crystal due to an incomplete plane of atoms. This is shown in diagram (b).

slip planes in metal crystals          the slip plane caused by a dislocation

> The 'neck' refers to the narrowing of the sample; this is sometimes called a 'waist'.

When slip occurs in a metal it can result in necking. If the width of the sample decreases at one point, the stress increases due to the reduction in cross-sectional area. The increased stress causes further stretching and narrowing at this point until the metal eventually fractures.

necking leading to fracture

## Treating metals

EDEXCEL A ▶ M3B

Work hardening is not effective with metals which have low melting points such as lead (mp 328°C) and zinc (mp 420°C).

A ductile metal can be made stronger by preventing slip from occurring. In high carbon steel the presence of carbon atoms between layers of iron atoms stops them from slipping over each other. Strangely, the introduction of further dislocations can also strengthen metals such as copper or mild steel. This happens in a process called **work-hardening**, where the metal is repeatedly worked by hammering, stretching or bending while it is cold. The increased strength is attributed to a bottle-neck of dislocations preventing them from moving past each other.

The size of the crystals or grains in a metal also affects its strength. Since dislocations cannot move past the grain boundaries, smaller grains lead to stronger metals. Heat treatment of a metal affects both the size of the grains and the concentration of dislocations. It can be of several types:

When steel is heated to red heat, the grains enlarge. Slow cooling allows excess carbon to diffuse out of the crystals and results in the metal being annealed. Rapid cooling results in a brittle material as the excess carbon does not have time to diffuse out of the crystals.

- **annealing** involves heating a metal until it is red hot followed by controlled cooling. This makes the metal more ductile and less prone to fracture.
- **quenching** involves heating the metal until it is red hot and then cooling it rapidly by putting it in cold water or oil. This makes the metal strong but brittle.
- **tempering** is the process of quenching followed by annealing at a lower temperature. This is done to produce a balance between strength and toughness.

## Failure

EDEXCEL A ▶ M3B

Lead pipes and roofing gradually become deformed under their own weight due to the effects of creep.

**Creep** can occur in metals and other materials subjected to a constant load, normally much lower than the yield stress, for a long time. Gradual deformation takes place due to a number of mechanisms, including:

- the movement of dislocations
- grain boundaries sliding relative to each other
- necking.

Creep affects structures such as the steel cables used to support suspension bridges, but it is more significant in metals used at temperatures that are above $0.3 \times$ the melting point of the metal in K. This is the case for lead used at room temperatures and also for metals used in engines. Creep-resistant alloys have been developed for use in applications where failure due to creep could be disastrous, for example, in the turbine blades used in aircraft jet engines.

**Fatigue** is another type of failure caused by a cumulative effect over time. It affects metals subjected to repeated cycles of stress such as that caused by stretching, rotation or vibration. Fatigue failure starts with a small crack appearing due to repeated working. Once a crack has formed, stress builds up around it due to the reduction in cross-sectional area. This causes the crack to enlarge until the material fractures.

The stress concentration at the tip of a crack increases as the crack grows.

The risks of fatigue failure can be reduced during both design and manufacture. In designing structures that are subjected to repetitive stress, sharp corners should be avoided as this is where cracks can develop. In the manufacturing process, care should be taken to avoid poor workmanship at places where materials are rivetted or bolted together leaving tiny cracks that then propagate.

## Polymers

EDEXCEL A   M3B       WJEC       M1
EDEXCEL B   M2

Amorphous describes any structure that is irregular, with no ordering. Glass has an amorphous structure.

Polymers consist of long chain molecules; natural rubber is a polymer whose molecules have no ordering. Such polymers are called amorphous. Perspex is a synthetic polymer whose structure is also amorphous. Nylon and polythene have both crystalline and amorphous regions, their structure is described as semi-crystalline. The diagram below shows how the chains in a semi-crystalline polymer are folded; the straight rows represent crystalline regions and the bends are amorphous.

Both nylon and polythene soften on heating because the bonds between the long chain molecules are weak; they are called thermoplastics. Because of this property they can be used to manufacture articles by moulding.

refolded chains in polythene

cross-linking makes rubber less stretchy

The rubber used in the manufacture of car tyres has a low sulphur content, about 5%. Hard rubbers contain up to 50% sulphur.

The rigidity of a polymer is increased by cross-linking, the formation of strong bonds between the molecules by introducing other substances. The vulcanised rubber used in car tyres contains sulphur which forms bonds between chains, making it less elastic. Adding more sulphur makes the rubber more rigid. As more cross-linking is employed, the polymer becomes a thermoset, one which hardens with increasing temperature.

Bakelite and melamine are thermosets commonly used for household objects.

Thermosetting polymers are used in the manufacture of electrical fittings such as lamp holders, sockets and kitchen worktops. Because they cannot be softened by heating, any moulding has to be done as the polymer is manufactured.

## Composite materials

EDEXCEL A   M3B

A composite is a mixture of two or more materials bonded together to obtain some overall desirable property.
Types of composite include:

- laminated materials
- fibre composites
- particle composites.

the grain in layers of plywood

Wood and bone are natural composite materials. Wood is a fibre composite, consisting of fibres of cellulose in a lignin cement.

A laminate consists of thin sheets of material joined together. Common examples are plywood and car windscreens. Wood is strong when subjected to a tensile force parallel to its grain, but if the force is across the grain it splits easily. The diagram above shows how plywood is manufactured by glueing sheets together so that the grains in adjacent sheets are at right angles to each other.

A windscreen is made in a similar way; two sheets of toughened glass are bonded to a sheet of clear plastic in a sandwich. If the glass shatters, the plastic holds the broken pieces together.

Chipboard is a **fibre composite** that consists of wooden chips glued together. The packing of the chips is random, so the wooden fibres lie in all directions. Unlike natural wood, chipboard does not have a grain so it is equally strong in all directions.

Fibre composites such as glass reinforced plastic (glass fibre) and carbon fibre reinforced plastics have replaced metals in some applications. They are lighter than metals and are not affected by fatigue. Some car bodies are made from glass fibre. Its low weight gives improved acceleration and fuel consumption but it is not as strong as the same thickness of steel, so it gives less protection to the driver and passengers.

> Glass fibre does not corrode like metals, so it is also suitable for applications such as the hulls of boats. Because glass fibre is non-magnetic, it is particularly suitable for minesweepers.

The diagram below shows two ways in which glass fibres can be arranged within the plastic. When the fibres are parallel the glass fibre is strong in the direction of the fibres but weak in the direction at right angles to them. A random arrangement of the fibres gives a material that is equally strong in all directions.

**Two ways of embedding glass fibres in plastic**

> In normal use glass breaks easily. This is because its surface is easily scratched and these scratches propagate. By protecting the surface from scratches, the ultimate tensile strength of glass is increased.

Carbon fibre reinforced plastic has the same tensile strength as steel but is much stiffer. It is still a very expensive material to manufacture so its use is limited to some parts of jet engines and such things as specialist tennis rackets.

Both types of fibre reinforced plastic rely on the intrinsic strength of carbon and glass. These materials are very brittle and become much weaker when the surface is scratched. The plastic protects the surface and prevents any cracks that do form from propagating along other fibres.

## Using concrete

EDEXCEL A ▸ M3B    WJEC ▸ M1

> The term aggregate applies to any filler material used in concrete. Sand is the finest aggregate used; coarse aggregates include gravel and crushed stone.

Concrete is a **particle composite** material made from a mixture of cement, water and aggregates. The cement bonds together the sand and small stones that make up the aggregate. This makes a cheaper material than using cement on its own.

sand and cement mixture

stones

**The structure of concrete**

Concrete is very strong when subjected to a compressive force, making it a suitable material for the pillars that support a road bridge. It breaks easily when in tension, so it is not suitable for the horizontal surfaces supported by the pillars, parts of which are in compression and parts of which are in tension.

To overcome this, concrete can be pre-stressed which ensures that the concrete remains under compression even when subjected to its normal tensile load. In the manufacture of pre-stressed concrete, the concrete is cast around steel bars that are held in tension.

Concrete that is always in compression in normal use is not pre-stressed, but it can be reinforced by casting the concrete around a steel mesh.

steel bars in tension

concrete

**Pre-stressed reinforced concrete**

After the concrete has hardened the tension is released, shrinking the steel and compressing the concrete. The steel both **reinforces** the concrete, sharing the load, and keeps it in compression even when it is stretched.

## Forces in structures

EDEXCEL A ▶ 3B      OCR A ▶ M1

The resultant force and resultant moment both have to be zero for the structure to be stable.

Resolving horizontal and vertical components will always give the same answers as taking moments.

If a beam or rod is **uniform** its mass is evenly distributed so that it is the same as if all the mass was concentrated at a single point at the centre of the beam. The weight is the force acting vertically downwards from this point.

The resultant force and resultant moment at any point in a stable structure are both zero. Either of these principles can be used to find the values of forces $F_1$ and $F_2$ in the ropes that support the beam shown in the diagram below.

- Both $F_1$ and $F_2$ could be found by resolving the vertical components of all the forces (which must balance) and the horizontal components of all the forces (which must also balance).
- Since $F_1$ does not have a moment about A, taking moments about this point gives a relationship between $F_2$ and the forces we know.
- Similarly, taking moments about B gives a relationship between $F_1$ and the known forces.

To take moments about A, $F_2$ needs to be resolved into horizontal and vertical components. The moment of the horizontal component is zero and that of the vertical component is in the anticlockwise direction. The equation is:

$$F_2 \cos 60° \times 4.0 \text{ m} = 200 \text{ N} \times 2.0 \text{ m} + 80 \text{ N} \times 1.0 \text{ m}$$
$$F_2 = 480 \text{ N m} \div (\cos 60° \times 4.0 \text{ m}) = 240 \text{ N}.$$

## Progress check

1 Natural rubber is an **amorphous polymer**.
   a Explain the meaning of the terms **amorphous** and **polymer**.
   b Explain why rubber becomes stiffer as it is stretched.

2 What effect do each of the following have on a metal?
   a Work-hardening.
   b Annealing.

3 Use the principle of moments to calculate the value of force $F_1$ in the diagram above.

**3** 262 N
**b** Annealing causes a metal to become less stiff and more ductile.
**2 a** Work-hardening causes a metal to become stiffer and more brittle.
A polymer is made up of long chain molecules.
**1 a** Amorphous means the structure is irregular.

# 1.5 Motion in one and two dimensions

**After studying this section you should be able to:**

The term 'curve' means the line drawn on the graph to show the relationship between the quantities. It could be straight or curved.

- *interpret graphs used to represent motion, and understand the physical significance of the gradient and the area between the curve and the time axis*
- *apply the equations that describe motion with uniform acceleration*
- *analyse the motion of a projectile*

**LEARNING SUMMARY**

## Speed and velocity

| | | | |
|---|---|---|---|
| AQA A | M2 | NICCEA | M1 |
| AQA B | M1 | OCR A | M1 |
| EDEXCEL A | M1 | OCR B | M2 |
| EDEXCEL B | M1 | WJEC | M1 |

Information about the speed of an object only states how fast it is moving; the velocity also gives the direction. For an object moving along a straight line, positive (+) and negative (–) are usually used to indicate movement in opposite directions. As long as it is moving, the distance travelled by the object is increasing, but its displacement can increase or decrease and, like velocity, can have both positive and negative values.

The fixed point is usually the starting point of the motion or the object's rest position.

- displacement is a vector quantity; it specifies both the distance and direction of an object measured from a fixed point.

The diagram shows a distance–time graph and a displacement–time graph for the same motion.

Check that you understand why distance has only positive values, but displacement has both positive and negative values.

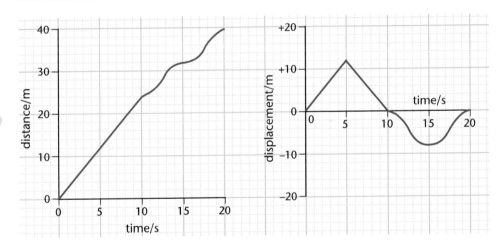

Information from the graphs can be used to calculate the average speed and average velocity over any time period, using the relationships:

The same symbols are used for speed and velocity. When an object is changing speed/velocity, $u$ is often used for the initial speed/velocity, with $v$ being used for the final speed/velocity. The symbol $c$ is usually reserved for the speed of light.

| | | | | |
|---|---|---|---|---|
| average speed | = distance travelled | ÷ time taken | $v = \Delta d \div \Delta t$ | **KEY POINT** |
| average velocity | = displacement | ÷ time taken | $v = \Delta s \div \Delta t$ | |

These show that:

- over the 20 s time period, the average speed is 2 m s$^{-1}$ and the average velocity is 0
- the average speed is constant for the first 10 s, after which it changes.

Calculations of the average speed or velocity over a period of time do not show any changes that have taken place. Instantaneous values are defined in terms of the rate of change of distance or displacement and are represented by the gradients of the graphs.

> instantaneous speed = rate of change of distance. It is represented by the gradient of a distance–time graph.
>
> instantaneous velocity = rate of change of displacement. It is represented by the gradient of a displacement–time graph.

*KEY POINT*

The gradient of a displacement–time graph can have either a positive or a negative value, so it shows the direction of motion as well as the speed.

## Speed–time and velocity–time graphs

| | | | |
|---|---|---|---|
| AQA A | M2 | NICCEA | M1 |
| AQA B | M1 | OCR A | M1 |
| EDEXCEL A | M1 | OCR B | M2 |
| EDEXCEL B | M1 | WJEC | M1 |

Any object that is changing its speed or direction is **accelerating**. Speeding up, slowing down and going round a corner at constant speed are all examples of **acceleration**. Acceleration is a measure of how quickly the velocity of an object changes.

> In physics, any change in velocity is an *acceleration*. This means that an object moving at constant speed, but changing direction, is accelerating.

> Instantaneous acceleration = rate of change of velocity. It is represented by the gradient of a velocity–time graph.
>
> average acceleration = change in velocity ÷ time          $a = \Delta v \div \Delta t$
> Acceleration is a vector quantity and is measured in m s$^{-2}$.

*KEY POINT*

Speed–time and velocity–time graphs both give information about the motion of an object that is accelerating. The graphs below both represent the motion of a ball after it has been thrown vertically upwards with an initial speed of 15 m s$^{-1}$.

> The 'negative' and 'positive' velocities show motion in opposite directions; in this case 'positive' is up and 'negative' is down.

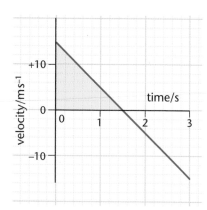

The speed–time graph shows that the speed decreases to zero as the ball reaches its maximum height and then increases. The velocity–time graph also shows the change in direction of the ball. In each case the gradient of the graph is numerically equal to the acceleration (as the ball is moving vertically this is free-fall acceleration), but the gradient of the velocity–time graph also shows that the direction of the acceleration (vertically downwards) is opposite to that of the initial velocity.

> Remember, the area of a triangle = ½ × base × height.

The distance that the ball travels in any time interval can also be obtained from the graphs. Since **distance travelled = average speed × time**, this is represented by the area between the curve and the time axis. The area that represents the distance travelled by the ball while moving upwards is shaded on each graph.

> The gradient of a velocity–time graph represents the acceleration.
>
> The area between the curve and the time axis of a velocity–time or a speed–time graph represents the distance travelled.

*KEY POINT*

Check that the shaded area represents a distance of 11.25 m in each case.

## Uniform acceleration

| | | | |
|---|---|---|---|
| AQA A | M2 | NICCEA | M1 |
| AQA B | M1 | OCR A | M1 |
| EDEXCEL A | M1 | OCR B | M2 |
| EDEXCEL B | M1 | WJEC | M1 |

The gradient of the velocity–time graph on page 40 has a constant value; it represents a constant or **uniform** acceleration. A number of equations link the variable quantities when an object is moving with uniform acceleration. They are:

- $v = u + at$
- $s = ut + \frac{1}{2}at^2$
- $v^2 = u^2 + 2as$
- $s = \frac{1}{2}(u + v)t$

The symbols have the meanings already defined:

$u$ = initial velocity
$v$ = final velocity
$s$ = displacement
$a$ = acceleration
$t$ = time

These equations involve a total of five variables but only four appear in each one, so if the values of three are known the other two can be calculated.

To work out the distance travelled by an object that changes direction, calculate the displacement both before and after the change in direction, and add these together.

When using the equations of uniformly accelerated motion:

- take care with signs; use + and − for vector quantities such as velocity and acceleration that are in opposite directions
- remember that $s$ represents displacement; this is not the same as the distance travelled if the object has changed direction during the motion.

## Motion in two dimensions

| | | | |
|---|---|---|---|
| AQA A | M2 | NICCEA | M1 |
| AQA B | M1 | OCR A | M1 |
| EDEXCEL A | M1 | OCR B | M2 |
| EDEXCEL B | M1 | WJEC | M1 |

People who play sports such as tennis and squash know that, no matter how hard they hit the ball, they cannot make it follow a horizontal path. The motion of a ball through the air is always affected by the Earth's pull.

There is a simple relationship between the total distances travelled after $t$, $2t$ etc. by an object falling vertically. Can you spot it?

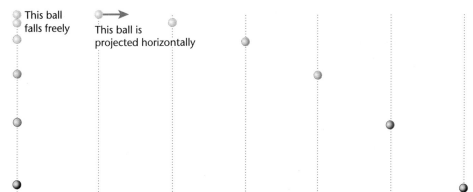

This ball falls freely

This ball is projected horizontally

The diagram shows the results of photographing the motion of a ball projected horizontally alongside one released so that it falls freely. The photographs are taken at equal time intervals.

The free-falling ball travels increasing distances in successive time intervals. This is because the vertical motion is accelerated motion, so the average speed of the ball over each successive time interval increases.

The ball projected horizontally travels equal distances horizontally in successive time intervals, showing that its horizontal motion is at **constant speed**. Vertically, its motion matches that of the free-falling ball, showing that **its vertical motion is not affected by its horizontal motion**.

The motion of any object can be resolved in two directions at right angles to each other. The two motions can then be treated separately.

> **KEY POINT**
>
> When an object has both horizontal and vertical motion, these are independent of each other.

This important result means that the horizontal and vertical motions can be analysed separately:

- for the horizontal motion at constant speed, the equation $v = s \div t$ applies
- for the vertical, accelerated, motion, the equations of motion with uniform acceleration apply.

## Solving projectile problems

| | | | |
|---|---|---|---|
| AQA A | M2 | NICCEA | M1 |
| AQA B | M1 | OCR A | M1 |
| EDEXCEL A | M1 | OCR B | M2 |
| EDEXCEL B | M1 | WJEC | M1 |

The common factor between the two independent motions of an object that is travelling both vertically and horizontally is $t$; the time for which it is travelling. It is important not to confuse the horizontal velocity and displacement with those that apply to the vertical motion.

Consider this problem.

A ball is thrown horizontally with a speed of 15.0 m s$^{-1}$ from a vertical height of 18.0 m. How far has it travelled horizontally when it reaches the ground? Take $g = 10.0$ m s$^{-2}$.

To solve this problem:

- apply the equations of motion with uniform acceleration to the vertical motion to find the time that the ball is in the air. Note that vertically $u = 0$.
- use the constant speed equation to calculate the distance travelled horizontally by the ball in this time.

Check that this gives an answer of 27 m.

> A common error in solving this type of problem is to take $u$, the initial vertical velocity, to be equal to the value of the horizontal velocity.

## Progress check

1 The diagram below shows a velocity–time graph. Calculate the distance travelled and the acceleration on each labelled section of the graph.

> Hint for Q2: what is the velocity of the pigeon at its maximum height?

2 A clay pigeon is fired upwards with an initial velocity of 23 m s$^{-1}$. What height does the pigeon reach? Take $g = 10$ m s$^{-2}$.

3 A tennis player is standing 5.5 m from the net. She hits the ball horizontally at a speed of 32 m s$^{-1}$. How far has the ball dropped when it reaches the net? Take $g = 10$ m s$^{-2}$.

3 0.15 m
2 26.45 m
c 80 m; 0 m s$^{-2}$
b 64 m; -3.1 m s$^{-2}$
1 a 36 m; 2 m s$^{-2}$;

# Sample question and model answer

A wooden pole is held in an upright position by two wires in tension. These forces are shown in the diagram below.

The pole is fixed at its base.

Always write down the formula that you intend to use, and each step in the working. It would be very easy in this case to use a wrong distance. Simply writing down a wrong answer always gets no marks, but showing how you arrive at that wrong answer allows credit to be awarded for the correct process.

(a) Calculate the moment of the 500 N horizontal force about X. [2]

moment = force × perpendicular distance to pivot
       = 500 N × 5.3 m       1 mark
       = 2650 N m            1 mark

Only the horizontal component of T has a turning effect; its moment must counterbalance that of the 500 N horizontal force.

(b) Calculate the value of the force T. [3]

The horizontal component of the force T must also have a moment of 2650 N m:       1 mark

Having written down the physical principle, the next step is to write it in the form of an equation. Tcos60° or Tsin30° is the horizontal component of T.

Tcos60° × 3.5 m = 2650 N m       1 mark

The final step is to calculate the value of T.
Although it is not essential to show the units of physical quantities at each step in the calculation, it is good practice and you MUST include the correct unit with your answer to each part.

T = 2650 N m ÷ (cos60° × 3.5 m) = 1514 N       1 mark

This problem cannot be solved by resolving T and the 500 N force as there is a force at X (you can see that this is not just a vertical reaction force by taking moments about Y – there must be a horizontal component with a moment that balances the horizontal component of T).

## Practice examination questions

Take the value of free-fall acceleration, *g*, to be 10 m s⁻².

**1** A microlight aircraft heads due north at a speed through the air of 18 m s⁻¹. A wind from the west causes the aircraft to move east at 6 m s⁻¹.
Draw a vector diagram and use it to work out the resultant velocity of the aircraft. [3]

**2** The diagram shows a light fitting held in place by two cables.

(a) Calculate the vertical component of the 450 N force. [2]

(b) Write down the weight of the light fitting. [1]

(c) Explain why a vector diagram that represents the forces on the light fitting is a closed triangle. [2]

(d) Draw the vector diagram and use it to find the value of the force *S*. [2]

**3** The diagram shows a garden tool being used to cut through a branch.

(a) Calculate the moment of the force on each handle. [2]

(b) Assuming equilibrium, calculate the value of the resistive force, *F*, that acts on each blade. [2]

(c) Suggest why a similar tool designed to cut through thicker branches has longer handles. [2]

4 The diagram shows the base of a steel girder that supports the roof of a building. The downward force in the girder is 4500 N and it is fixed to a square steel plate, each side of which is 0.30 m long.

(a) Calculate the values of the horizontal and vertical components of the force in the girder. [2]

(b) Calculate the pressure that the steel plate exerts on the ground. [3]

(c) Explain why the girder is fixed to a steel plate rather than being fixed directly to the ground. [2]

5 A large aquarium has a horizontal cross-section of 10.5 m × 4.2 m. It is filled with water to a depth of 3.5 m. The density of water is 1000 kg m⁻³.

(a) Calculate the weight of the water in the aquarium. [3]

(b) Calculate the pressure on the base of the aquarium due to the weight of the water. [3]

(c) Explain why the glass used for the sides of the aquarium needs to be as strong as the base. [2]

(d) A similar aquarium has a horizontal cross-section of 10.5 m × 2.1 m and is filled with water to the same depth.
How does the pressure on the base compare to that calculated in b?
Give the reason for your answer. [2]

6 (a) How does a ductile material behave differently from a brittle one when subjected to a stretching force? [2]

(b) The graph shows the relationship between the stress on a sample of cast iron and the strain that it causes. The curve ends when the sample breaks.

(i) Write down the value of the breaking stress of the cast iron. [1]
(ii) For what range of stresses does the cast iron follow Hooke's law? [1]
(iii) Calculate the Young modulus for the cast iron in the region where it follows Hooke's law. [2]
(iv) A second sample of the same material has twice the diameter of the original sample.
Explain whether the stress–strain graph for this sample would be the same as that shown in the diagram. [2]

## Practice examination questions (continued)

7   The diagram is a velocity–time graph for a train travelling between two stations.
A positive velocity represents motion in the forwards direction.

(a)   Calculate the acceleration of the train during the first 80 s.          [3]

(b)   During which other times shown on the graph is the train increasing
speed?                                                                [1]

(c)   For what length of the time was the train:
(i)   not moving?                                                     [1]
(ii)  travelling at a constant non-zero speed?                        [1]

(d)   Calculate the total distance travelled by the train.                   [2]

(e)   At the end of the time shown on the graph, how far was the train from
its starting point?                                                  [1]

8   In a game of cricket, a ball leaves a bat in a horizontal direction from a height
of 0.39 m above the ground.
The speed of the ball is 35 m s$^{-1}$.
(a)   Calculate the time interval between the ball leaving the bat and reaching the
ground.                                                              [3]

(b)   What horizontal distance does the ball travel in this time?            [2]

(c)   Fielders are not allowed within 15 m of the bat.
What is the maximum speed that the ball should leave the bat, travelling
in a horizontal direction, for the player not to be caught out?       [2]

9   A motorist travelling at the legal speed limit of 28 m s$^{-1}$ (60 mph) takes his
foot off the accelerator as he passes a sign showing that the speed limit is
reduced to 14 m s$^{-1}$ (30 mph). The car decelerates at 2.0 m s$^{-2}$.
(a)   For what time interval is the motorist exceeding the speed limit?      [2]

(b)   How far does the car travel in that time?                              [2]

10  A fountain is designed so that the water leaves the nozzle and rises vertically to a
height of 3.5 m.
(a)   Calculate the speed of the water as it leaves the nozzle.              [3]

(b)   For how long is each drop of water in the air?                         [2]

# 1.6 Force and acceleration

**After studying this section you should be able to:**

- *recall and use the relationship between resultant force, mass and acceleration*
- *explain the principles of seat belts, air bags and crumple zones*
- *interpret and explain data relating to vehicle stopping distances*

**LEARNING SUMMARY**

## Getting going

| | | | |
|---|---|---|---|
| AQA A | M2 | NICCEA | M1 |
| AQA B | M1 | OCR A | M1 |
| EDEXCEL A | M1 | OCR B | M2 |
| EDEXCEL B | M1 | WJEC | M1 |

Changing motion requires a force. Starting, stopping, getting faster or slower and changing direction all involve forces. One way to start an object moving is to release it and allow the Earth to pull it down. The Earth's pull causes the object to accelerate downwards at the rate of $10 \text{ m s}^{-2}$.

> The force that pushes a cycle or other vehicle forwards is called the driving force, or motive force.

To set off on a cycle, you push down on the pedal. The chain then transmits this force to the rear wheel. The wheel pushes on the road and, provided that there is enough friction to prevent the wheel from slipping, the push of the road on the wheel accelerates the cycle forwards.

However, as every cyclist knows, this acceleration is not maintained. The cyclist eventually reaches a speed at which the driving force no longer causes the cycle to accelerate. This is due to the **resistive** forces acting on the cycle. The main one is **air resistance**, although other resistive forces act on the bearings and the tyres.

Air resistance or **drag** also affects the motion of an object released and allowed to fall vertically; in fact it opposes the motion of anything that moves through the air. This is due to the air having to be pushed out of the way. The faster the object moves, the greater the volume of air that has to be displaced each second and so the greater the resistive force.

The diagram shows the directions of the forces acting on a cyclist travelling horizontally and a ball falling vertically.

> The single arrow drawn acting on the front of the cyclist represents all the resistive forces acting against the motion.

The effect of the resistive forces is to reduce the size of the resultant, or unbalanced, force that causes the acceleration. How this affects the speed of the cycle and ball is shown by the graph.

The decreasing gradient of the speed–time graph shows that the acceleration decreases as the resultant force becomes smaller. At a speed where the resistive force is equal to the driving force, the resultant force is zero and the object travels at a constant speed.

> When the forces acting on a moving object are balanced, the constant speed is called its terminal speed or terminal velocity.

To summarise:
- an object accelerates if there is a resultant, or unbalanced, force acting on it
- the acceleration is proportional to the resultant force and in the direction of the resultant force.

## What else affects the acceleration?

| AQA A | M2 | NICCEA | M1 |
|-------|----|--------|-----|
| AQA B | M1 | OCR A | M1 |
| EDEXCEL A | M1 | OCR B | M2 |
| EDEXCEL B | M1 | WJEC | M1 |

> Galileo's insistence that heavy objects do not fall faster than lighter ones cost him his job as professor of mathematics at the University of Pisa.

Galileo appreciated that heavy objects have the same free-fall acceleration as lighter ones, provided that they are not affected by air resistance. You can confirm this by releasing two different-sized coins side-by-side; at low speeds, air resistance has little effect so you see and hear them reaching the ground together.

In vertical motion the **mass** of an object affects the force causing the acceleration; the heavier the object, the greater the force pulling it down. Doubling the mass doubles the pulling force (since $W = m \times g$) **but does not affect the acceleration.**

This is not the case in horizontal motion, where increasing the mass of an object has no effect on the force causing it to accelerate, but does affect the acceleration. As the number of passengers on a bus increases, its acceleration away from the bus stop decreases.

> If the mass of an object is doubled, its acceleration is halved for the same pulling force.

Results of experiments using a constant force to accelerate different masses show that:

the acceleration of an object is inversely proportional to its mass

> The symbol Σ, meaning 'sum of' is used here to emphasise that the relationship applies to the resultant force on an object, and not to individual forces.

The dependence of acceleration on both the resultant force and the mass is summarised by the relationship:

> **KEY POINT**
>
> resultant force = mass × acceleration
> $$\Sigma F = ma$$
> Where the unit of force, the newton, is defined as the force required to cause a mass of 1 kg to accelerate at 1 m s$^{-2}$.

## Forces and braking

| OCR A | M1 |
|-------|-----|

Vehicles that travel on wheels rely on friction between the tyres and the road surface to provide the driving force. Ice on a road surface and wet leaves on railway lines can reduce the friction force, resulting in the wheels spinning round as they slide over the surface.

Friction is also used in braking systems to bring cycles and motor vehicles to a halt. Two common types of brake used are **drum brakes** and **disc brakes**. These are shown in the diagram.

> Most modern vehicles use disc brakes because of their greater efficiency.

disc and drum brakes

> A deceleration is a negative acceleration. In this case the acceleration is in the opposite direction to the vehicle's velocity, so it has the opposite sign.

When the driver applies the brakes, high friction material in the shoe or pad is pressed against the drum or disc. The friction force exerts a torque in the opposite direction to the torque that drives the wheel. During braking the tyre pushes **forwards** on the road surface, causing the road surface to push **backwards** on the tyre. The resultant force in the backwards direction causes the vehicle to decelerate.

the braking force on a wheel

Braking forces, like driving forces, rely on friction. The size of the friction force depends on the nature of both the road and the tyre surface. The greater the friction force, the shorter the distance required to stop the vehicle.

Grand prix drivers match their tyres to the road conditions. Excessive friction wastes fuel and reduces the acceleration and top speed. However, having fitted tyres suitable for dry road conditions, a sudden shower of rain can suddenly send the cars out of control!

Tyre **tread** is designed to prevent a layer of water building up between the tyre and the road. This reduces the friction force, reducing the effectiveness of steering, drive and braking. If there is insufficient tread on a tyre it cannot push the water out of the way quickly enough, as the channels in the tyres are too small for all the water to pass through.

## Car safety

OCR A    M1

If there is not enough friction, a passenger appears to be 'thrown' in the opposite direction to the change in velocity.

When the velocity of a car changes, that of the people inside it has to change too. During normal driving, the force to accelerate the driver and passengers comes from the seat; it pushes the person forwards to cause an increase in speed and the friction force between the person and the seat is sufficient to slow the person down or cause a change in direction.

During hard braking or in a collision, friction is not enough to match the car's deceleration. Viewed from the outside of the car, a passenger appears to have been thrown forwards. Despite what the television advertisements say, this is **not** the case! The photograph shows what can happen to a passenger in a collision. **Seat belts** prevent this from happening. However, a rigid seat belt that caused a passenger to decelerate at the same rate as the car could prove fatal in a head-on collision.

To be effective a seat belt should:
- restrain the passenger and prevent collisions with the inside of the car
- allow the passenger to come to rest over a longer time period than that taken by the car.

Excessive force from a seat belt can break the ribs and damage internal organs.

To achieve these conflicting aims, seat belts are designed so that they stretch sufficiently to allow the passenger to carry on moving for a short time after the car has stopped, but not so much that would result in the passenger hitting the windscreen or the seat in front.

There is less space in front of the driver than there is in front of the passengers because of the steering wheel. In modern cars this is designed to collapse on impact.

Seat belts can still inflict severe injuries during a collision. The amount of space to allow stretching, particularly in front of a driver, is limited so the restraining force from the belt can be large. There is also a design conflict in deciding on the width of the belt; wide belts exert less pressure than narrow ones but are less comfortable to wear.

49

**Airbags** are designed to restrain a driver and passengers without any risk of causing physical damage to the person. The bag only operates during a rapid deceleration such as a head-on collision. This releases a gas which causes the bag to inflate and surround the driver or passenger like a cushion. The force exerted on the person is similar to that from a seat belt, but because it is over a much larger area the pressure is smaller.

Modern cars, and railway carriages, are also designed so that parts of them crush on impact. Although the part of the vehicle containing passengers is rigid to give them protection, the front and rear are crumple zones.

the crumple zones on a car

crumple zone

rigid passenger cage

crumple zone

As the name implies, the crumple zone should squash during a collision. This has two effects:

- it increases the time it takes to stop the vehicle, reducing the deceleration and the force
- it absorbs much of the kinetic energy of the moving vehicle, so that it does not bounce backwards, causing further injury to passengers.

## Forces and towing

OCR A  M1

The term 'acceleration' is used here with the everyday meaning, i.e. speeding up.

Towing a caravan or trailer affects the steering, acceleration and braking of a vehicle.

When accelerating, there is more mass to be accelerated, so the maximum acceleration is reduced. The force that accelerates the towed vehicle is transmitted through the tow bar, which is in tension. As well as pulling forwards on the towed vehicle, the tow bar also pulls backwards on the car, reducing the resultant force that causes acceleration.

Tension in the tow bar is maintained when the car and towed vehicle are travelling at a steady speed, but now the force is reduced, being equal to the resistive force acting on the towed vehicle.

During braking, the car pushes back on the vehicle being towed, which in turn pushes forwards on the car. This reduces the car's deceleration, so braking distances are increased. The diagram shows the forces in the tow bar when the car and vehicle being towed are accelerating and braking.

acceleration

braking

caravan pulls car

car pulls caravan

caravan pushes car

car pushes caravan

## Stopping distances

OCR A  M1

In the highway code there is a chart that shows the stopping distances in good conditions for a car travelling at different speeds. There are two components to the stopping distance at a particular speed, thinking distance and braking distance.

Thinking distance is how far the vehicle travels while the driver is reacting. The two factors that determine this are:
- reaction time
- vehicle speed.

A driver's reaction time depends on a number of things, including the state of the driver and the nature of the hazard, but a driver should be able to react to an unexpected hazard within 0.6 s.

During the driver's reaction time the vehicle continues at a constant speed, so distance travelled = speed × time. Assuming a constant reaction time, this means that:

**thinking distance is proportional to speed.**

This is not the case with braking distance. As with reaction time, a number of factors could affect this, including the nature and condition of the road surface, the tyres and the brakes. The following calculations to work out the braking distances from speeds of 10 m s⁻¹ and 20 m s⁻¹ assume a deceleration of 8.3 m s⁻².

| initial speed/m s$^{-1}$ | 10 | 20 |
|---|---|---|
| time taken to brake/s | 1.2 | 2.4 |
| average speed during braking/m s$^{-1}$ | 5 | 10 |
| braking distance/m = average speed × time | 6 | 24 |

This shows that doubling the speed quadruples the braking distance. Not only is the average speed during braking doubled, so is the time it takes to stop, so:

**braking distance is proportional to (speed)².**

The diagram shows how stopping distance is related to speed:

The overall effect of speed on stopping distance is approximately 'doubling the speed trebles the stopping distance'.

## Progress check

1 a The total mass of a cyclist and her cycle is 85 kg.
Calculate the acceleration as she sets off from rest if the driving force is 130 N.

  b The driving force remains constant at 130 N. What is the size of the resistive force when she is travelling at terminal velocity?

2 A fully-laden aircraft weighs 300 000 kg. It accelerates from rest to a take-off speed of 70 m s⁻¹ in 20 s.

  a Calculate the acceleration of the aircraft.

  b Calculate the size of the resultant force needed to cause this acceleration.

  c Explain why the force from the engines must be greater than the answer to **b.**

3 A car is travelling at the legal maximum speed of 28 m s⁻¹ on a single carriageway. The driver notices a hazard in the road ahead.

  a If the driver's reaction time is 0.5 s and the car decelerates at 7 m s⁻² during braking, calculate the stopping distance.

  b Calculate the size of the force required to decelerate the car if its total mass is 850 kg.

1 a 1.53 m s⁻²
  b 130 N
2 a 3.5 m s⁻²
  b 1.05 × 10⁶ N
  c the force also has to act against air resistance and friction.
3 a 70 m
  b 5950 N

# 1.7 Newton I and III

**After studying this section you should be able to:**

- *identify the forces acting on an object that is not changing its velocity*
- *for a given force, describe the other force that makes up the third law 'pair'*

**LEARNING SUMMARY**

### Newton's laws of motion

With his three laws of motion, Newton thought that he could describe and predict the movement of every object in the Universe. Like many other 'laws' in physics, they give the right answers much of the time. The paths of planets, moons and satellites are all worked out using Newton's laws.

The ideas behind the three laws have already been used informally in this chapter. This section looks at formal statements of the first and third laws and how they apply to everyday situations.

## The first law

| | | | |
|---|---|---|---|
| AQA A | M2 | NICCEA | M1 |
| AQA B | M1 | OCR A | A2 |
| EDEXCEL A | M1 | OCR B | M1 |
| EDEXCEL B | M1 | WJEC | A2 |

TEXT M2
TEXT M1
TEXT A2

The phrase 'uniform motion' means moving in a straight line at constant speed; i.e. moving at constant velocity.

> **KEY POINT**
>
> Newton's first law states that:
>
> *An object maintains its state of rest or uniform motion unless there is a resultant force acting on it.*

At first, this seems contrary to everyday experience. Give an object a push and it slows down before coming to rest. This is not 'continuing in a state of uniform motion'. Newton realised that in this case the unseen resistive forces of friction and air resistance together act in opposition to the motion; as there is no longer a driving force after the object has been pushed, there is a resultant backwards-directed force acting on it.

friction and air resistance create a resultant force on this sliding object

Motion without resistive forces is difficult to achieve on Earth: there is always air resistance or friction from a surface that a moving object rests on. The nearest that we can get to modelling motion with no resistive forces is to study motion on ice or an air track.

The converse of Newton's first law is also true:

> if an object is at rest or moving at constant velocity the resultant force on it is zero

The closed figure is a triangle if there are three forces acting; a quadrilateral for four forces, etc.

This means that where there are two forces acting on an object that satisfies these conditions, they must be equal in size and opposite in direction. If there are three or more forces acting then the vector diagram is a closed figure.

A normal reaction force acts on all four wheels; the arrow shown on the diagram represents the sum of these forces.

You have already studied examples of a vehicle moving at constant velocity where the driving force and resistive force are equal in size and act in opposite directions. The diagram shows an example of a situation where three forces sum to zero; it illustrates a vehicle parked on a hill at rest and a vector triangle that shows the forces.

## The third law

| | | | |
|---|---|---|---|
| AQA A | M2 | NICCEA | M1 |
| AQA B | M1 | OCR A | M1 |
| EDEXCEL A | M1 | OCR B | M1 |
| EDEXCEL B | M1 | WJEC | M1 |

The laws were originally written in Latin. The third law is directly translated as 'to every action there is an equal and opposite reaction'. This was widely misinterpreted as meaning that two equal and opposite forces act on the same object, resulting in zero acceleration.

This is both the simplest in its statement and the most misunderstood of the three laws. The statement given here is not a direct translation of the original, but it helps to remove some of the misunderstanding.

> **KEY POINT**
>
> Newton's third law can be stated as:
>
> If object A exerts a force on object B, then B exerts a force equal in size and opposite in direction on A.

According to the third law, forces do not exist individually but in pairs. However, it is important to remember that:

- the forces are of the same type, i.e. both gravitational or electrical
- the forces act on different objects
- the third law applies to all situations.

## Some examples

| | | | |
|---|---|---|---|
| AQA A | M2 | NICCEA | M1 |
| AQA B | M1 | OCR A | M1 |
| EDEXCEL A | M1 | OCR B | M1 |
| EDEXCEL B | M1 | WJEC | M1 |

Pluto and its moon Charon are so close together in terms of their mass that they spin round like a pair of dancers, each of them orbiting a point in space between the planet and its moon.

This type of situation is where the greatest misunderstanding of Newton's third law occurs. Remember, third law pairs of forces act on **different** objects.

If you step off a wall the Earth pulls you down towards the ground. According to the third law, you also pull the Earth up with an equal-sized force. So why doesn't the Earth accelerate upwards to meet you instead of the other way round? The answer is it does, but if you apply $F = ma$ to work out the Earth's acceleration, it turns out to be minimal.

the Earth pulls the person

the person pulls the Earth

Charon pulls Pluto

Pluto pulls Charon

Newton's third law: pairs of forces that are equal in size and opposite in direction.

Planets pull moons, and according to the third law, moons pull planets with an equal-sized force. Why does the moon go round the planet instead of the other way round?

Again, the answer is, it does. In fact, they both rotate around a common centre of mass. In the case of the Earth and its Moon, the centre of mass is so close to the centre of the Earth that to all intents and purposes the Moon orbits the Earth.

The diagram shows another pair of forces that are equal in size and opposite in direction. This is an application of the first law – the vase is in a state of rest so there is no resultant force acting on it.

the Earth pulls the vase

the table pushes the vase

These forces are also equal in size and opposite in direction – an application of Newton's first law, not third.

## Progress check

1 A tightrope walker stands in the middle of a rope.
   a What exerts the upward force on the tightrope walker?
   b Explain why it is not possible for the rope to be perfectly horizontal.

2 Explain how you can tell from a vector diagram that an object is in a state of rest or uniform motion.

3 Identify the other forces that make up the third law pairs in the diagram of the vase above.

3 The vase pulls the Earth.
The vase pushes the table.

2 The arrows that represent the vectors form a closed figure; i.e. the last arrow finishes where the first one starts.

1 a The tension in the rope.
   b A rope can only pull in its own direction. For the tension to have a vertical component, the rope must be inclined to the horizontal.

# 1.8 Momentum and the second law

**After studying this section you should be able to:**

- apply the principle of conservation of momentum to collisions in one dimension
- state the relationship between the change of momentum of an object and the resultant force on it
- explain the significance of the impulse of a force

**LEARNING SUMMARY**

## Momentum

| | | | |
|---|---|---|---|
| AQA A | M2 | NICCEA | M1 |
| AQA B | A2 | OCR A | A2 |
| EDEXCEL A | M1 | OCR B | A2 |
| EDEXCEL B | A2 | WJEC | A2 |

A collision between two objects need not involve physical contact. Imagine two positive charges approaching each other; the repulsive forces could cause them to reverse in direction without actually touching.

Which is more effective at demolishing a brick wall, a 1 kg iron ball moving at 50 m s⁻¹ or a 1000 kg iron ball moving at 1 m s⁻¹? Both the mass and the velocity need to be taken into account to answer this question.

Newton realised that what happens to a moving object involved in a collision depends on two factors:

- the mass of the object
- the velocity of the object.

He used the concept of momentum to explain the results of collisions between objects.

> **KEY POINT**
>
> momentum = mass × velocity
> $$p = mv$$
> Momentum is a vector quantity, the direction being the same as that of the velocity. It is measured in N s or kg m s⁻¹; these units are equivalent.

A ten-pin bowling ball has a mass of several kilogrammes, so even at low speeds it has much more momentum than a tennis ball (mass 0.06 kg) travelling as fast as you can throw it.

Everything that moves has momentum and exerts a force on anything that it interacts with. This applies as much to the light that is hitting you at the moment as it does to a collision between two vehicles.

No matter how hard you throw a tennis ball at a garden wall you will not knock the wall down; nor will you be very successful if you use a tennis ball to play ten-pin bowling!

## Conservation of momentum

| | | | |
|---|---|---|---|
| AQA A | M2 | NICCEA | A2 |
| AQA B | A2 | OCR A | A2 |
| EDEXCEL A | M1 | OCR B | A2 |
| EDEXCEL B | A2 | WJEC | A2 |

The phrase 'equal and opposite' is used as a shorthand way of writing 'equal in size and opposite in direction'.

When a force causes an object to change its velocity, there is also a change in momentum. When two objects interact or collide, they exert equal and opposite forces on each other and so the momentum of each one changes. This is illustrated below where the blue ball loses momentum and the black ball gains momentum.

| before | | after | |
|---|---|---|---|
| 5.0 m s⁻¹ | 3.0 m s⁻¹ | 4.2 m s⁻¹ | 6.2 m s⁻¹ |
| 0.4 kg | 0.1 kg | | |

Like the forces they exert on each other during the collision, the changes in momentum of the balls are *equal in size and opposite in direction*.

It follows that:

- the combined momentum of the balls is the same before and after they collide.

This is an example of the principle of conservation of momentum.

To satisfy the conditions for conservation of momentum to apply, the only forces acting on the objects must be the ones between the objects themselves.
This is why conservation of momentum is usually demonstrated on an air track or other low friction surface.

The principle of conservation of momentum states that:
When two or more objects interact the total momentum remains constant provided that there is no external resultant force.

In the context of the colliding balls, an external resultant force could be due to friction or something hitting one of the balls. Either of these would result in a change in the total momentum.

## Different types of interaction

### Objects moving in opposite directions

When two objects approach each other and collide head-on, the result of the collision depends on whether they stick together or whether one or both rebound.

The diagram shows two 'vehicles' approaching each other on an air track. The vehicles join together and move as one.

To make two air track vehicles stick together, one has some Plasticine stuck to it and the other is fitted with a pin that sticks into the Plasticine.

0.40 m s$^{-1}$  →    0.30 m s$^{-1}$ ←    velocity = ?

0.30 kg        0.20 kg              0.50 kg

**before**                         **after**

The principle of conservation of momentum can be used to work out the combined velocity, $v$, after the collision.

Taking velocity from left to right as being positive:

total momentum before collision =
 0.3 kg × +0.4 m s$^{-1}$ + 0.2 kg × –0.3 m s$^{-1}$ = 0.06 N s.

This must equal the momentum after the collision, so 0.5 kg × $v$ = 0.06 N s,

i.e. $v$ = 0.12 m s$^{-1}$. As this is positive, it follows that the 'vehicle' is moving from left to right.

### Rebound

When a light object collides with a heavier one, the light object may rebound, reversing the direction of its momentum. The diagram shows an example of such a collision.

A rebound collision can be modelled on an air track by fitting the vehicles with repelling magnets.

**before**                              **after**

10 m s$^{-1}$ →      0 m s$^{-1}$        $v$ →        3.3 m s$^{-1}$ →

2 kg              10 kg

The figures here are rounded. The 3.33 is a rounded value of 3⅓.

Application of the principle of conservation of momentum in this case gives:

2 kg × 10 m s$^{-1}$ + 10 kg × 0 m s$^{-1}$ = 2 kg × $v$ + 10 kg × 3.33 m s$^{-1}$

so 2 kg × $v$ = 20 Ns – 33.3 Ns  i.e. $v$ = –6.7 m s$^{-1}$

The negative value means the object is moving from right to left.

### Recoil

In some situations two objects are stationary before they interact, having a combined momentum of zero. After they interact, both objects move off but the combined momentum is still zero. Examples of this include:

- a stationary nucleus undergoes radioactive decay, giving off an alpha particle
- a person steps off a boat onto the quayside
- two ice skaters stand facing each other; then one pushes the other
- two air track vehicles fitted with repelling magnets are held together and then released.

In each case, both objects start moving from rest. For the combined momentum to remain zero, each object must gain the same amount of momentum but these

push   0.6 m s$^{-1}$   $v = ?$

In recoil situations, each object gains the same amount of momentum, but in opposite directions. It follows that the one with the smaller mass ends up with the faster speed.

momentum gains must be in *opposite directions*. The ice skater who does the pushing **recoils** as an equal and opposite force is exerted by the one who is pushed.

In the example shown in the above diagram, an 80 kg man pushes a 50 kg child. Their combined momentum is zero both before and after this interaction.

Taking velocity from left to right as positive, this means that:

80 kg × − 0.6 m s$^{-1}$ + 50 kg × $v$ = 0

so $v$, the final velocity of the child, is 0.96 m s$^{-1}$ to the right.

## Elastic or inelastic

| | | | |
|---|---|---|---|
| AQA A | M2 | OCR A | A2 |
| EDEXCEL B | A2 | WJEC | A2 |
| NICCEA | A2 | | |

In all three examples above, both momentum and energy are conserved. This is always the case. A collision where the total kinetic energy before and after the interaction is the same is called an **elastic** collision. Where there is a gain or loss of kinetic energy as a result of the interaction the collision is called **inelastic**.

> in an elastic collision: momentum, kinetic energy and total energy are conserved
> in an inelastic collision: momentum and total energy are conserved, but the amount of kinetic energy changes

Of the examples given, only the second is an elastic collision.

## Momentum and the second law

| | | | |
|---|---|---|---|
| AQA A | M2 | NICCEA | M1 |
| AQA B | A2 | OCR A | A2 |
| EDEXCEL A | M1 | OCR B | A2 |
| EDEXCEL B | A2 | WJEC | A2 |

Newton's second law establishes the relationship between the resultant force acting on an object and the change of momentum that it causes. Important results that follow from the second law include $F = ma$, studied in force and acceleration, and the **impulse of a force**, to be met later in this section.

The expression $\Delta p/\Delta t$ can be read as: change in momentum ÷ time taken for the change.

> **KEY POINT**
>
> Newton's second law states that:
> The rate of change of momentum of an object is proportional to the resultant force acting on it and acts in the direction of the resultant force.
> $$\Delta p/\Delta t \propto F$$

The definition of the newton (see force and acceleration) fixes the proportionality constant at 1. This enables the second law to be written as:

> **Force = rate of change of momentum**
> $$F = \Delta p / \Delta t$$
> **KEY POINT**

In this form, the second law is useful for working out the force in situations such as jet and rocket propulsion where the change in momentum each second can be calculated easily.

A rocket carries its own oxygen supply, so that it can fire the engines when it is travelling in a vacuum. A spacecraft travelling where it is not affected by any gravitational fields or resistive forces maintains a constant velocity, so it only needs to fire the rocket engines to change speed or direction.

The principle of rocket propulsion can be seen by blowing up a balloon and letting it go; the air squashed out of the neck of the balloon gains momentum as it leaves, causing the balloon and the air that remains inside to gain momentum **at the same rate but in the opposite direction**. The diagram shows the principle of rocket propulsion.

*Jet and rocket engines use the same principle; hot gases are ejected from the back of the engine to provide a force in the forwards direction. The difference is that jet engines are designed to operate within the Earth's atmosphere, so they can take oxygen from their surroundings.*

exhaust gases gain momentum in this direction

rocket gains momentum in this direction

conservation of momentum and rocket propulsion

*The unit kg m s$^{-2}$ is equivalent to the N, so a rate of change of momentum of 4 500 000 kg m s$^{-2}$ is another way of stating: a force of 4 500 000 N.*

In this example, the exhaust gases are emitted at a rate of 3000 kg s$^{-1}$. The speed of the exhaust gases relative to the rocket is 1500 m s$^{-1}$. The rate of change of momentum of these gases is therefore 4 500 000 kg m s$^{-2}$, i.e. 4 500 000 kg m s$^{-1}$ each second. This is the size of the force on both the gases and the rocket.

## Impulse

| AQA B | A2 | NICCEA | M1 |
| EDEXCEL A | M1 | OCR B | A2 |
| EDEXCEL B | A2 | | |

*A modern jet engine is capable of producing a force of 1 MN (1 × 10$^6$ N). Four of these engines are used to accelerate a jumbo jet to the speed required for take-off.*

Jet and rocket engines are often working over a long period of time, so the second law written as **force = rate of change of momentum** can be readily applied when the time for which the force acts is not a fixed quantity. A manufacturer of aircraft engines, for example, uses the maximum force required to design the engine to achieve the required rate of output of exhaust gases.

In situations where the time for which the force acts is known, or is short, the concept of **impulse** can be more readily applied. The impulse of a force is the change in momentum that it causes. Using the concept of impulse, the second law can be written as:

> **impulse** = force × time
> = change in momentum
> Impulse, like momentum, is measured in the equivalent units of N s or kg m s$^{-1}$.
> **KEY POINT**

*When a child is taught how to catch a hard ball, he is shown how to pull back the hands while catching the ball. This extends the time required to stop the ball, so a smaller force is exerted on the hands.*

When tennis players or snooker players hit a ball, they are aware that the effect of the force that they exert depends on the time of contact with the ball as well as the size of the force. By maintaining contact with the ball over a longer period of time, a greater change in momentum can be caused by a given force.

The change in momentum of a ball during a collision is called the impulse

In the example of a ball hitting a wall and rebounding, the force is not constant. It becomes greater as the ball becomes more compressed and decreases again as the ball returns to its original shape. The diagram below shows how the force on a ball changes during a collision with a wall.

Remember, the area between the curve on a graph and the x-axis represents the product of the quantities plotted on the x- and y- axes.

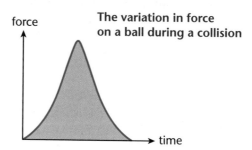

**The variation in force on a ball during a collision**

The impulse, or change in momentum of the ball, is represented by the area between the curve and the time axis, shown shaded on the graph.

In practice sessions, long-jumpers and high-jumpers often land on a bed made of pieces of foam rubber. This gives a long 'stopping' time, so there is no chance of physical injury due to the impulse as they land.

Although the concept of impulse is often used to describe the effect of a force that acts over a short time, it can be applied to any interaction between objects. When a pole-vaulter lands on the ground, her momentum is typically 400 to 500 N s. This is the size of the impulse needed to stop her. On a hard concrete surface she could be brought to rest in a tiny fraction of a second, needing a back-breaking force. As **impulse = force × time**, the same change in momentum can be achieved by a smaller force acting over a longer time.

The cushions lengthen the time of the impulse, so a small force is exerted.

## Progress check

1   In an air track collision, a 0.10 kg vehicle travelling at 0.20 m s$^{-1}$ from left to right approaches a 0.30 kg vehicle travelling at 0.30 m s$^{-1}$ in the opposite direction. After the head-on collision, the vehicles stick together. Calculate their combined velocity.

2   A rocket burns fuel at the rate of 3000 kg each second. The exhaust gases have a speed, relative to the rocket, of 2200 m s$^{-1}$.
   a   Calculate the rate of change of momentum of the exhaust gases.
   b   The mass of the rocket is 500 000 kg. Calculate the acceleration while the fuel is being burned. Assume that no other forces act on the rocket.
   c   Suggest why the acceleration of the rocket increases as more fuel is burned.

3   A tennis ball with no initial horizontal velocity leaves a tennis racquet with a horizontal speed of 58 m s$^{-1}$. The mass of the ball is 0.060 kg.
   a   Calculate the impulse given to the ball.
   b   If the time of contact with the racquet is 0.025 s, calculate the average force exerted on the ball.

# 1.9 Energy to work

*After studying this section you should be able to:*

- *identify situations where a force is working*
- *state and use the relationships between force, distance, energy, work, power and time*
- *calculate the efficiency of an energy transfer*

**LEARNING SUMMARY**

## Working

| | | | |
|---|---|---|---|
| AQA A | M2 | NICCEA | A2 |
| AQA B | A2 | OCR A | M1 |
| EDEXCEL A | M1 | OCR B | M2 |
| EDEXCEL B | M1 | WJEC | A2 |

Every event requires **work** to make it happen. Even if you relax and try to do nothing, your body is working just to keep you alive. Any force that causes movement is doing work. Pushing a supermarket trolley is working, as is throwing or kicking a ball. However, holding some weights above your head is not working; it may cause your arms to ache, but the force on the weights is not causing any movement!

How much work a force does depends on:

- the size of the force
- the direction of movement
- the distance that an object moves.

> The phrase, 'distance moved in the direction of the force', is used here instead of 'displacement' to emphasise that the force and displacement it causes are measured in the same direction.

**KEY POINT**

The work done by a force that causes movement is defined as:

work = average force × distance moved in the direction of the force

$$W = F \times s$$

Work is measured in joules (J) where 1 joule is the work done when a force of 1 N moves its point of application 1 m in the direction of the force.

Here are some examples of forces that are working and forces that are not working.

The shelf is not moving, so no work is being done.

The tension in the string is not working as there is no movement **in the direction of the force.**

> A pylon that supports electricity transmission cables is not working, but a wind that causes the cables to move is!

This force causes movement **in the direction of the force.**

The horizontal component of the force on the log is doing the work here.

> In circular motion, the direction of the velocity is along a tangent to the circle.

In the case of the stone being whirled in a horizontal circle, the stone's velocity is always at **right angles** to the force that maintains the motion, so the force is not working.

Although the force pulling the log and the movement it causes are not in the same direction, the force on the log has a component in the same direction as the log's movement. In calculating the work done the horizontal component of the force is multiplied by the horizontal distance moved by the log.

## Work and energy transfer

| | | | |
|---|---|---|---|
| AQA A | M2 | NICCEA | A2 |
| AQA B | A2 | OCR A | M1 |
| EDEXCEL A | M1 | OCR B | M2 |
| EDEXCEL B | M1 | WJEC | A2 |

The physical processes of working and energy transfer are inseparable. Work requires an energy source, for example, a fuel, an electric cell or a wound-up spring. As the work is done, energy leaves the source and ends up elsewhere, this is what is meant by energy transfer.

> An object has energy if it can exert a force that causes movement of the point of application of the force.

> **KEY POINT**
> Energy is the ability to do work.
> The energy transfer from a source is equal to the amount of work done.
> Like work, energy is measured in joules.

Here are some examples of everyday energy transfers.

### A bus accelerates away from a bus stop and then maintains a steady speed

> A common error is to describe the energy transfer of a vehicle moving at a constant speed as 'energy from the fuel and oxygen is transferred to kinetic energy'.

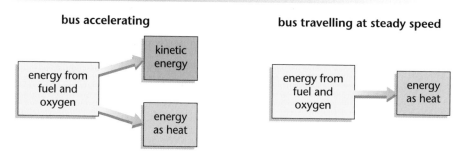

As the bus speeds up, some of the energy from the fuel and oxygen is transferred to the kinetic energy of the bus, i.e. the energy it has due to its movement. The rest of the energy is transferred to the surroundings. This takes place in a number of ways: the exhaust gases transfer energy as heat directly into the surrounding air and some energy is also transferred to heat wherever there are resistive forces.

Once the bus is maintaining a steady speed there is no increase in its kinetic energy, so all the energy transferred from the fuel and oxygen is going to the surroundings.

### An electric motor lifts a load at a steady speed

> The phrase 'gravitational potential energy' is often abbreviated to 'gpe' or just 'potential energy', although strictly 'potential energy' can be any form of stored energy, for example in a spring or a fuel.

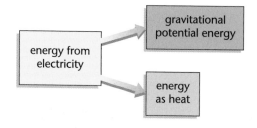

The load gains gravitational potential energy as it is lifted, i.e. the energy it has due to its position above the Earth's surface. All the movement is at a constant speed, so there is no change in the kinetic energy of the system. As with the bus travelling at a steady speed, work done against resistive forces causes heating, and this energy is transferred as heat to the surroundings.

## Energy and a filament lamp

| | | | |
|---|---|---|---|
| AQA A | M2 | OCR A | A2 |
| EDEXCEL A | M1 | WJEC | A2 |

When a filament lamp is switched on, some energy is absorbed by the filament as it reaches its operating temperature. This takes a fraction of a second. Some energy is absorbed by the glass envelope and the gas inside. These take a little longer to reach a steady temperature, perhaps a few seconds.

Once the lamp has reached its steady state, of the 60 J of energy passing into the lamp each second from the electricity supply, typically 3 J passes out as light. This is shown in the diagram. The Sankey diagram gives a visual indication of the relative proportions of energy transferred into a desirable output and wasted.

**Energy flow through filament lamp**

a Sankey diagram

The 'wasted' energy is transferred to the surroundings in two main ways:
- from the glass envelope to the surrounding air by conduction and convection
- from all parts of the lamp as non-visible electromagnetic radiation (mainly infra-red).

The light and the infra-red radiation both cause heating when they are absorbed, so all the energy input to the lamp ends up as heat!

These examples illustrate some important points about energy transfer:
- energy due to movement is called kinetic energy
- energy due to position above the Earth's surface is called gravitational potential energy, often abbreviated to potential energy
- work usually involves some transfer of energy to the surroundings.

## Conservation of energy

| | | | |
|---|---|---|---|
| AQA A | M2 | NICCEA | M1 |
| AQA B | M1 | OCR A | M1 |
| EDEXCEL A | M1 | OCR B | M2 |
| EDEXCEL B | M1 | WJEC | A2 |

The examples of energy transfer identify some ways in which energy flows into and out of a process where work is done. Like mass and charge, energy is a conserved quantity.

> **KEY POINT**
>
> The principle of conservation of energy states that:
> energy cannot be created or destroyed.

This simple statement means that energy is never **used up** although often the energy output from a working process is in a form which is difficult to harness to do more work. The energy output from a filament lamp, for example, is effectively **lost** as heat in the surroundings. Like the energy from the bus, it raises the temperature of its surroundings by such a small amount that makes it difficult to recover.

In many energy transfer processes the amount of energy at each stage cannot easily be quantified; it is difficult to put a figure on the total kinetic and potential energy of the moving parts in an engine for example. There are simple formulas for calculating the kinetic energy and gravitational potential energy of individual objects.

> **KEY POINT**
>
> kinetic energy = ½ × mass × (speed)²
> $$E_k = \tfrac{1}{2} mv^2$$
> change in gravitational potential energy = weight × change in height
> $$\Delta E_p = mg\Delta h$$

The formula for calculating gravitational potential energy gives the *change* in energy because ground level varies and so does not provide an accurate reference point for zero energy. This is not the case with kinetic energy as no movement relative to the Earth's surface is the same the world over.

> The principle of conservation of energy applies to all energy transfers.

When a streamlined object is falling freely in the atmosphere at a low speed, the resistive forces are small and so little energy is transferred to the surroundings. In this case the principle of conservation of energy can be applied to the transfer of gravitational potential energy to kinetic energy. When a parachutist is falling at a constant speed conservation of energy still applies, but the energy transfer is from gravitational potential energy to heat in the atmosphere.

### Energy transfer of falling objects

**A ball falling freely towards the Earth**

gravitational potential energy → kinetic energy

**A parachutist falling at constant speed**

gravitational potential energy → heat in the surrounding air

## Efficiency

| | | | |
|---|---|---|---|
| AQA B | A2 | NICCEA | A2 |
| EDEXCEL A | M2 | OCR A | A2 |
| EDEXCEL B | M1 | WJEC | A2 |

> An energy-efficient lamp has an efficiency of 0.25 (25%), compared to 0.05 (5%) for a filament lamp.

A filament lamp is a good example of an everyday device that does not do its job very well. Of the 60 J of energy input to the lamp each second, only 3 J is output as light. It has a very low efficiency!

Fluorescent lamps are much more **efficient** at doing the same job, but until recently they were only available in the form of long tubes. Modern energy-efficient lamps are fluorescent lamps that are designed as direct replacements for filament lamps. Although their running costs are much lower and they last much longer, the initial cost is higher. To achieve the same light output, an energy-efficient lamp needs only 20% of the energy input from electricity that a filament lamp needs. It is more efficient at doing the same job.

> **KEY POINT**
>
> Efficiency is defined as:
>
> useful energy output ÷ total energy input
>
> It is usually expressed as a fraction or decimal, which can be multiplied by 100 to give a percentage efficiency.

## Power

| | | | |
|---|---|---|---|
| AQA A | M2 | NICCEA | A2 |
| AQA B | A2 | OCR A | M1 |
| EDEXCEL A | M1 | OCR B | M2 |
| EDEXCEL B | M1 | WJEC | A2 |

> Do not confuse power with speed. A double-decker bus may be slower than many cars, but it can transport sixty passengers a given distance much faster than any car!

In choosing or designing a machine to do a particular job, important factors to consider include the **power input** and the **power output**. You would expect a hairdryer with a high power output to dry your hair faster than one with a low power output, but a vacuum cleaner with a high power input may not clean the carpet any faster than one with a lower power input, it may just be less efficient!

Steam trains have a very high power input, but a very low efficiency at transferring this to output power.

Power is a measure of the work done or energy transferred each second.

> **KEY POINT**
>
> Power is the rate of working or energy transfer.
>
> Average power can be calculated as work done ÷ time taken, $P = \Delta W \div \Delta t$
>
> Power is measured in watts (W), where 1 watt is a rate of working of 1 joule per second ($J\ s^{-1}$)

## Power and velocity

| | | | |
|---|---|---|---|
| AQA A | M2 | NICCEA | A2 |
| AQA B | A2 | OCR A | M1 |
| EDEXCEL A | M1 | OCR B | M2 |
| EDEXCEL B | M1 | | |

The expression $(F \times s) \div t$, representing **work done** ÷ **time taken** is also equal to **force** × **velocity**. This gives an alternative way of calculating power in situations where the force and speed in the direction of the force (the component of the velocity in that direction) are known:

> **KEY POINT**
>
> Power = force × velocity
> $$P = Fv$$

This expression can be applied to the motive power of a vehicle, where $F$ is the driving force and $v$ is the speed of the vehicle.

For example, if a force of 600 N is used to tow a trailer at a speed of 12 m s$^{-1}$, then the motive power required is 600 N × 12 m s$^{-1}$ = 7200 W.

## Power and efficiency

| | | | |
|---|---|---|---|
| AQA B | A2 | EDEXCEL B | M1 |
| EDEXCEL A | M2 | NICCEA | A2 |

Where the power input and output of a machine are known, it is not necessary to calculate the work input and output in order to calculate the efficiency. The input and output powers can be used directly. This gives the alternative relationship for efficiency:

> **KEY POINT**
>
> efficiency = useful power output ÷ total power input

## Progress check

1 A 240 N force is used to drag a 50 kg mass a distance of 5.0 m up a slope (see diagram). The mass moves through a vertical height of 1.8 m. Take $g = 10$ m s$^{-2}$.

Calculate:

a The work done on the mass.

b The gravitational potential energy gained by the mass.

c The efficiency of this way of lifting the mass through a vertical height of 1.8 m.

d The time taken to drag the mass up the slope by a motor that has an output power of 150 W.

2 A car pulls a trailer with a force of 750 N, transmitted through the tow bar. The tow bar is inclined at 20° to the horizontal. Calculate the work done by the car in pulling the trailer through a horizontal distance of 200 m.

3 A car has a maximum speed of 65 m s$^{-1}$. At this speed, the motive power is 90 kW. Calculate the value of the driving force.

3 1385 N
2 1.4 × 10$^5$ J
d 8.0 s
c 0.75
b 900 J
1 a 1200 J

# 1.10 Drag and terminal velocity

**After studying this section you should be able to:**

- explain the difference between laminar and turbulent flow
- calculate the drag force acting on a vehicle and explain how it varies with the speed of the vehicle
- explain why objects reach a terminal velocity

LEARNING SUMMARY

## Streamlines and turbulence

EDEXCEL A    M1    OCR A    M1
EDEXCEL B    M2

In a wind tunnel the air moves over a stationary object such as a car body to model the movement of the car through the air. This is a valid model, as it is the relative speed of the air and the object that determines the pattern of flow.

To be as fuel-efficient as possible, a car, train or aircraft needs to be designed to minimise the resistive forces acting on it. To investigate the resistive or **drag** forces due to movement through the air, a car body is placed in a **wind tunnel**.

Here the car body is held stationary and the air moves around it; vapour trails show the air flow over the car body.

The diagram below shows the air flow at a low speed over a car body. The flow is said to be **laminar** or **streamlined**.

In laminar flow:

- the air moves in layers, with the layer of air next to the car body being stationary and the velocity of the layers increasing away from the car body
- particles passing the same point do so at the same velocity, so the flow is regular
- the drag force is caused by the resistance of the air to layers sliding past each other

Viscosity is a measure of the resistance of a fluid to flowing. Syrup and tar are viscous fluids; hydrogen has a very low viscosity.

- more **viscous** air has a greater resistance to relative motion and exerts a bigger drag force
- the drag force is proportional to the speed of the car relative to the air.

The air flow over a car travelling at low speed

turbulence occurs at higher speeds

As the speed of the air passing over the car body is increased, the flow pattern changes from laminar to **turbulent**. This is shown in the diagram above.

The changeover from laminar flow to turbulent occurs at a speed known as the **critical velocity**. Turbulent flow causes much more drag than laminar flow.

In turbulent flow:

- the air flow is disordered and irregular
- the drag force depends on the **density** of the air and not the viscosity
- the drag force is proportional to the (speed)$^2$ of the car relative to the air.

## Terminal velocity

AQA A    M1
AQA B    M1
EDEXCEL B    M2
OCR A    M1
OCR B    M2
WJEC    M1

If an object is falling through a fluid, e.g. air or water, as its speed increases the drag on it will also increase. Eventually a speed is reached where the upward force will equal the weight of the object. As there is no net force on the object the acceleration will be zero. The object will fall at a constant velocity. This velocity is called the terminal velocity.

Similarly for a car, or other vehicle, when it reaches a speed where the sum of all the forward forces, i.e. the thrust = the sum of all the resistive forces, i.e. the drag, the vehicle will move at a terminal velocity.

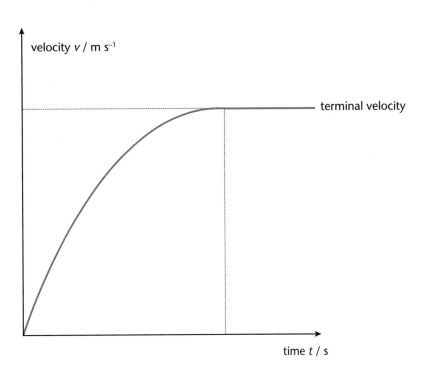

## Progress check

**1** State **two** differences between laminar flow and turbulent flow of air over an object such as a car body.

**2** Suggest why it is dangerous for an aircraft to attempt to take off when there is a layer of ice on its wings.

**3** The aerofoils that form the wings of an aircraft have a lift coefficient, $C_L = 50$. The surface area of the underside of each wing is 10 m$^2$ and the aircraft leaves the ground when its speed through the air is 60 m s$^{-1}$. Calculate the lift force required for take-off if the density of the air is 1.1 kg m$^{-3}$.

**1** any two from: laminar flow is regular; in laminar flow the layers of air slide over each other; in laminar flow the drag force is due to resistance of the layers of the air to sliding over each other; in laminar flow the drag force is proportional to the relative speed of the object and the air; in laminar flow the drag is proportional to the viscosity of the air.

**2** the ice increases the turbulence, reducing the lift.

**3** $1.98 \times 10^5$ N

## *Sample question and model answer*

In an emergency stop, a driver applies the brakes in a car travelling at 24 m s⁻¹. The car stops after travelling 19.2 m while braking.

(a) Calculate the average acceleration of the car during braking. [3]

The final velocity, $v$, is implicit in the question, although not clearly stated. As the car brakes to a halt, the final velocity must be 0.

The initial velocity, $u$, final velocity, $v$, and displacement, $s$, are known.
The unknown quantity is the acceleration, $a$.
The appropriate equation is $v^2 = u^2 + 2as$                1 mark
Rearrange to give $a = (v^2 - u^2)/2s$                1 mark
$= (0 - 24^2)/(2 \times 19.2)$
$= -15$ m s⁻²                1 mark

Note that the acceleration is negative, as the speed in the direction of the velocity is decreasing.

(b) Sketch a graph that shows how the speed of the car changes during braking, assuming that the acceleration is uniform. Mark the value of the intercept on the time axis.
Show on your graph the feature that represents the distance travelled during braking. [3]

The graph shows a straight line from 24 m s⁻¹ to 0 m s⁻¹                1 mark
The intercept on the time axis is 1.6 s                1 mark
The area between the graph line and the time axis represents the distance travelled.                1 mark

(c) Calculate the size of the force required to decelerate the driver, of mass 75 kg, at the same rate as the car. [3]

Again, the negative sign shows that the direction of the force is opposite to the direction of motion.

force $=$ mass $\times$ acceleration                1 mark
$= 75$ kg $\times -15$ m s⁻²                1 mark
$= -1125$ N                1 mark

(d) Suggest what is likely to happen to a driver who is not wearing a seat belt. The car does not have air bags. [2]

The cue word 'suggest' indicates that you are not meant to have studied this example as part of your course, but should be able to reason it from your understanding of forces and motion.

The driver will carry on moving (1 mark) as there is nothing to stop her/him until (s)he hits the steering wheel and windscreen (1 mark).

## Practice examination questions

Take the value of free-fall acceleration, $g$ to be 10 m s$^{-2}$.

1   An aircraft has a total mass, including fuel and passengers, of 70 000 kg.
    Its take-off speed is 60 m s$^{-1}$ and it needs to reach that speed before the end
    of the runway, which is 1500 m long.
    (a)  Calculate the minimum acceleration of the aircraft.                                  [3]
    (b)  Calculate the average force needed to achieve this acceleration.                     [3]
    (c)  Explain how the resultant force on the aircraft is likely to change during
         take-off.                                                                            [2]

2   A car pulls a trailer with a force of 150 N. This is shown in the diagram.

150 N

(a)  According to Newton's third law, forces exist in pairs.
     (i)   What is the other force that makes up the pair?                                    [1]
     (ii)  Write down the size and direction of this force.                                   [1]

(b)  The mass of the trailer is 190 kg. As it sets off from rest, there is a resistive
     force of 20 N.
     (i)   Calculate the size and direction of the resultant force on the trailer.            [1]
     (ii)  Calculate the initial acceleration of the trailer.                                 [3]

(c)  The force from the car on the trailer is maintained at 150 N.
     When the car and trailer are travelling at constant speed:
     (i)   what is the size of the resultant force on the trailer?                            [1]
     (ii)  write down the size and direction of the resistive force on the trailer.           [1]

3   The diagram shows the forces acting on a child on a playground slide. Air
    resistance is negligible.

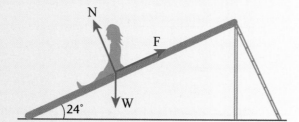

(a)  The mass of the child is 40 kg.
     (i)   Calculate the size of the force W.                                                 [1]
     (ii)  Describe the force that, along with W, makes up the 'equal and
           opposite pair' of forces described by Newton's third law.                          [1]

(b)  The size of the friction force is 90 N.
     (i)   Calculate the component of W parallel to the slide.                                [2]
     (ii)  Calculate the acceleration of the child.                                           [3]

(c)  The slide is 5.5 m long. If the child maintains this acceleration, calculate
     her speed at the bottom.                                                                 [3]

## *Practice examination questions* (continued)

**4** The diagram shows two 'vehicles' on an air track approaching each other. After impact they stick together.

0.45 kg          0.30 kg

The speed of the heavier vehicle is 0.60 m s⁻¹.

(a) If the lighter vehicle is travelling at 0.42 m s⁻¹, calculate the velocity of the vehicles after the collision. [3]

(b) If both vehicles are stationary after the collision, what is the speed of the lighter vehicle before the collision? [2]

**5** The diagram shows a neutron of mass $1.7 \times 10^{-27}$ kg about to collide inelastically with a stationary uranium nucleus of mass $4.0 \times 10^{-25}$ kg. During the collision, the neutron will be absorbed by the uranium nucleus.

neutron

uranium nucleus

(a) Calculate the velocity of the uranium nucleus immediately after the neutron has been absorbed. [3]

(b) Collisions between neutrons and uranium nuclei can also be elastic. State, and explain briefly, how the speed of the uranium nucleus after impact would be different in the case of an elastic collision. **Do not perform any further calculations.** [3]

(c) Using the data at the beginning of the question, calculate the kinetic energy of the neutron before it collides with the uranium nucleus. [3]

*AEB AS-A PHYS 6 Jan 1999 Q 2*

**6** An astronaut on a space walk pushes against the side of the spacecraft.

(a) Explain how this causes the velocity of both the astronaut and the spacecraft to change. [2]

(b) The astronaut can change his velocity by firing jets of nitrogen gas. Explain how this changes the velocity of the astronaut. [2]

**7** A ball of mass 0.080 kg hits a wall at a speed of 32 m s⁻¹ and rebounds at a speed of 22 m s⁻¹.

(a) Calculate the change in momentum of the ball. [2]

(b) If the impact between the ball and the wall lasts for 0.15 s, calculate the average force on the ball. [3]

(c) Explain whether momentum is conserved in this interaction. [2]

## Practice examination questions *(continued)*

**8** A crane lifts a 3 tonne (3000 kg) load through a vertical height of 18 m in 24 s. Calculate:

(a) the work done on the load [3]

(b) the gain in gravitational potential energy of the load [1]

(c) the power output of the crane [2]

(d) the power input to the crane if the efficiency is 0.45. [2]

**9** In a hydroelectric power station, water falls at the rate of $2.0 \times 10^5$ kg s$^{-1}$ through a vertical height of 215 m before driving a turbine.

(a) Calculate the loss in gravitational potential energy of the water each second. [3]

(b) Assuming that all this energy is transferred to kinetic energy of the water, calculate the speed of the water as it enters the turbines. [3]

(c) The water leaves the turbines at a speed of 8 m s$^{-1}$. Calculate the maximum input power to the turbines. [2]

(d) The electrical output is 250 MW. Calculate the efficiency of this method of transferring gravitational potential energy to electricity. [2]

**10** The diagram shows a child's toy. The spring is compressed a distance of 0.15 m using a force of 25 N. When the release mechanism is operated the ball rises into the air.

spring release mechanism

(a) Calculate the energy stored in the spring. [2]

(b) The mass of the ball is 0.020 kg. Calculate:

(i) the maximum speed of the ball. [3]

(ii) the maximum vertical distance that the ball travels. [3]

# Electricity

**The following topics are covered in this chapter:**

- Ohm's law, resistance and resistivity
- Energy transfer in circuits
- Resistance in circuits
- Alternating current and the oscilloscope
- Electromagnetism

# 2.1 Ohm's Law, resistance and resistivity

**After studying this section you should be able to:**

- recall and use the formula for resistance
- recall and use the formula for resistivity
- state Ohm's law and describe its limitations
- sketch the current–voltage characteristics of a filament lamp, a wire at constant temperature and a diode

**LEARNING SUMMARY**

## Resistance

| | | | |
|---|---|---|---|
| AQA A | M3 | NICCEA | M1 |
| AQA B | M1 | OCR A | M2 |
| EDEXCEL A | M2 | OCR B | M1 |
| EDEXCEL B | M1 | WJEC | M2 |

Electromagnetism, heating and electrolysis are all due to currents passing in circuits. Our everyday appliances rely on these three effects of a current.

What determines the current in a component or a circuit? The answers include the voltage; which is a measure of the energy available to move the charge round the circuit, and how freely the charge can flow; which depends on the structure of the conducting material.

**Resistance** is a measure of the opposition to current. The greater the resistance of a component, the smaller the current that passes for a given voltage.

Just as different objects can weigh different amounts or the same amount, according to their size and the material they are made of, different components can have the same resistance or different resistances. However, if you change the shape of a piece of material its weight does not change but its resistance does.

> $1\ \Omega$ is equivalent to $1\ \text{V A}^{-1}$, so the voltage across a $1\ \Omega$ resistor is 1 V for each amp of current passing in it.

> **KEY POINT**
>
> Resistance is defined and calculated using the formula:
>
> $$Resistance = voltage \div current \quad or \quad R = V \div I$$
>
> The unit of resistance is the ohm ($\Omega$).

## Different components

| | | | |
|---|---|---|---|
| AQA A | M3 | NICCEA | M1 |
| AQA B | M1 | OCR A | M2 |
| EDEXCEL A | M2 | OCR B | M1 |
| EDEXCEL B | M1, M2 | WJEC | M2 |

> The same piece of equipment is normally used as both a variable resistor and a potential divider. When used as a variable resistor, connections are made to the slider and one of the fixed terminals. All three terminals are used in a potential divider circuit.

There are three key variables that affect the current passing in a conductor. They are:
- the voltage
- the material that the conductor is made from
- the physical dimensions of the conductor.

The diagram shows how a **potential divider** can be used to investigate how the current passing in a component depends on the potential difference, or voltage, across it. Movement of the sliding contact enables the voltage to be varied from zero up to the maximum of the supply voltage.

This section considers some other factors that can determine the size of the current.

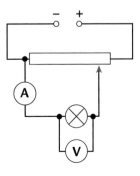

The graphs represent typical results obtained for a metal wire kept at a constant temperature, a filament lamp and a diode.

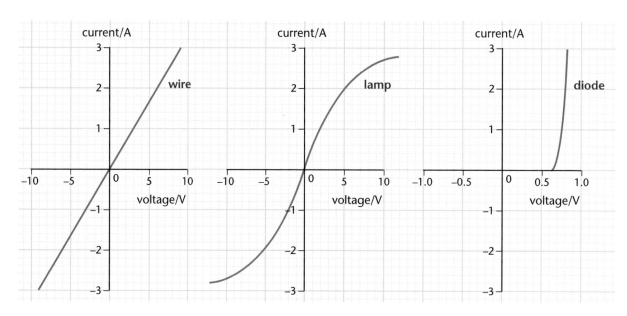

The text in *italics* is **Ohm's Law**, one that he formulated based on his own experiments. Because it only applies to metallic conductors kept at a constant temperature, it is of very limited use in predicting the values of electric currents.

**KEY POINT**

The current–voltage graph for the **wire** is a straight line through the origin, showing that the *current is directly proportional to the voltage.*

A **filament lamp** is an everyday example of a metallic conductor that does change in temperature during use.

In a short time after it is switched on, the temperature of the filament increases from room temperature to about 2000°C. The current–voltage graph shows how this change in temperature affects the relationship between these quantities. Like the one for the wire at constant temperature, this graph is symmetrical. Both the wire and the lamp filament show the same pattern of behaviour no matter which way round the voltage is applied.

Calculations based on this graph show that as the current in the filament increases, so does its resistance. This is due to the higher temperature causing an increase in the amplitude of the lattice vibrations, which in turn increases the frequency of the collisions that impede the electron flow.

The arrow shows the direction in which the diode allows current to pass.

The graph showing the characteristics of the **diode** also shows the property that makes diodes different from other components; they only allow current to pass in one direction.

This direction is shown by the arrow on the diode symbol. With the voltage in the reverse direction, no current passes. This property makes diodes useful in rectification; the process in which a direct current is obtained from an alternating one.

Other diodes in common use include the light-emitting diode (l.e.d.); used as an indicator and the laser diode; used to read information from compact discs.

Even with the voltage in the forwards direction, no current passes until a certain minimum voltage has been reached. The minimum voltage depends on the material that the diode is made from, and is about 0.6 V for a general purpose silicon diode. Once this has been achieved, small increases in voltage cause relatively large increases in current as the resistance of the diode decreases.

## Resistivity

| | | | |
|---|---|---|---|
| AQA A | M3 | NICCEA | M1 |
| AQA B | M1 | OCR A | M2 |
| EDEXCEL A | M2 | OCR B | M1 |
| EDEXCEL B | M2 | WJEC | M2 |

Resistors used in electrical and electronic circuitry are made in a variety of ways. Some are made from a precise length of wire wound on a ceramic base. These tend to be expensive and their use is limited to applications where there is a large current. Resistors used in low-current electronic devices are made from a thin film of conducting material coated onto a small ceramic tube.

> The large wire-wound resistor is able to dissipate heat at the rate of 10 W. This would burn out the small carbon-film resistor.

What factors determine the resistance of either type of resistor?

- The physical dimensions.
  Long wires have more resistance than short wires, so the longer the sample of material, the greater the resistance.

- Cross-sectional area.
  Doubling the cross-sectional area of a sample also doubles the number of charge-carriers available to carry the current, halving the resistance.

> **KEY POINT**
>
> The resistance of a sample of material is directly proportional to its length and inversely proportional to its cross-sectional area.
>
> This statement can be written as $R \propto l/A$.

> Good conductors have a low resistivity; that of copper being $1.7 \times 10^{-8}\ \Omega$ m. The resistivity of carbon is about one thousand times as great, $3.0 \times 10^{-5}\ \Omega$ m.

The other factor that determines the resistance is the material that the resistor is made of; its resistive properties are measured by its resistivity, symbol $\rho$. When this is taken into account the formula becomes $R = \dfrac{\rho l}{A}$. This formula can be used to calculate the resistance when the dimensions and material of a resistor are known or to calculate the dimensions required to give a particular value of resistance.

Resistance is a property of an individual component in a circuit, but resistivity is a property of a material. It is measured in units of $\Omega$ m.

## Progress check

1. Use the resistance formula to work out the following:
   a. The voltage across a lamp filament if the current passing is 2.4 A and the filament resistance is 7.3 $\Omega$.
   b. The resistance of an infra-red heater if the current is 9.5 A when it operates from the mains 240 V supply.
   c. The current passing in a 6.8 $\Omega$ resistor when it is connected to a 9 V battery.

> Note that $1\ \text{mm} = 1 \times 10^{-3}\ \text{m}$.

2. Use the resistivity formula to work out the following:
   a. The resistivity of silicon. A 1 mm cube of silicon has a resistance of $2.20 \times 10^{6}\ \Omega$ when measured across opposite faces.

> First re-arrange the resistivity equation to make *l* the subject.

   b. The length of carbon rod required to make a 3.9 $\Omega$ resistor.
   The resistivity of carbon is $3.00 \times 10^{-5}\ \Omega$ m and the rod has a cross-sectional area of $1.26 \times 10^{-6}\ \text{m}^2$.

3. An electrical heating element is to be made from nichrome wire. The current in the heating element is to be 5.00 A when the voltage across it is 12.0 V.
   a. Calculate the resistance of the heating element.
   The resistivity of nichrome is $1.10 \times 10^{-6}\ \Omega$ m.
   The cross-sectional area of the wire used is $7.85 \times 10^{-7}\ \text{m}^2$.
   b. Calculate the length of wire required.

1 a 17.5 V   b 25.3 Ω   c 1.32 A
2 a 2200 Ω m   b 0.164 m
3 a 2.4 Ω   b 1.71 m

# 2.2 Energy transfer in circuits

*After studying this section you should be able to:*

- *recall and use the relationship between charge flow and current*
- *recall and use the relationship between potential difference, energy transfer and charge*
- *explain the difference between electromotive force and potential difference*
- *use the relationships between electrical power, current, voltage and resistance*

**LEARNING SUMMARY**

## Charge and current

| | | | |
|---|---|---|---|
| AQA A | M3 | NICCEA | M1 |
| AQA B | M1 | OCR A | M2 |
| EDEXCEL A | M2 | OCR B | M1 |
| EDEXCEL B | M1 | WJEC | M2 |

In a metal, current is due to the movement of free electrons from negative to positive.

In a conducting non-metal, positive and negative ions flow in opposite directions.

To conduct electricity, a material must have charged particles that can move. In metals, these are negatively-charged **free electrons**; while both positively-charged and negatively-charged **ions** are responsible for conduction in gases and non-metallic liquids.

- An overall flow of charge in a particular direction is an **electric current**.
- Current = rate of flow of charge.

The relationship between charge flow and current can be written as $I = \Delta q/\Delta t$, where $\Delta q$ is the amount of charge that flows in a short time $\Delta t$.

> **This relationship is used to define the size of the unit of charge, the coulomb is the charge that flows past a point when a steady current of 1 A passes for 1 second.**
>
> **KEY POINT**

## Current and drift velocity

| | | | |
|---|---|---|---|
| AQA B | M1 | NICCEA | M1 |
| EDEXCEL A | M2 | WJEC | M2 |

The free electrons in a non-conducting metal are in constant motion. The motion of any one electron is random in both speed and direction, and constantly subject to change due to collisions, but a typical speed in a metal at room temperature is of the order of $1 \times 10^6$ m s$^{-1}$.

The random motion of a free electron

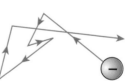

drift velocity is slow compared to the random motion of the electrons

Applying an electric force in the form of a voltage does not cause the free electrons to suddenly dash for the positive terminal. The random motion continues, but in addition the body of electrons moves at low speed in the direction negative to positive.

In a metal, a typical drift velocity is of the order of 1 mm s$^{-1}$, but in a semiconductor of similar dimensions carrying a similar current it is likely to be higher than this because the concentration of charge carriers is much less.

The charge carried by one electron, $e = -1.60 \times 10^{-19}$ C. 1 C of charge is equivalent to that carried by $6.25 \times 10^{18}$ electrons.

The overall speed of the body of electrons is called the **drift velocity**. There are three key variables that affect the drift velocity:

- current
- charge carrier concentration, $n$, defined as the number of charge carriers (free electrons in the case of a metal) per unit volume
- the cross-sectional area of the specimen, $A$.

These lead to the relationship for a metal $I = nAev$, where $e$ is the electronic charge.

A similar expression applies to non-metallic conductors. As the charge on each ion can vary between non-metals, the expression becomes $I = nAqv$, where $q$ is the ionic charge.

## Charge and energy transfer

| AQA A | M3 | NICCEA | M1 |
| AQA B | M1 | OCR A | M2 |
| EDEXCEL A | M2 | OCR B | M1 |
| EDEXCEL B | M1 | WJEC | M2 |

Charge is not used up or changed into anything else as it moves round a circuit. The job it does is that of transferring energy. It gains energy from the source of electricity (mains, battery or power supply) and transfers this energy to the circuit components as it flows through them.

In the case of a filament lamp, the energy transfer from the charge is due to collisions of the electrons moving through the filament. With a motor or other electromagnetic device the charge loses energy as it has to work against repulsive forces in moving through the armature.

### How much energy?

Energy transfer is a continuous process as the charge flows round a circuit. There are two separate physical processes:

> The charge carriers gain energy as they are accelerated by the source; they lose energy in collisions.

- energy transfer to the charge from the source
- energy transfer from the charge to the components

In each case, rather than measure the energy transfer for a single charge carrier, the transfer to and from each coulomb of charge is measured.

**Electromotive force (e.m.f.)** measures the energy transfer from the source.

A 6 V battery transfers 6 J of energy to each coulomb of charge that it moves around a circuit.

> Electromotive force is a mis-named quantity. It describes energy transfer rather than force.

> **KEY POINT**
> Electromotive force, symbol $E$, is defined as being the energy transfer from the source in driving unit charge round a complete circuit. It is measured in volts (V), where 1 volt = 1 joule/coulomb (1 J C$^{-1}$).

Electromotive force measures the work done on the charge by the source. As the charge flows in a circuit, a small amount of work is done in the connecting wires, transferring energy to heat, but most work is done in passing through the components.

**Potential difference (p.d.)** measures the work done by the charge in the circuit.

> **KEY POINT**
> The potential difference between two points, symbol $V$, is defined as being the energy transfer per unit charge from the charge to the circuit. Like e.m.f., it is measured in V.

- Electromotive force and potential difference are both defined in terms of the work done per unit charge, the difference being whether the work is done on the charge or by the charge.
- The definitions can be written as $E = W/q$ and $V = W/q$.
- The unit of e.m.f. and p.d. is the volt, where *one volt is the potential difference between two points if 1 J of energy is transferred when 1 C of charge passes between the points.*

> The same instrument – a voltmeter – is used to measure both e.m.f. and p.d.

This voltmeter is measuring the e.m.f. of the cell; the energy transfer to each coulomb of charge

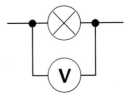

This voltmeter is measuring the p.d. across the lamp; the energy transfer from each coulomb of charge

## Electrical power

| | | | |
|---|---|---|---|
| AQA A | M3 | NICCEA | M1 |
| AQA B | M1 | OCR A | M2 |
| EDEXCEL A | M2 | OCR B | M1 |
| EDEXCEL B | M1 | WJEC | M2 |

> Unless stated otherwise, the term 'rate of' refers to 'rate of change with time', so power is the work done or energy transfer per second.

**Power** is the *rate of working or energy transfer.*

> **KEY POINT**
>
> The relationship between electrical power, current and potential difference is
> *power = current × potential difference or P=IV.*
> Power is measured in watts (W), where 1 W = 1 J s⁻¹.

This relationship can be combined with the resistance equation to give:

- $P = I^2R$
- $P = V^2/R$

These three equivalent relationships can all be used to calculate the value of either a constant or a varying power.

A battery-operated lamp transfers energy at a constant rate as there is no change in the current, potential difference or brightness of the filament. The power of a mains-operated lamp is constantly changing as all three of these variables change in step with each other. The graphs compare the power of a battery-operated lamp with that of a mains lamp.

> The smaller the value of $\Delta t$, the better the approximation that $E = IV\Delta t$ is to the area between the graph line and the time axis.

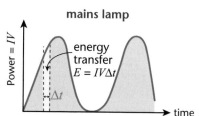

> The SI unit of energy is the joule. For domestic purposes, energy is measured in kilowatt-hours (kW h), where 1 kW h is the energy transfer by a 1 kW appliance in 1 hour.

The total energy transfer in time $\Delta t$ where the power is constant is equal to:

> **KEY POINT**
>
> $$\Delta E = power \times time = IV\Delta t.$$

For a varying power the energy transfer is represented by the area between the curve and the time axis of a graph of power against time. The shorter the value of $\Delta t$, the better the approximation of the above expression to this area.

## Progress check

> To answer i and ii, use the definitions of current and potential difference.

1  Charge flows through a lamp filament at the rate of 15 C in 60 s. In the same time, 1500 J of energy is transferred to the filament.
   **a**  Calculate:
   **i**   the current in the filament
   **ii**  the potential difference across the filament
   **iii** the power of the lamp
   **iv**  the resistance of the filament.
   **b**  Explain how the power of the filament is affected when the current is doubled.

2  When the current in a cell is 1.5 A it supplies energy at the rate of 18 W. Calculate the e.m.f. of the cell and the total resistance of the circuit.

3  A silicon diode has a cross-sectional area of 4.0 mm² (4.0 × 10⁻⁶ m²). The concentration of charge carriers is 2.5 × 10²⁷ m⁻³ and each charge has a charge of 1.60 × 10⁻¹⁹ C. Calculate the drift velocity when the current in the diode is 1.5 A.

3  9.4 × 10⁻⁴ m s⁻¹
2  12 V and 8 Ω.
1  **a** i 0.25 A  ii 100 V  iii 25 W  iv 400 Ω  **b** The power becomes four times as great, since P∝I².

# 2.3 Resistance in circuits

**After studying this section you should be able to:**

- calculate the effective resistance of a number of resistors connected in series or parallel
- understand that a source of electricity may contribute to the resistance of a circuit and the effects of this
- apply the principles of conservation of charge and conservation of energy to circuit calculations
- describe how the resistance of a metallic conductor, a thermistor and a light-dependent resistor change with environmental conditions

## Series and parallel

| | | | |
|---|---|---|---|
| AQA A | M3 | NICCEA | M1 |
| AQA B | M1 | OCR A | M2 |
| EDEXCEL A | M2 | OCR B | M1 |
| EDEXCEL B | M1 | WJEC | M2 |

Resistors can be combined in series, parallel or a combination of the two. There are simple formulas for calculating the effective resistance of two or more resistors connected together.

The effective resistance of a number of resistors in series is always greater than the largest value resistor in the combination.

### Series

Adding another resistor in series in a circuit always decreases the current. This means that the effective resistance in the circuit has increased.

> **KEY POINT**
> The formula for calculating the effective resistance of a number of resistors in series is:
> $$R = R_1 + R_2 + R_3.$$

This formula can be used for any number of resistors connected in series by adding on extra terms.

To find the effective resistance of a 4.7 Ω, a 6.8 Ω and a 10 Ω resistor connected in parallel
$1/R = 1/4.7 + 1/6.8 + 1/10$
$= 0.460$
$R = 1/0.460 = 2.2\ \Omega$
*A common error is to forget to carry out the final stage of the calculation, in this case giving an answer of:*
$R = 0.460\ \Omega$

### Parallel

When an additional resistor is added in parallel to a circuit, the current always increases. The additional resistor opens up another current path, without affecting the current in any of the existing paths. This results in less resistance in the circuit.

The effective resistance of a number of resistors in parallel is always smaller than the smallest value resistor in the combination.

> **KEY POINT**
> The formula for calculating the effective resistance of a number of resistors in parallel is:
> $$1/R = 1/R_1 + 1/R_2 + 1/R_3.$$

As with the formula for resistors in series, any number of terms can be added to this expression.

## Internal resistance

| | | | |
|---|---|---|---|
| AQA A | M3 | NICCEA | M1 |
| AQA B | M1 | OCR A | M2 |
| EDEXCEL A | M2 | OCR B | M1 |
| EDEXCEL B | M1 | WJEC | M2 |

All parts of a circuit have resistance and energy is needed to move the charge carriers through them. A cell of e.m.f. 1.5 V, for example, does 1.5 J of work on each coulomb of charge that completes a circuit.
This work is done on moving the charge through:

- the connecting wires
- the circuit components
- the cell itself, due to the cell's own internal resistance, symbol $r$.

The internal resistance of a cell or power supply acts as an extra resistor in series with the rest of the circuit.

The greater the current in the cell, the more work is done against the internal resistance, so less can be done in the external circuit.

How does internal resistance affect the circuit?

For example, a cell of e.m.f ($E$) 1.5 V and internal resistance ($r$) 0.2 Ω causes a current ($I$) of 0.5 A in an external resistor ($R$). The work done per coulomb of charge against the internal resistance = $I \times r$ = 0.5 A $\times$ 0.2 Ω = 0.1 V. This leaves 1.4 V for the external circuit.

> The greater the current in the cell, the greater the reduction in potential difference.

- *The effect of internal resistance is to reduce the potential difference across the external circuit.*

When the only component connected to a cell is a high-resistance voltmeter such as a digital voltmeter, the reading is the cell's e.m.f. This is because when it is connected to the voltmeter alone, virtually no current passes in the cell so no work is being done against internal resistance. However, when the cell is connected to a circuit, a voltmeter placed across its terminals reads less than the e.m.f.

> A typical digital voltmeter has a resistance of 10 MΩ ($1.0 \times 10^7$ Ω), so the current passing in it is very small, causing negligible reduction in the p.d. across the terminals.

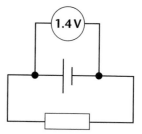

The e.m.f. of the cell is 1.5 V. The cell supplies 1.5 J of energy to move each coulomb of charge around a complete circuit.

0.1 J of energy is needed to move each coulomb of charge through the cell, leaving 1.4 J to move the charge round the rest of the circuit.

The difference between the e.m.f. and the 'terminal p.d.' is the work done within the cell against the internal resistance.

> **KEY POINT**
>
> The relationship between e.m.f. and terminal p.d. is:
> e.m.f. = terminal p.d. + p.d. across internal resistance
> $$E = V + Ir$$

## Conservation rules

| | | | |
|---|---|---|---|
| AQA A | M3 | NICCEA | M1 |
| AQA B | M1 | OCR A | M2 |
| EDEXCEL A | M2 | WJEC | M2 |
| EDEXCEL B | M1 | | |

> Mass, charge and energy are fundamental conserved quantities. They cannot be created or destroyed.

In a circuit, the flow of **charge** transfers **energy** from the source to the components. Neither of these physical quantities becomes used up in the process. Both are **conserved**. **Kirchoff's laws** are re-statements of these fundamental laws in terms that apply to electric circuits.

> **KEY POINT**
>
> The **first law** states that the total current that enters a junction is equal to the total current that leaves the junction.

This is illustrated in the diagram and is simply a statement that charge is conserved at the junction.

> The **second law** is about conservation of energy. It states that **around any closed loop (i.e. complete series path), the total e.m.f. is equal to the sum of the potential differences**, $E = \Sigma IR$.
>
> **KEY POINT**

Remember that e.m.f. represents energy transfer **to** the charge and p.d. represents energy transfer **from** the charge, so this law is stating that when a quantity of charge makes a complete circuit, the energy transfers each way are balanced. The second law can be applied to a series circuit or any series path within a parallel circuit.

> There are three closed loops in this parallel circuit. Can you identify the other two?

## Resistance and the environment

| | | | |
|---|---|---|---|
| AQA B | M1 | NICCEA | M1 |
| EDEXCEL A | M1 | OCR A | M2 |
| EDEXCEL B | M1 | OCR B | M1 |
| | | WJEC | M2 |

Any change in environmental conditions, for example moisture level, temperature, pressure, stress and illumination can cause a change in resistance of a suitable sensor. The consequent change in current or potential difference can be used to monitor the forces in a structure or switch control circuits that perform tasks, such as the automatic watering of greenhouse crops and the maintaining of a constant temperature in an incubator.

The current in a **metallic conductor** is due to a flow of electrons, and resistance to the current is caused by anything that impedes that flow. Collisions between the free electrons and the metal structure become more frequent at higher temperatures, so the resistance increases. The change in resistance with temperature for a metallic conductor is small and approximately linear.

> Because of their large change in resistance with temperature, thermistors are useful in applications such as heat-sensitive cameras, which need to be able to detect small changes in temperature.

Much bigger changes in resistance occur when the temperature of a **thermistor** changes. Thermistors are made out of semi-conducting material; as the temperature rises the number of charge-carriers increases and the material becomes a better conductor. The effect of this swamps any increase in resistance due to increased collisions, and the resistance falls sharply with increasing temperature.

> These graphs show the variation in resistance with environmental conditions for a metallic conductor, a thermistor and a light dependent resistor.

A **light-dependent resistor** (LDR) also shows large changes in resistance when the light level changes. Typically the resistance can vary from millions of ohms in total darkness to a few hundred ohms in daylight.

In a potential-divider circuit the voltage is shared in proportion to the resistance. If the resistance of one component increases, so does the voltage across it.

Thermistors and LDRs are widely used in control circuits, to switch on heaters and lamps for example. They are normally used in a potential-divider arrangement with a fixed resistor. In this arrangement the supply voltage is **shared** between the sensor (thermistor or LDR) and the fixed or variable resistor. If the resistance of a sensor changes, so does the potential difference across it; this can be used to switch an electronic circuit that controls other devices.

## Progress check

**1** Calculate the effective resistance of each of the following combinations of resistors.

In **b**, work out the resistance of the parallel combination first.

a            b            c

Remember that the p.d. across the terminals is equal to that across the external resistor.

**2** A cell has an e.m.f. of 1.6 V and internal resistance 0.5 Ω.
Calculate the current in the cell and the p.d. across the terminals when it is connected to:
**a** a 0.5 Ω resistor
**b** a 3.5 Ω resistor.

**3** A battery of e.m.f 4.5 V and internal resistance 0.6 Ω is connected to a 3.3 Ω resistor in series with a 5.1 Ω resistor.
Calculate:
**a** the current in the circuit
**b** the p.d. across each resistor
**c** the p.d. across the terminals of the cell.

**4** The diagram shows a thermistor used in a potential divider as part of a temperature-controlled switch. The internal resistance of the power supply is small and can be neglected.

3 V

1000 Ω

to electronic
switch

The electronic switch turns on a heater when the p.d. across the thermistor rises to 0.6 V.
**a** Calculate the p.d. across the thermistor when its resistance is 200 Ω.
**b** Explain how the heater becomes switched off.
**c** Calculate the resistance of the thermistor when the heater switches off.

The quickest way to answer **a** is to use ratios.

**1 a** 1.33 Ω   **b** 7.67 Ω   **c** 12.5 Ω
**2 a** 1.6 A; 0.8 V   **b** 0.4 A; 1.4 V
**3 a** 0.50 A and 2.55 V   **c** 4.2 V
**4 a** 0.5 V   **b** The temperature of the thermistor rises, causing its resistance and the p.d. across it to fall.
**c** 250 Ω.

# 2.4 Alternating current and the oscilloscope

*After studying this section you should be able to:*

- *describe how to use an oscilloscope as a voltmeter and to measure time intervals and frequencies*
- *explain the meaning of root mean square, peak and peak-to-peak values of a sinusoidal waveform*

## Alternating and direct current

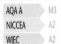

The current from a battery is a **direct current** (d.c.), while that from the mains is an **alternating current** (a.c.).

- An alternating current changes direction.
- A direct current may vary in value, but maintains the same direction.

A direct current and an alternating current. Which is which?

## Controlling an oscilloscope

An oscilloscope is a high-resistance voltmeter that can display how a voltage changes over a period of time. An electron beam strikes the screen and causes it to fluoresce, making a bright dot on the screen.

The oscilloscope has two key controls:

- the **sensitivity** or **gain** controls the vertical deflection of the dot
- the **time base** controls the speed of horizontal movement of the dot.

The **sensitivity** control is marked in 'volts/cm' or 'volts/div', this is the voltage required to cause a 1 cm vertical deflection.

Because of its high resistance, an oscilloscope can only be used as a voltmeter to study the waveform of a voltage. It cannot be used in place of an ammeter to study the waveform of a current.

The **time base** control is marked in 'time/cm' or 'time/div', the setting being the time it takes for the dot to move horizontally one centimetre. The shorter this time interval, the faster the dot moves across the screen. With all but the slowest settings, the dot moves so fast that it appears to be a horizontal line.

time base control

sensitivity control

These diagrams compare the traces obtained for a 3.0 V direct voltage input with different sensitivity and time base settings.

sensitivity = 2 V/cm
time base on

sensitivity = 1 V/cm
time base on

sensitivity = 1 V/cm
time base off

## *Measuring time and frequency*

AQA A — M3
NICCEA — A2
WJEC — A2

The time base setting on the oscilloscope can be used to measure short time intervals and the frequency of an alternating voltage.

- A time interval is measured by multiplying the appropriate horizontal distance on the screen by the time base setting.
- A frequency is measured by calculating (1/time) for one complete cycle of a wave.

> Remember to convert times in ms or $\mu$s into s before calculating a frequency.

In the example shown in the diagram, the time base is set to 1 ms cm$^{-1}$, or $1.0 \times 10^{-3}$ s cm$^{-1}$. One complete cycle of the wave occupies a distance of 4.0 cm on the screen, so the time taken for one cycle is $4.0 \times 10^{-3}$ s.
Since *frequency = 1/time period* or $f = 1/t$, the frequency of the wave is
$1 \div 4.0 \times 10^{-3}$ s = 250 Hz.

## *Measuring alternating voltage*

AQA A — M3
NICCEA — A2
WJEC — A2

Alternating voltage from the mains varies in the form of a sine wave. This means that it not only changes direction, it is also continually changing its value. Two measurements that can be made from an oscilloscope trace are:

- the **peak value**, $V_0$, the maximum or minimum value of the voltage
- the **peak-to-peak value**, the difference between the maximum and minimum values.

> Normally the peak-to-peak value of an alternating voltage is twice the peak value.

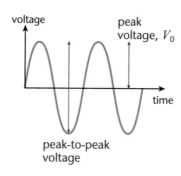

The peak value of an alternating current or voltage can be a misleading measurement, since for most of the time the current or voltage has a lower value than this. To make a fair comparison with a direct current or voltage, the **root mean square** or **r.m.s** value is quoted. The r.m.s value of an alternating current or voltage has the same heating effect as a direct current or voltage of the same value. This means that a lamp lights with the same brightness when connected to 12 V d.c. and 12 V r.m.s a.c.

The relationship between the r.m.s value of an alternating current or voltage and the peak value is:

> Mains voltage is described as 240 V. This is an r.m.s value, the peak value being $240 \times \sqrt{2}$ = 339 V.

$$I_{r.m.s} = \frac{I_0}{\sqrt{2}} \qquad V_{r.m.s} = \frac{V_0}{\sqrt{2}}$$

**KEY POINT**

This relationship is only valid for alternating currents and voltages that are sinusoidal, i.e. that follow the pattern of a sine wave.

## Progress check

The diagram shows an oscilloscope display of an alternating voltage.

The sensitivity is set to 10 V/cm and the time base to 5 ms/cm. Calculate:
i the peak value of the voltage
ii the r.m.s. value of the voltage
iii the frequency of the voltage.

i 25 V ii 17.7 V iii 33.3 Hz

# 2.5 Electromagnetism

**After studying this section you should be able to:**

- sketch the magnetic field patterns due to a current in a straight wire, a circular coil and a solenoid
- use Fleming's left-hand rule to predict the direction of the force on a current-carrying conductor placed in a magnetic field
- define magnetic field strength (flux density) and the tesla
- recall and use F = Bll

**LEARNING SUMMARY**

## Magnetic field patterns

| | | | |
|---|---|---|---|
| AQA A | A2 | NICCEA | A2 |
| AQA B | A2 | OCR A | M2 |
| EDEXCEL A | A2 | OCR B | A2 |
| EDEXCEL B | A2 | WJEC | A2 |

At AS/A level you study three types of field: magnetic fields, gravitational fields and electric fields. Take particular note about the similarities and differences between them.

A field is a name given to a region where forces are exerted on objects with a certain property. A gravitational field describes a region where there are forces on objects that have mass. An electric field affects objects that have charge. A magnetic field affects:

- permanent magnets
- magnetic materials
- electric currents.

You should be familiar with magnetic field patterns drawn around permanent magnets. Although the force at any point is in a straight line, the force lines joined together to make a series of curved lines. Arrows on the magnetic field patterns show the direction of the force on the north-seeking pole of a magnet.

Similarly, every electric current has its own magnetic field. The diagrams show the field patterns due to the currents passing in:

- a long straight wire
- a circular coil
- a solenoid (a long coil).

Notice that the magnetic field in the centre of the solenoid is uniform, while on the outside it resembles that of a bar magnet.

## Magnetic forces

| | | | |
|---|---|---|---|
| AQA A | A2 | NICCEA | A2 |
| AQA B | A2 | OCR A | M2 |
| EDEXCEL A | A2 | OCR B | A2 |
| EDEXCEL B | A2 | WJEC | A2 |

Electric motors make use of the force that acts on a current-carrying conductor placed at right angles to a magnetic field. Provided that it is not parallel to the field (in which case there is no force), any electric current in a magnetic field experiences a force.

- The maximum force occurs when the current and magnetic field are at right angles to each other.
- The direction of the force is at right angles to both the field and the current.
- Fleming's left-hand rule shows the relationship between the directions of the current, magnetic field and force.

field

N force S

current

The force on a wire loop.

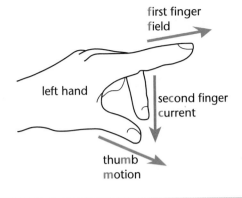

first finger
field

left hand

second finger
current

thumb
motion

1 point your first finger in the direction of the magnetic field.

2 point your second finger in the direction of the current.

3 your thumb points in the direction of motion.

## The size of the force

| AQA A | A2 | NICCEA | A2 |
| AQA B | A2 | OCR A | M2 |
| EDEXCEL A | A2 | OCR B | A2 |
| EDEXCEL B | A2 | WJEC | A2 |

The size of the force that acts on a current-carrying conductor in a magnetic field depends on:

- the direction of the current relative to the magnetic field
- the size of the current
- the length of conductor in the magnetic field
- the strength of the magnetic field.

> **KEY POINT**
> These factors are used to define the strength of a magnetic field as:
> the force per unit (current × length) perpendicular to the field.
> Magnetic field strength is also known as *flux density*. Its symbol is *B* and the unit is the tesla (T).

With the current and magnetic field directions shown in the diagram, the force is into the paper and the definition of magnetic field strength can be written as:

$B = F/Il$ or $F = BIl$.

You must always specify the directions of the quantities when using this formula to define magnetic field strength.

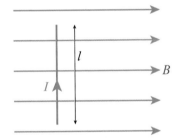

> **KEY POINT**
> This equation is also used to define the size of the tesla as:
> **the magnetic field strength that exerts a force of 1 N on a (current × length) of 1 A m placed at right angles to the field.**

The force between two parallel conductors like those in Q2 below is now used as the basis for the definition of the unit of current, the ampère.

## Progress check

1 a Draw a clear, labelled diagram that shows how magnetic field strength is defined in terms of the force on a current-carrying conductor. Your diagram should show the directions of any vector quantities.
  b The magnetic field strength in the centre of a solenoid is $2.0 \times 10^{-4}$ T. Calculate the size of the force on a wire of length 0.15 m carrying a current of 2.5 A placed:
   i along the axis of the solenoid
   ii perpendicular to the axis of the solenoid.

2 Two parallel wires carry currents in opposite directions. Draw a diagram showing the magnetic field due to each wire and the direction of the force on each wire.

2 The diagram should show that each wire is repelled from the other wire.
1 a see text b i 0 ii $7.5 \times 10^{-5}$ N

## Sample question and model answer

(a) Show that the unit of resistivity is the Ω m. [1]

Since $\rho$ = AR/l, the units of $\rho$ must equal those of AR/l.
Units of AR/l = m² × Ω ÷ m = Ω m.

This technique relies on the fact that a relationship between physical quantities must be homogenous, i.e. the units on each side are the same.

(b) A cable consists of seven straight strands of copper wire each of diameter 1.35 mm as shown in the diagram. [4]

strand of copper wire

Calculate:

(i) the cross-sectional area of **one strand** of copper wire.

Cross-sectional area = ¼πd²
= ¼ × π × (1.35 × 10⁻³)² = 1.43 × 10⁻⁶ m²   1 mark

It is important here to convert the diameter from mm to m in order to maintain a consistent set of units. Many AS candidates, faced with this type of calculation, work out the answer in mm² and then apply a wrong conversion factor to change it to m².

(ii) the resistance of a 100 m length of the cable, given that the resistivity of copper is $1.6 \times 10^{-8}$ Ω m.

The resistance of one strand,
R = $\rho l$/A (1 mark)
= 1.6 × 10⁻⁸ Ω m × 100 m ÷ 1.43 × 10⁻⁶ m² = 1.12 Ω   1 mark
The resistance of 100 m of the cable is ¹/₇ of this, i.e. 0.16 Ω.   1 mark

The emphasis here is on reading the question thoroughly. A common error is to calculate the resistance of one strand only, ignoring the fact that the resistance of the cable is emphasised in the question.

(c) (i) If the cable in part (b) carries a current of 20 A, what is the potential difference between the ends of the cable? [2]

V = IR = 20 A × 0.16 Ω = 3.2 V.   1 mark

(ii) If a single strand of the copper wire in part (b) carried a current of 20 A, what would be the potential difference between its ends?

20 A × 1.12 Ω = 22.4 V.   1 mark

The calculations in **c** are straightforward applications of the resistance equation. Do not worry if your answers to **b** were wrong. You still gain full marks in **c** for correct working with the wrong resistance values.

(d) State **one** advantage of using a stranded rather than a solid core cable with copper of the same total cross-sectional area. [1]

A stranded cable is more flexible than a solid one and is less prone to snap due to repeated bending.

*NEAB Mechanics and Electricity (PH01), June 1999 Q 2*

The fixed wiring that connects the sockets in a house back to the consumer unit is done with solid cables as these are not subjected to repeated flexing. Connections between an appliance and a mains socket use stranded cables.

## Practice examination questions

1 (a) Write down the formula that relates the resistivity of a material to the resistance of a particular sample. [1]

(b) Explain the difference between *resistance* and *resistivity*. [2]

(c) The diagram shows a cuboid made from carbon.

The resistivity of carbon is $3.00 \times 10^{-5}$ Ω m.
Calculate the resistance of the cuboid, measured between the faces that are 10 mm apart. [3]

2 In the circuit shown in the diagram, the internal resistance of the cell can be neglected.

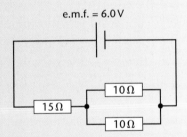

Calculate:
(a) the effective resistance of the two resistors in parallel [2]

(b) the current in the 15 Ω resistor [3]

(c) the current in each 10 Ω resistor [2]

(d) the potential difference across each 10 Ω resistor. [2]

3 (a) (i) Draw a sketch graph to show how the resistance of a thermistor varies with increasing temperature. [2]

(ii) State TWO ways in which this graph differs from a graph that shows how the resistance of a metallic conductor varies with increasing temperature. [2]

(iii) Explain why Ohm's law does not apply to either graph. [2]

(b) The diagram shows a thermistor in a series circuit with a 100 Ω resistor.

## Practice examination questions (continued)

The table gives some data about the resistance of the thermistor.

| temperature/°C | thermistor resistance/$\Omega$ |
| --- | --- |
| 5 | 500 |
| 20 | 300 |
| 40 | 75 |

(i) Calculate the potential difference across the thermistor at a temperature of 5°C. [3]

(ii) Explain how the potential difference across the thermistor changes as the temperature rises. [2]

(iii) The potential difference across the resistor is used to switch off a heater when the temperature reaches 40°C. Calculate the potential difference at which the switch operates. [3]

(iv) The circuit is adapted to switch off the heater when the temperature reaches 20°C. Calculate the value of the fixed resistor required. [3]

4  A cell has an e.m.f. of 1.60 V. When connected to a digital voltmeter of resistance 10 M$\Omega$ ($1.0 \times 10^7$ $\Omega$), the reading on the voltmeter is 1.60 V. When the cell is connected to a moving coil voltmeter of resistance 1 k$\Omega$ ($1.0 \times 10^3$ $\Omega$) the reading on the voltmeter is 1.55 V.

(a) Explain why there is a difference in the voltmeter readings. [3]

(b) Calculate the internal resistance of the cell. [3]

5  A 12 V car battery is used to light four 6 W parking lamps connected in parallel.

(a) Calculate the current in the battery. [1]

(b) How much charge flows through the battery in one minute? [3]

(c) Calculate the effective resistance of the four lamps. [3]

6  (a) A 4.7 $\Omega$ resistor has a power rating of 10 W.

(i) Calculate the maximum current that should pass in the resistor. [3]

(ii) Calculate the maximum potential difference across the resistor. [2]

(iii) The resistor is manufactured from constantan wire of cross-section $2.0 \times 10^{-7}$ m$^2$ and resistivity $5.0 \times 10^{-7}$ $\Omega$ m.
Calculate the length of wire used to make the resistor. [3]

(b) Low-power resistors are often made from carbon. The resistivity of carbon is $3.0 \times 10^{-5}$ $\Omega$ m. What is the advantage of making resistors from carbon? [2]

## Practice examination questions *(continued)*

**7** The diagrams show two circuits that can be used to investigate the current–voltage characteristics of component X.

circuit A          circuit B

(a) What name is given to circuit A? [1]

(b) Describe what circuit A is able to do. [1]

(c) What is the advantage of using circuit A rather than circuit B to investigate the characteristics of a diode? [2]

(d) The table shows data obtained by varying the voltage across a diode.

| voltage/V | 0.65 | 0.71 | 0.76 | 0.78 | 0.80 |
|---|---|---|---|---|---|
| *current*/A | 0.06 | 0.20 | 0.44 | 0.68 | 1.00 |
| *resistance*/$\Omega$ | | | | | |

(i) Complete the table. [2]

(ii) Draw a graph of resistance against voltage. [4]

(iii) Use the graph to estimate the voltage at which the resistance of the diode is 1 $\Omega$. [1]

(iv) Suggest why the data only span a narrow voltage range. [1]

**8** In a simple lighting circuit, a 60 W lamp is connected to a 12 V battery using copper cable of cross section 1.25 mm$^2$ ($1.25 \times 10^{-6}$ m$^2$). The lamp filament is made of tungsten of cross section $3.0 \times 10^{-4}$ mm$^2$ ($3.0 \times 10^{-10}$ m$^2$). The number of free electrons per unit volume for copper is $8.0 \times 10^{28}$ m$^{-3}$ and that for tungsten is $3.4 \times 10^{28}$ m$^{-3}$. The electronic charge $e = -1.60 \times 10^{-19}$ C.

(a) Calculate the drift velocity of the electrons in the copper cable. [3]

(b) Give TWO reasons why the drift velocity of the electrons in the tungsten is much greater than that of electrons in the copper. [2]

(c) Suggest why the tungsten becomes heated while the copper remains cool. [2]

**9** A cell of e.m.f. 3.0 V and internal resistance 0.5 $\Omega$ is connected to a 3.0 $\Omega$ resistor placed in series with a parallel combination of a 6.0 $\Omega$ and a 12.0 $\Omega$ resistor.

Calculate:

(a) the current in the circuit [3]

(b) the p.d. across the terminals of the cell. [2]

# Particle physics

## The following topics are covered in this chapter:

- A model gas
- Internal energy
- Radioactive decay

- Forces and particles
- Stability and the nucleus

## 3.1 A model gas

### After studying this section you should be able to:

- understand the concept of absolute zero of temperature and be able to convert temperatures between the Celsius and Kelvin scales
- recall and use the ideal gas equation of state $pV = nRT$
- explain the relationship between the kelvin temperature and the mean kinetic energy of the particles of an ideal gas

**LEARNING SUMMARY**

### Gas pressure

| | | | |
|---|---|---|---|
| AQA A | M2 | NICCEA | A2 |
| AQA B | A2 | OCR A | A2 |
| EDEXCEL A | M2 | OCR B | A2 |
| EDEXCEL B | A2 | WJEC | A2 |

As you will see later in this section, the average speed of the particles is different for different gases.

Brownian motion was first observed by Robert Brown while studying pollen grains suspended in water. The explanation of Brownian motion is due to Einstein.

The kinetic model pictures a gas as being made up of large numbers of individual particles in constant motion. At normal pressures and temperatures the particles are widely-spaced compared to their size. The motion of an individual particle is:

- **rapid** – a typical average speed at room temperature is 500 m s$^{-1}$
- **random** in both speed and direction – these are constantly changing due to the effects of collisions.

Gases exert **pressure** on the walls of their container and any other objects that they are in contact with. This pressure acts in all directions and is due to the forces as particles collide and rebound.

Evidence for this movement of gas particles comes from **Brownian motion**; the random, lurching movement of comparatively massive particles such as smoke specks when suspended in air. This movement can only be attributed to the bombardment by much smaller air particles which are too small to be seen by an optical microscope and must therefore be moving very rapidly.

Gas particles are widely spaced. Their movement is random in both speed and direction.

Experiments show that increasing the temperature of a gas also increases the pressure that it exerts. At higher temperatures the average speed of the particles increases. This affects the pressure in two ways:

- it increases the average force of the collisions
- collisions are more frequent.

The graph shows how the pressure of a gas depends on its Celsius temperature when the volume of

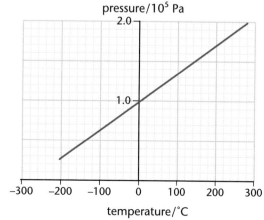

the gas is kept constant. Although the graph is a straight line, it does not show that pressure is directly proportional to Celsius temperature since the line does not go through the origin.

This is not surprising since 0°C does not represent 'no temperature'; it is merely a reference point that is convenient to use because it is easily reproduced.

The graph does however give an indication of whereabouts 'no temperature' is on the Celsius scale. If temperature is a measure of particle movement, then the minimum possible temperature corresponds to the minimum amount of particle movement. By extrapolating the graph back to the corresponding pressure, zero, it gives a figure for the minimum temperature at –273°C.

The extrapolation assumes that the substance remains a gas as it is cooled. Real gases would reach their boiling point and liquefy before the minimum temperature was reached.

Knowledge of the zero of temperature allows it to be measured on an absolute scale. On an absolute scale of measurement:

- the zero is the minimum amount of a quantity – it is not possible for a smaller measurement to exist
- a measurement of 40 units is twice that of 20 units.

Note that the kelvin is not called a degree and does not have a ° in the unit.

> **KEY POINT**
>
> The absolute scale of temperature is called the Kelvin scale; its unit is the kelvin (K) and its relationship to the Celsius scale is:
> $$T/\text{K} = \theta/°\text{C} + 273$$

The diagram shows the relationship between Celsius and kelvin temperatures.

When Celsius temperatures are converted to kelvins, the experimental results show that:

The value of the constant depends on the amount of gas and its volume.

> **KEY POINT**
>
> For a fixed mass of gas at constant volume:
> pressure ∝ absolute temperature or $p/T$ = constant

Remember: the relationship is only valid when temperature is measured in kelvin.

## The ideal gas equation

| AQA A | M2 | NICCEA | A2 |
| AQA B | A2 | OCR A | A2 |
| EDEXCEL A | M2 | OCR B | A2 |
| EDEXCEL B | A2 | WJEC | A2 |

When experimenting with a fixed mass of gas the three variables are pressure, volume and temperature. Fixing one of these enables the relationship between the other two to be established. In addition to the relationship between pressure and temperature above:

The relationship $pV$ = constant is known as Boyle's law.

> **KEY POINT**
>
> For a fixed mass of gas at constant temperature, pressure is inversely proportional to volume:
> $$\text{pressure} \propto 1/\text{volume} \quad \text{or} \quad pV = \text{constant}$$
> For a fixed mass of gas at constant pressure, the volume is proportional to absolute temperature:
> $$\text{volume} \propto \text{temperature} \quad \text{or} \quad V/T = \text{constant}$$

These relationships are based on experiments and are valid for all gases provided that the gas is neither at a high pressure nor close to its boiling point. They lead to the concept of an **ideal gas** as one that obeys Boyle's law, i.e. a gas for which $pV$ = constant.

Halving the volume of a gas leads to twice the rate of collisions – doubling the pressure. This is an example of Boyle's Law.

> The concept of an ideal gas is useful because the behaviour of all gases is close to ideal provided that certain conditions about temperature and pressure are met.

Real gases do not obey Boyle's law when the particles are close enough together so that they occupy a significant proportion of the gas's volume and exert appreciable forces on each other.

For an ideal gas the three gas laws can be combined into the single equation:

$pV/T$ = constant

> 1 mole of gas consists of the Avagadro constant, $N_A$, of particles. The value of the Avagadro constant is $6.02 \times 10^{23}$ mol$^{-1}$.

In this case the value of the constant depends only on the amount of gas. For 1 mole of any gas under ideal conditions it has the value of 8.3 J mol$^{-1}$ K$^{-1}$. This constant is called the **molar gas constant** and has the symbol $R$. Doubling the number of particles by considering two moles of gas has the effect of doubling the value of the constant, so:

> **KEY POINT**
>
> The equation of state for an ideal gas is:
> $$pV = nRT$$
> where $n$ is the number of moles of gas and $R$ is the molar gas constant.

## An algebraic model

| | | | |
|---|---|---|---|
| AQA A | M2 | NICCEA | A2 |
| AQA B | A2 | OCR A | A2 |
| EDEXCEL A | M2 | OCR B | A2 |
| EDEXCEL B | A2 | WJEC | A2 |

If a gas is sealed in a container, then increasing its temperature results in an increase in the pressure of the gas. This can be attributed to an increase in the average speed of the particles. The kinetic model can be extended to establish the relationship between the temperature of a gas and the speed of the particles. First of all it is necessary to establish just what is meant by 'ideal gas behaviour'.

The kinetic theory of gases involves some basic assumptions about the particles of a gas:

- a gas consists of a large number of particles in a state of rapid, random motion
- gas pressure is a result of collisions between the particles and the container walls.

In addition, for an ideal gas it is assumed that:

- the volume occupied by the particles is negligible
- intermolecular forces are negligible
- all collisions are elastic, so there is no loss in kinetic energy.

With these assumptions, it can be shown that **the pressure of an ideal gas is proportional to the mean of the squares of the particle speeds**:

> Note that working out $< c^2 >$ involves:
> - listing all the individual particle speeds
> - squaring them
> - calculating the mean of the squares.

> **KEY POINT**
>
> $$pV = 1/3\ Nm < c^2 >$$
> Where $N$ = total number of particles
> $m$ = mass of each particle
> $< c^2 >$ = mean of the squares of the particle speeds

Alternatively, since $Nm/V$ = mass ÷ volume = density:

> **KEY POINT**
>
> $$p = 1/3\ \rho < c^2 >$$
> Where $\rho$ = density of the gas

If the mass and volume of an ideal gas are fixed, the gas pressure is proportional to its temperature, $p/T$ = constant. The above results show that, as pressure is proportional to $< c^2 >$ it must be proportional to the mean kinetic energy of the particles.

These factors taken together give the relationship between kinetic energy and temperature:

> **KEY POINT**
>
> the mean kinetic energy of the particles of an ideal gas is proportional to the kelvin temperature
>
> $$\tfrac{1}{2}m < c^2 > \propto T$$

This does not mean that hydrogen molecules move, on average, four times as fast as those of oxygen at a particular temperature. The root mean square speed, $\sqrt{} < c^2 >$, can be used as a typical speed but not the mean speed of the particles.

This means that doubling the kelvin temperature of a gas doubles the mean kinetic energy and the mean square speed of its particles. Remember that these relationships apply to all ideal gases, irrespective of the gas. So at a particular temperature the mean kinetic energy of the molecules in a sample of hydrogen is the same as those in a sample of oxygen. Since an oxygen molecule has sixteen times the mass of a hydrogen molecule, the mean square speed of the hydrogen molecules must be sixteen times as great as that of oxygen molecules.

## The Boltzmann constant

| | | | |
|---|---|---|---|
| AQA A | M2 | NICCEA | A2 |
| AQA B | A2 | OCR B | A2 |
| EDEXCEL B | A2 | WJEC | A2 |

The algebraic model can be used to put a figure on the mean kinetic energy of the particles of an ideal gas at any temperature.

For one mole of gas the number of particles, $N$ is equal to Avagadro's number, $N_A$. So for one mole, combining the equations $pV = RT$ and $pV = 1/3 N_A m < c^2 >$ gives:

$$\tfrac{1}{2}m < c^2 > = \tfrac{3}{2}RT/N_A \text{ or } N_A \times \tfrac{1}{2}m < c^2 > = \tfrac{3}{2}RT$$

This means that the total kinetic energy of one mole of an ideal gas is equal to $\tfrac{3}{2}RT$, since $N_A$ is the number of particles and $\tfrac{1}{2}m < c^2 >$ is their mean kinetic energy.

The Boltzmann constant, $k$, relates the mean kinetic energy of the particles in an ideal gas directly to temperature:

> **KEY POINT**
>
> $$\tfrac{1}{2}m < c^2 > = \tfrac{3}{2}RT/N_A$$
> $$= 3/2\ kT$$
>
> Where $k = R/N_A$ is the Boltzmann constant and has the value $1.38 \times 10^{-23}$ J K$^{-1}$

To work out the mass of a molecule of a gas, divide the molar mass by the Avagadro constant.

So, for example, at a temperature of 20°C (293 K) the mean kinetic energy of the particles of an ideal gas is equal to $3/2 \times 1.38 \times 10^{-23}$ J K$^{-1} \times 293$ K $= 6.07 \times 10^{-21}$ J. This may seem a tiny amount of energy, but remember that it refers to a single gas particle which has a very small mass.

## Progress check

Assume that $R = 8.3$ J mol$^{-1}$ K$^{-1}$

1 2.0 mol of an ideal gas is sealed in a container of volume $5.5 \times 10^{-2}$ m$^3$.
  a Calculate the pressure of the gas at a temperature of 18°C.
  b At what Celsius temperature would the pressure of the gas be doubled?

2 1.0 mol of oxygen has a mass of 32 g. Calculate the mean square speed of the molecules at 0°C and a pressure of $1.01 \times 10^5$ Pa.

3 A room measures 3.5 m × 3.2 m × 2.2 m.
  It contains gas at a pressure of $1.03 \times 10^5$ Pa
  and a temperature of 25°C. Calculate the
  number of moles of gas in the room.

1 a $8.8 \times 10^4$ Pa
  b 309°C
2 $2.1 \times 10^5$ m$^2$ s$^{-2}$
3 1026

# 3.2 Internal energy

**After studying this section you should be able to:**

- *explain the meaning of* specific heat capacity *and* specific latent heat *and use the appropriate relationships*
- *describe the principle of operation of a* heat engine *and a* heat pump
- *calculate the maximum efficiency of a heat engine*

**LEARNING SUMMARY**

## Energy of individual particles

| | | | |
|---|---|---|---|
| AQA A | M2 | NICCEA | A2 |
| AQA B | A2 | OCR A | A2 |
| EDEXCEL A | M2 | OCR B | A2 |
| EDEXCEL B | A2 | WJEC | A2 |

All materials are made up of particles, and particles have energy. The particles of an ideal gas have only kinetic energy since forces between the particles are negligible. However, this cannot be true of a solid or a liquid or the particles would not stay in that phase, and the particles in real gases exert both attractive and repulsive forces on each other during collisions.

Imagine a collision between two particles in a real gas:

- as the particles approach, the attractive forces cause an increase in the speed and kinetic energy
- the particles then get closer and the speed decreases, being momentarily zero before they start to separate
- the process is then reversed.

> The speed of colliding particles is momentarily zero as their direction of travel is reversed.

There is an interchange between potential (stored) energy and kinetic energy during the collision, as shown in the diagram.

particles speed up due to short-range attractive forces

particles slow down due to close-range repulsive forces

> As the separation between two particles decreases, the net force changes from attractive to repulsive. The equilibrium position is where these forces are equal in size, so the resultant force is zero.

The particles of real gases, like those of solids and liquids, have both potential and kinetic energy. The distribution of energy between kinetic and potential is constantly changing and so is said to be random.

The total amount of energy of the particles in an object is known as its **internal energy**. At a constant temperature the internal energy of an object remains unchanged, but the contributions of individual particles to that energy change due to the transfer of energy during interactions between particles.

## Being specific

| | | | |
|---|---|---|---|
| AQA A | M2 | NICCEA | A2 |
| AQA B | A2 | OCR A | A2 |
| EDEXCEL A | M2 | OCR B | A2 |
| EDEXCEL B | M1 | WJEC | A2 |

Changing the temperature of an object involves a change in its internal energy, the total potential and kinetic energy of the particles. One way of doing this is to place the object in thermal contact with a hotter or colder object, so that there is a net flow of energy from hot to cold (see diagram).

> Two objects are in thermal contact if they can exchange heat. Energy flows both ways between a warm and a cold object in thermal contact, but there is more energy flow from the warm object to the cold one than the other way.

 a cold dish in a hot oven

 warm food placed in a refrigerator

**The energy flow between objects at different temperatures**

The energy transfer required to change the temperature of an object depends on:

- the temperature change
- the mass of the object
- the material the object is made from.

These are all taken into consideration in the concept of specific heat capacity. The term 'specific' means 'for each kilogram', so any physical measurement that is described as 'specific' refers to 'per kilogram of material'.

> When using this relationship the temperatures can be in either Celsius or kelvin since the temperature change is the same in each case.

> **KEY POINT**
>
> The specific heat capacity of a material, $c$, is defined as:
> The energy transfer required to change the temperature of 1 kg of the material by 1°C.
> $$\Delta E = mc\,\Delta\theta$$
> where $c$, specific heat capacity, is measured in $J\ kg^{-1}\ °C^{-1}$ or $J\ kg^{-1}\ K^{-1}$

## Changing phase

| AQA A | M2 | EDEXCEL A | M2 |
| AQA B | A2 | OCR A | A2 |
|  |  | OCR B | A2 |

If you try to squash a solid or a liquid, the particles are pushed closer together and they repel each other. Stretching has the opposite effect; when the separation of the particles is increased the forces are attractive. At increased separations the particles have increased potential energy and this energy has to be supplied for any process that involves expansion to take place.

> The term 'phase' means whether the substance is a solid, liquid or gas.

When a substance changes phase there is a change in the potential energy of the particles. For most substances there is a small increase in potential energy when changing from solid to liquid and a much bigger change from liquid to gas.

> The increase in particle separation during a change of phase from solid to liquid is small, but that during a change of phase from liquid to gas is large.

*There is a small increase in potential energy of the particles when a substance changes from solid to liquid, and a much larger increase in changing from liquid to gas.*

The energy absorbed or released during a change of phase is called latent heat. As with heat capacity, the term 'specific latent heat' refers to 1 kilogram of material.

> This definition applies to a change of phase where energy is removed from the substance, as well as a change where energy is supplied.

> **KEY POINT**
>
> The specific latent heat of a material, $l$, is defined as:
> The energy required to change the phase of 1 kg of the material without changing its temperature.
> $$E = ml$$

Note that there are two values of the specific latent heat for any material:

- the specific latent heat of fusion refers to a change of phase between solid and liquid
- the specific latent heat of vaporisation refers to a change of phase between liquid and gas

On each horizontal part of the curve the substance exists in two phases. Can you identify them?

The diagram shows a typical temperature–time graph as a solid is heated at a constant rate and passes through the three phases. Where the curve is horizontal it shows that energy is being absorbed with no change in temperature. This corresponds to a change of phase. Note that much more energy is absorbed during the change from liquid to gas than during the change from solid to liquid.

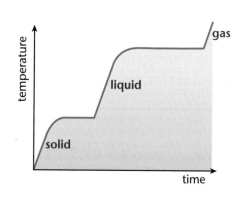

## Heat and work

| AQA B | A2 | OCR A | A2 |
|---|---|---|---|
| EDEXCEL A | M2 | WJEC | A2 |

Placing an object in thermal contact with one that is hotter or colder, for example an immersion heater or the inside of a freezer, is one way of changing its temperature. Anyone who has ever pumped up a bicycle tyre is aware that compressing the air in the pump also causes it to become warmer.

Compressing a gas causes heating and expanding a gas causes cooling. Whether the gas is heated or cooled depends on whether work is done **on** it or **by** it.

Other everyday examples of work causing a change in temperature include:

- striking a match, where work done against friction forces results in heating of the match head
- refrigeration, where the rapid expansion of a liquid into a vapour causes cooling.

The **first law of thermodynamics** states the internal energy of a gas depends only on its state, i.e. the conditions of temperature, pressure and volume and the amount of gas, and not how it reached that state. Transferring energy as heat to a gas has the same effect as doing work on it – they both cause an increase in the internal energy, and therefore the temperature.

In other words, the first law states that work and heat are equivalent.

> **KEY POINT**
>
> The first law of thermodynamics can be written as:
> increase in internal energy of gas = heat transferred to gas + work done on gas
>
> $$\Delta U = Q + W$$
>
> Note that the values of $\Delta U$, $Q$ and $W$ are negative if the internal energy decreases or heat is removed from the gas or work is done by the gas.

## Heat engines

| EDEXCEL A | M2 |
|---|---|

Internal combustion engines drive cars, buses and some trains. In these engines:

- the burning of fuel supplies heat to a gas, increasing its internal energy
- the gas does work as it pushes the pistons, decreasing the internal energy.

In an internal combustion engine, the pistons are driven by the hot gases produced when fuel burns. Energy that is not removed from the gases is wasted in the exhaust gases and passes to the atmosphere, which is at a lower temperature.

Rockets, jet engines and steam engines are all examples of heat engines.

This illustrates what is meant by a **heat engine**; a device that uses the energy flow from a **hot source** to a **cold sink** to do work. The principle of operation of a heat engine is shown in the diagram.

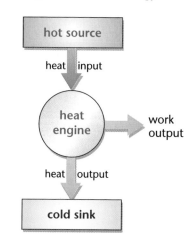

It is not possible to extract all the internal energy from the steam in a steam turbine. Can you explain why?

The steam turbines that drive the generators in power stations are also heat engines. Steam enters the turbines at high temperature and pressure and leaves at much lower temperature and pressure. Of the internal energy lost by the steam:

- some does work on the turbines
- some is lost to the surroundings.

The turbines are designed to extract as much energy from the steam as possible; the greater the proportion of the available energy extracted, the more efficient the process is.

If there are no energy losses, then the maximum efficiency of a heat engine is given by the relationship:

Refer to page 62 for the definition of efficiency.

> **KEY POINT**
>
> maximum efficiency = $(T_H - T_C) \div T_H$
> where $T_H$ is the temperature of the hot source and $T_C$ is the temperature of the cold sink. Both temperatures are measured in kelvin.

For example, if steam enters a turbine at 250°C (523 K) and leaves at 120°C (393 K) then the maximum efficiency is (523 K – 393 K) ÷ 523 K = 0.25. In practice, real heat engines do not operate at their maximum efficiency since there are always other energy losses.

## Heat pumps

EDEXCEL A    M2

Some ovens and convector heaters use fans to speed up the rate of energy transfer.

When two objects at different temperatures are placed in thermal contact, the net energy flow is always from hot to cold. No other energy source is needed to drive this flow, although sometimes fans are used to speed up the natural process. In refrigeration, the direction of energy flow is from cold to hot. To achieve this, an external source of energy is needed to drive a **heat pump**.

The principle of a refrigerator is shown in the diagram.

The pipes at the back of a refrigerator are painted black. How does this help to remove energy?

A volatile liquid is one that evaporates easily.

- Volatile liquid evaporates as it is forced through a narrow jet inside the refrigerator.
- This evaporation causes cooling and the cold vapour absorbs energy from its surroundings.
- The vapour returns to the outside where it is compressed, causing it to become heated.
- Energy is dissipated to the surroundings as the hot vapour passes through cooling pipes, where it condenses.

This method of heating a building can be used effectively where there is a convenient river to extract the energy from.

Heat pumps are also used to heat buildings by extracting energy from the surroundings. The energy available for heating is more than the energy needed to drive the pump, so this method is both cost-efficient and energy-efficient. Unfortunately, the time of year when most energy is needed is also that when there is least available.

### Progress check

1 Hot water enters a central heating radiator at a temperature of 52°C and leaves it at a temperature of 39°C. The specific heat capacity of water is 4200 J kg⁻¹ K⁻¹.

   a Calculate the energy transfer from each kg of water that passes through the radiator.

   b If the rate of flow of water through the radiator is 3.0 kg min⁻¹, calculate the power output of the radiator.

2 The specific latent heat of vaporization of nitrogen is $2.0 \times 10^5$ J kg⁻¹. Calculate the energy required to change 0.65 kg of liquid nitrogen into vapour without changing its temperature.

3 Some air is trapped in a sealed syringe. The piston is pushed in and does 450 J of work on the air. While this is happening 300 J of heat flows out of the air.

   a Calculate the change in internal energy of the air.

   b Explain whether the air ends up hotter or colder than it was to start with.

1 a $5.46 \times 10^4$ J   b 2.73 kW
2 $1.30 \times 10^5$ J
3 a +150 J   b hotter as the internal energy has increased.

# 3.3 Radioactive decay

*After studying this section you should be able to:*

- *explain how the results of alpha-scattering experiments give evidence for the atomic model*
- *describe the effects of radioactive decay on the nucleus*
- *calculate the half-life of a radioactive isotope from an activity–time graph*

**LEARNING SUMMARY**

## Evidence for the atomic model

| | | | |
|---|---|---|---|
| AQA A | M1 | NICCEA | A2 |
| AQA B | M2 | OCR A | A2 |
| EDEXCEL A | M1 | OCR B | A2 |
| EDEXCEL B | A2 | WJEC | A2 |

An alpha particle is a positively-charged particle consisting of two protons and two neutrons. The experiments were carried out in a vacuum so that the alpha particles were not scattered by air particles.

The simple atomic model pictures the atom as a tiny, positively-charged nucleus surrounded by negatively-charged particles (electrons) in orbit. Evidence for this model comes from the scattering of alpha particles as they pass through a thin material such as gold foil. The results of these experiments, first carried out under the guidance of Rutherford in 1911, can be summarised as:

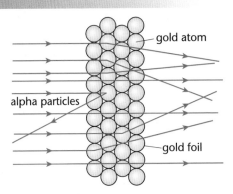

- most of the alpha particles travel straight through the foil with little or no deflection
- a small number are deflected by a large amount
- a tiny number are scattered backwards.

Rutherford concluded that the atoms of gold are mainly empty space, with tiny regions of concentrated charge. This charge must be the same sign as that of alpha particles (positive) to explain the back-scattering as being due to the repulsion between similar-charged objects.

## The nucleus

| | | | |
|---|---|---|---|
| AQA A | M1 | NICCEA | A2 |
| AQA B | M2 | OCR A | A2 |
| EDEXCEL A | M1 | OCR B | A2 |
| EDEXCEL B | A2 | WJEC | M2 |

You will not find El in the periodic table – it is fictitious.

Later experiments showed that there are two types of particle in the nucleus, **protons** and **neutrons**.

> The nucleus of an element is represented as $_Z^A$El.
> $Z$ is the atomic number, the number of protons.
> $A$ is the mass number, the number of nucleons (protons and neutrons).
>
> **KEY POINT**

The element is fixed by $Z$, the number of protons. A neutral atom has equal numbers of protons in the nucleus and electrons in orbit. Different atoms of the same element can have different values of $A$ due to having more or fewer neutrons. As this does not affect the number of electrons in a neutral atom, the chemical properties of these atoms are the same. They are called **isotopes** of the element. The most common form of carbon, for example, is carbon-12, $_6^{12}C$, which has six protons and six neutrons in the nucleus. Carbon-14, $_6^{14}C$, has the same number of protons but two extra neutrons. These two forms of carbon are isotopes of the same element.

The phrase 'relative to the value' means compared to the actual amount of charge, ignoring the sign.

The charges and relative masses of atomic particles are shown in the table. The masses are in atomic mass units (u), where 1u = 1/12 the mass of a carbon-12 atom = $1.661 \times 10^{-27}$ kg. The charges are relative to the value of the electronic charge, $e = 1.602 \times 10^{-19}$ C.

| atomic particle | mass | charge |
|---|---|---|
| proton | 1.01 | +1 |
| neutron | 1.01 | 0 |
| electron | $5.49 \times 10^{-4}$ | −1 |

## Radiation all around us

| | | | |
|---|---|---|---|
| AQA A | A2 | NICCEA | A2 |
| AQA B | M2 | OCR A | A2 |
| EDEXCEL A | M1 | WJEC | A2 |
| EDEXCEL B | M2 | | |

The term 'background radiation' is also used to describe the microwave radiation left over from the 'Big Bang'. This is a different type of radiation to nuclear radiation.

Radioactive decay occurs when an atomic nucleus changes to a more stable form. This is a random event that cannot be predicted. The emissions from these nuclei are collectively called radioactivity or radiation.

We are subjected to a constant stream of radiation called background radiation. Most of this is natural in the sense that it is not caused by the activities of people living on Earth. Sources of background radiation include:

- the air that we breathe – radioactive radon gas from rocks can concentrate in buildings
- the ground and buildings – all rocks contain radioactive isotopes
- the food that we eat – the food chain starts with photosynthesis. Radioactivity enters the food chain in the form of carbon-14, an unstable isotope of carbon that is continually being formed in the atmosphere
- radiation from space, called cosmic radiation
- medical and industrial uses of radioactive materials.

## The emissions

| | | | |
|---|---|---|---|
| AQA A | A2 | NICCEA | A2 |
| AQA B | M2 | OCR A | A2 |
| EDEXCEL A | M1 | OCR B | A2 |
| EDEXCEL B | M2 | WJEC | A2 |

The range of a beta-plus particle is effectively zero, since it is annihilated when it collides with an electron.

Alpha radiation is the most intensely ionising and can cause a lot of damage to body tissue. Although it cannot penetrate the skin, alpha emitters can enter the lungs during breathing.

Radioactive emissions are detected by their ability to cause ionisation, creating charged particles from neutral atoms and molecules by removing outer electrons. This results in a transfer of energy from the emitted particle, which is effectively absorbed when all its energy has been lost in this way. The four main emissions are alpha ($\alpha$), beta-plus ($\beta^+$), beta-minus ($\beta^-$) and gamma ($\gamma$). Of these, alpha radiation is the most intensely ionising and has the shortest range, with the exception of beta-plus, while gamma radiation is the least intensely ionising and has the longest range.

- In alpha emission the nucleus emits a particle consisting of two protons and two neutrons. This has the same make-up as a helium nucleus.
- Beta-minus emission occurs when a neutron decays into a proton, emitting an electron in the process.
- In beta-plus emission a proton changes to a neutron by emitting a positron.
- A gamma emission is short-wavelength electromagnetic radiation.

Some properties of these radiations are shown in the table.

| radioactive emission | nature | charge/e | symbol | penetration | causes ionisation | affected by electric and magnetic fields |
|---|---|---|---|---|---|---|
| alpha | two neutrons and two protons | +2 | $^4_2\text{He}$ or $^4_2\alpha$ | absorbed by paper or a few cm of air | intensely | yes |
| beta-minus | high-energy electron | −1 | $^0_{-1}\text{e}$ or $^0_{-1}\beta$ | absorbed by 3 mm of aluminium | weakly | yes |
| beta-plus | positron (antielectron) | +1 | $^0_{+1}\text{e}$ or $^0_{+1}\beta$ | annihilated by an electron | | yes |
| gamma | short-wavelength electromagnetic radiation | none | $^0_0\gamma$ | reduced by several cm of lead | very weakly | no |

Notice that the electron emitted in beta-minus decay and the positron emitted in beta-plus decay have been allocated the atomic numbers −1 and +1. This is because of the effect on the nucleus when these particles are emitted.

## Balanced equations

| AQA A | A2 | NICCEA | A2 |
| AQA B | M2 | OCR A | A2 |
| EDEXCEL A | M1 | OCR B | A2 |
| EDEXCEL B | A2 | WJEC | A2 |

Technetium-99 is an artificial isotope that emits gamma radiation only. It is used as a tracer in medicine. The gamma radiation can be detected outside the body and there is less risk of cell damage than with an isotope that also emits alpha or beta radiation.

When a nucleus decays by alpha or beta emission, the numbers of protons and neutrons are changed. Gamma emission does not change the make-up of the nucleus, but corresponds to the nucleus losing excess energy. Gamma emission often occurs alongside alpha and beta emissions, though some artificial radioactive isotopes emit gamma radiation only.

The changes that take place due to alpha and beta emissions are:

- **alpha** – the number of protons decreases by 2 and the number of neutrons also decreases by two
- **beta-minus** – the number of neutrons decreases by one and the number of protons increases by one
- **beta-plus** – the number of neutrons increases by one and the number of protons decreases by one.

In writing equations that describe nuclear decay, both charge (represented by $Z$) and the number of nucleons (represented by $A$) are conserved. The table summarises these changes and gives examples of each type of decay.

Check that the equations given as examples are balanced in terms of charge and number of nucleons.

| particle emitted | effect on A | effect on Z | example |
|---|---|---|---|
| alpha | −4 | −2 | $^{226}_{88}\text{Ra} \rightarrow {}^{222}_{86}\text{Rn} + {}^{4}_{2}\text{He}$ |
| beta-minus | unchanged | +1 | $^{14}_{6}\text{C} \rightarrow {}^{14}_{7}\text{N} + {}^{0}_{-1}\text{e}$ |
| beta-plus | unchanged | −1 | $^{11}_{6}\text{C} \rightarrow {}^{11}_{5}\text{B} + {}^{0}_{+1}\text{e}$ |

## Rate of decay and half-life

| AQA A | A2 | NICCEA | A2 |
| AQA B | M2 | OCR A | A2 |
| EDEXCEL A | M1 | OCR B | A2 |
| EDEXCEL B | A2 | WJEC | A2 |

> **KEY POINT**
>
> The activity, or rate of decay, of a sample of radioactive material is measured in **becquerel** (Bq). An activity of 1 Bq represents a rate of decay of 1 $\text{s}^{-1}$.

Radioactive decay is a random process and the decay of an individual nucleus cannot be predicted. However, given a sample containing large numbers of undecayed nuclei, then statistically the rate of decay should be proportional to the number of undecayed nuclei present. Double the size of the sample and, on average, the rate of decay should also double.

Unlike chemical reactions, radioactive decay is not affected by changes in temperature.

There are only two factors that determine the rate of decay of a sample of radioactive material. They are:

- the radioactive isotope involved
- the number of undecayed nuclei.

> **KEY POINT**
>
> The relationship between the rate of decay, or activity, of a radioactive isotope is:
>
> $$\text{activity} = \lambda N$$
>
> where the activity is measured in becquerel, $N$ is the number of undecayed nuclei present and $\lambda$ is the **decay constant** of the substance. $\lambda$ has units of $\text{s}^{-1}$.

When the activity of a radioactive isotope is plotted against time, the result is a curve that shows the activity decreasing as the number of undecayed nuclei decreases. A decay curve is shown in the diagram. The shape of the curve is the same for all radioactive substances, but the activities and time scales depend on the size of the sample and its decay constant.

The graph shown is a plot of activity against time. A plot of number of undecayed nuclei against time would be identical but with a different scale on the *y*-axis.

The curve shows **exponential decay**. In an exponential decay curve:

- the rate of change of *y*, represented by the gradient of the curve, is proportional to *y*
- in equal intervals of *x* the value of *y* always changes in the same ratio.

This second point means that it always takes the same time interval for the activity to decrease to a given fraction of any particular value. The time for the activity to halve is a measure of the **half-life** of the substance.

Note that although half-life is defined in terms of the number of undecayed nuclei, it is usually measured as the average time for the activity of a sample to halve.

> **KEY POINT**
>
> The half-life of a radioactive isotope, $t_{\frac{1}{2}}$ is the average time taken for the number of undecayed nuclei of the isotope to halve.

Because radioactive decay is a random process, the results of any experiment to measure activity do not fit the curve exactly and there is some variation in the change of activity when identical samples of the same material are compared. This is why the term *average* is used in the definition of half-life. The values of half-lives range from tiny fractions of a second to many millions of years. Starting with $N$ nuclei of a particular isotope, the number remaining after one half-life has elapsed is $N/2$, after two half-lives $N/4$ and after n half-lives it is $N/2^n$.

There is a relationship between the half-life and the decay constant of any particular radioactive isotope. The shorter the half-life, the greater the rate of decay and the decay constant.

When using this relationship, $\lambda$ and $t_{\frac{1}{2}}$ must be measured in consistent units, e.g. $\lambda$ in $s^{-1}$ and $t_{\frac{1}{2}}$ in s.

> **KEY POINT**
>
> The half-life of a radioactive isotope and its decay constant are related by the equation:
> $$\lambda t_{\frac{1}{2}} = \ln 2 = 0.69$$

## Progress check

1 a What feature of the results of alpha-particle scattering experiments shows that the nucleus is positively charged?

   b Explain how the results of these experiments suggest that the nucleus occupies a very small proportion of the volume of the atom.

2 Strontium-90, $_{38}^{90}Sr$, decays to yttrium, symbol Y, by beta-minus emission. Write a balanced nuclear equation for this decay.

3 Use the graph on this page to determine the values of the half-life and the decay constant for a radioactive isotope that has this decay curve.

3 half-life = 60 s
decay constant = $1.15 \times 10^{-2}\ s^{-1}$

2 $_{38}^{90}Sr \rightarrow {}_{39}^{90}Y + {}_{-1}^{0}e$

1 a The fact that some alpha particles are back-scattered shows that they are repelled from the nucleus, so they must have the same charge.

   b Most alpha particles pass through thin gold foil without being deviated.

# 3.4 Forces and particles

**After studying this section you should be able to:**

- distinguish between the groups of fundamental and other particles
- interpret Feynman diagrams
- explain forces in terms of exchange particles
- describe the processes of pair production and pair annihilation

## Holding the nucleus together

| | | | |
|---|---|---|---|
| AQA A | M1 | NICCEA | A2 |
| EDEXCEL A | M3C | OCR B | A2 |
| EDEXCEL B | A2 | | |

*On the nuclear scale, gravitational forces are so weak that they can be ignored.*

The forces that we experience every day can be classified as one of two types:

- **gravitational forces** affect all objects that have mass – they are very weak and have an infinite range
- **electromagnetic forces** also have an infinite range and are much stronger than gravitational forces – they include all forces due to static or moving charges.

The nucleus is very concentrated in terms of both mass and charge. The electrostatic force between the protons is immense and the gravitational attraction is tiny, so if these were the only forces acting, the nucleus would be unstable. In addition to the forces above, there are two other fundamental forces.

*The strong force decreases very rapidly with increasing separation of the particles.*

- The **strong nuclear force** affects protons and neutrons but not electrons. Its limited range, $1 \times 10^{-15}$ m, means that it only acts between nucleons that are very close, i.e. next to each other in the nucleus.
- The **weak nuclear force** affects all particles. It has an even shorter range than the strong force, $1 \times 10^{-18}$ m, and is responsible for beta decay.

It is the strong nuclear force that keeps the nucleus together. The strong attraction between the nucleons balances the electrostatic repulsion.

## Fundamental particles

| | | | |
|---|---|---|---|
| AQA A | M1 | NICCEA | A2 |
| AQA B | M2 | OCR B | A2 |
| EDEXCEL A | M1 | WJEC | A2 |
| EDEXCEL B | A2 | | |

*In an inelastic collision, kinetic energy is not conserved.*

Electrons are fundamental particles, they cannot be split up. Protons and neutrons are themselves made up of other particles. Evidence for the structure of protons and neutrons comes from **deep inelastic scattering** experiments. In these experiments very-high-energy electrons are fired at nucleons. The electrons are not affected by the strong force and they penetrate the nucleons in an inelastic collision, resulting in the electrons being scattered through a range of angles, with some being back-scattered in the same way that alpha particles can be back-scattered by gold foil.

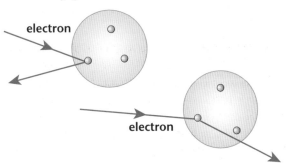

deep inelastic scattering experiments show that
nucleons contain small regions of intense charge

This gives evidence that a nucleon contains small, dense regions of charge that are themselves fundamental particles. These particles are called **quarks**. The diagram above illustrates some of the findings of deep inelastic scattering. In the diagram below each nucleon is shown as containing three quarks.

 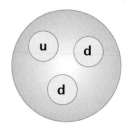

**The quarks contained in nucleons**

The symbol *e* is used throughout this section to represent the amount of charge on an electron, ignoring the sign. Positive (+) and negative (–) show the sign of a charge.

Of the six types of quark, two occur in protons and neutrons. These are the **up** and **down** quarks. The up quark, symbol u, has a charge of $+2e/3$ and that on the down quark, symbol d, is $-1e/3$. The diagram above shows the quarks in a proton and a neutron.

## Particles and their antis

| | | | |
|---|---|---|---|
| AQA A | M1 | EDEXCEL B | A2 |
| AQA B | M2 | NICCEA | A2 |
| EDEXCEL A | M3C | | |

Where a particle has zero charge, for example the neutron, its antiparticle also has zero charge.

For every type of particle there is an antiparticle. An antiparticle:

* has the same mass as the particle, but has the opposite charge if it is the antiparticle of a charged particle
* annihilates its particle when they collide, the collision resulting in energy in the form of gamma radiation or the production of other particles. This process is called **pair annihilation**
* can be created, along with its particle, when a gamma ray passes close to a nucleus. This process is called **pair production**.

**Anti-electrons**, or **positrons**, are emitted in some types of radioactive decay and can be created when cosmic rays interact with the nuclei of atoms. Like all antiparticles, they cannot exist for very long on Earth before being annihilated by the corresponding particles.

The symbols β⁻ and β⁺ are normally used to refer to electrons and positrons given off as a result of radioactive decay.

> **using symbols for particles**
> An electron is represented by the symbol e⁻ or β⁻.
> The positron can be represented by either of the symbols e⁺ and β⁺.
> The proton and antiproton are written as p⁺ and p⁻.
> More generally, a line drawn above the symbol for a particle refers to an antiparticle, so u and ū (read as u-bar) refer to the up quark and its antiquark.

**KEY POINT**

The greek letter ν is pronounced as 'new'.

The **antineutrino**, $\bar{\nu}$, is a particle of antimatter emitted along with an electron when an nucleus undergoes β⁻ decay. It is a particle that has almost no mass and no charge. Its corresponding particle the **neutrino**, ν, is emitted with a positron in β⁺ decay.

## The quark family

| | | | |
|---|---|---|---|
| AQA A | M1 | EDEXCEL B | A2 |
| AQA B | M2 | NICCEA | A2 |
| EDEXCEL A | M3C | OCR B | A2 |

There are six quarks which, together with their antiquarks, make up the family. The name of each quark describes its type, or **flavour**. Some of the properties of quarks are:

The kaons (see later in this section) were described as 'strange' particles when their lifetime turned out to be much longer than expected.

* charge – this is $+2e/3$ or $-1e/3$ for quarks and is always conserved
* baryon number – like charge, this is conserved in all interactions
* strangeness – this property describes the strange behaviour of some particles that contain quarks with strangeness value that is non-zero. It is conserved in strong and electromagnetic interactions but can be changed in weak interactions.

The values of these properties are shown in the table.

The six quarks can be thought of as three pairs, u–d, c–s and t–b.

| quark | symbol | charge/e | baryon number, B | strangeness, S |
|---|---|---|---|---|
| up | u | +2/3 | 1/3 | 0 |
| down | d | −1/3 | 1/3 | 0 |
| charm | c | +2/3 | 1/3 | 0 |
| strange | s | −1/3 | 1/3 | −1 |
| top | t | +2/3 | 1/3 | 0 |
| bottom | b | −1/3 | 1/3 | 0 |

The properties of antiquarks are similar to those of the corresponding quark but with the opposite sign, so the antiquark $\bar{s}$ has charge +1/3, baryon number −1/3 and strangeness +1.

## The hadron families

AQA A    M1    EDEXCEL B    A2
AQA B    M2    NICCEA    A2
EDEXCEL A    M3C

The intermediate vector bosons, responsible for the weak interaction, form another group of particles.

The **hadrons** and **leptons** are two groups of particles. Hadrons are affected by both the strong and the weak nuclear force, but leptons are affected by the weak force only.

Hadrons are not fundamental particles; they are made up of quarks:
* **baryons** are made up of three quarks
* **mesons** consist of a quark and an antiquark.

The familiar baryons are the proton, whose quark structure is uud, and the neutron, udd. There are other baryons which have different combinations of quarks but all are unstable and have short lifetimes. All baryons have a baryon number +1, as each quark contributes a baryon number of +1/3.

Two families of mesons are the **pi-mesons**, or **pions**, and the **kaons**. Each family consists of three particles with charges +1, −1 and 0 along with their antiparticles. Because a meson consists of a quark and an antiquark, it has baryon number 0.

The table shows the structure and some properties of the kaons and pions.

| particle | structure | charge/e | baryon number, B | strangeness, S |
|---|---|---|---|---|
| $\pi^0$ | $u\bar{u}$ or $d\bar{d}$ | 0 | 0 | 0 |
| $\pi^+$ | $u\bar{d}$ | +1 | 0 | 0 |
| $\pi^-$ | $\bar{u}d$ | −1 | 0 | 0 |
| $K^0$ | $d\bar{s}$ | 0 | 0 | +1 |
| $K^+$ | $u\bar{s}$ | +1 | 0 | +1 |
| $K^-$ | $\bar{u}s$ | −1 | 0 | −1 |

## The leptons

AQA A    M1    EDEXCEL B    A2
AQA B    M2    NICCEA    A2
EDEXCEL A    M3C

Remember, a fundamental particle cannot be split up into other, simpler particles.

Leptons are fundamental particles. There are three families of leptons:
* the electron and its antiparticle the positron, e⁻ and e⁺, together with the neutrinos given off in β⁻ and β⁺ decay. These neutrinos are called the electron-antineutrino, $\bar{v}_e$, and the electron-neutrino, $v_e$, to distinguish them from the neutrinos associated with the other leptons
* the muons, μ⁻ and μ⁺, together with their neutrinos $v_\mu$ and $\bar{v}_\mu$. Muons have 200 times the mass of electrons, but are unstable, decaying to an electron and a neutrino
* the tauons, τ⁻ and τ⁺ and their neutrinos $v_\tau$ and $\bar{v}_\tau$. The tauons are the most massive leptons, having twice the mass of a proton. Like the muons, they are unstable.

Cosmic rays create pions in the upper atmosphere due to collisions with nuclei. These pions decay, charged pions producing muons and their neutrinos and uncharged pions decaying to gamma radiation.

Muons are produced by cosmic rays. Their properties are similar to those of electrons and positrons but they are much more massive. Tauons are the elephants of the lepton tribe; they are produced by the interaction of high-energy electrons and positrons.

## Exchanging particles

AQA A     M1     OCR B     A2
EDEXCEL A     M3C     WJEC     A2

The concept of a **field** is used to describe the effects of gravitational and electromagnetic forces and how they vary with distance. This is a useful concept that enables the size and direction of forces to be calculated, but it does not explain what causes them.

The interaction between particles that results in attractive and repulsive forces is due to a continual exchange of other particles. These exchange particles have a very short lifetime and owe their existence to borrowed energy, so they are often referred to as virtual particles.

There are different exchange particles associated with each of the four fundamental forces. For electromagnetic forces the exchange particles are photons, short bursts of electromagnetic radiation.

> A photon is a 'packet' of electromagnetic radiation. If you are unfamiliar with the concept of a photon, refer to chapter 4.

When two electrons approach each other they exchange virtual photons. The closer the electrons are to each other, the shorter the wavelength of the virtual photons exchanged.

**The exchange of virtual photons between two interacting electrons**

This exchange of virtual photons results in the electrons moving away from each other in the same way as two ice skaters on a collision course would move away from each other if they were to keep throwing things at each other!

> Richard Feynman was an American physicist. He is famous for his books *Lectures on Physics* and for his way of explaining physics concepts by using diagrams.

The diagram above represents the exchange of virtual photons between two interacting electrons.

Interactions between particles can be represented by **Feynman diagrams**. In these diagrams:

- straight lines are used for the interacting particles
- wavy lines represent the exchange particles
- the straight lines extend beyond the diagram, showing the existence of these particles before and after the interaction
- the wavy lines are contained within the diagram, showing that the exchange particles are short-lived.

The diagram opposite is a Feynman diagram for the interaction between two electrons shown in the diagram above. The arrows do not represent directions of travel; arrows pointing in show the particles before the interaction and those pointing out show the particles after the interaction. The symbol γ is used to represent the exchange of virtual photons.

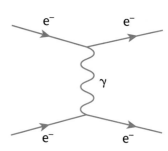

> The same symbol, g, is used for both gluons and gravitons.

The exchange particles for gravitational forces are called **gravitons**, symbol g, though these have not yet been observed.

# The strong force

| AQA A | M1 | NICCEA | A2 |
| AQA B | M2 | OCR B | A2 |
| EDEXCEL A | M3C | WJEC | A2 |
| EDEXCEL B | A2 | | |

It is the strong force that holds nucleons together in a nucleus and holds quarks together in a nucleon. The exchange particles between quarks are called **gluons**, symbol g. The quarks that make up a proton or neutron are constantly exchanging gluons as they interact with each other.

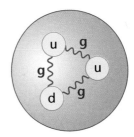

The exchange of gluons between the quarks in a proton

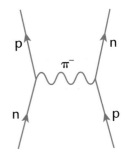

**Interactions between nucleons**

> Since there is no change in charge in the interactions between two neutrons or two protons, only neutral particles can be exchanged.

Pions, or pi-mesons, are responsible for the strong force between nucleons. Two protons or two neutrons exchange pi-zero, $\pi^0$, particles, but any one of the three pions can be exchanged during an interaction between a proton and a neutron. Feynman diagrams for some of the possible interactions are shown above.

Notice that charge and baryon number are conserved at each junction of the diagrams.

# The weak force

| AQA A | M1 | EDEXCEL B | A2 |
| AQA B | M2 | NICCEA | A2 |
| EDEXCEL A | M3C | OCR B | A2 |

> The intermediate vector bosons form another group of particles, separate from the hadrons and leptons.

Weak interactions involve the exchange of one of three particles called **intermediate vector bosons**. Like the pi-mesons, their symbols, W+, W- and Z0 indicate the charge on each boson. The W+ boson carries charge $+e$ and the W- boson carries charge $-e$.

- The W+ boson transfers charge $+e$; it is exchanged in an interaction between a neutrino and a neutron, resulting in an electron and a proton.
- The W- boson is responsible for β- decay of a neutron.
- Weak interactions where there is no transfer of charge involve the Z0 boson, for example a **collision** between an electron and an antineutrino.

Feynman diagrams for these interactions are shown below.

> Can you interpret these Feynman diagrams? Try to describe the interaction that each one represents.

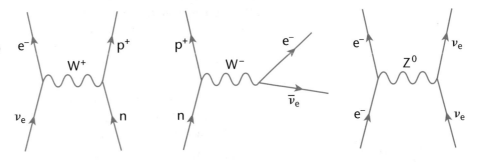

In two of these examples the type or flavour of a quark in the neutron has been changed from d to u by the weak interaction involving a charge-carrying boson. In β⁻ decay a down quark emits a W⁻ particle which decays to an electron and its antineutrino. This is shown in the diagram below.

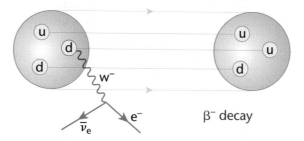

β⁻ decay

## Progress check

1 Describe the differences between:
   a   Hadrons and leptons.
   b   Baryons and mesons.

2 The diagram represents the decay of a meson.

   a   Which meson is decaying?
   b   What is the exchange particle?
   c   What particles are produced as a result of the decay process?

3 What change takes place in a proton in β⁺ decay?

3   A u quark changes to a d quark, changing the quark structure of the nucleon from uud to udd.
       c   The μ⁺ muon and its neutrino.
       b   The intermediate vector boson W⁺.
2   a   π⁺.
       b   Baryons contain 3 quarks. Mesons contain a quark and an antiquark.
           Leptons are fundamental particles.
1   a   Hadrons are made up of quarks.

# 3.5 Stability and the nucleus

**After studying this section you should be able to:**

- use the N–Z curve to explain how the type of decay depends on the position of an unstable nucleus relative to the region of stability
- explain why energy is released in nuclear decay
- describe the similarities and differences in the energy spectra for $\alpha$, $\beta$ and $\gamma$ emission

*LEARNING SUMMARY*

## How big?

AQA A · A2 · NICCEA · A2
EDEXCEL A · M3C

Atoms and nuclei are not hard, rigid particles with a fixed shape and size. Evidence from alpha-scattering experiments shows that the diameter of a nucleus is approximately one ten-thousandth ($1 \times 10^{-4}$) that of the atom and the nuclear volume is $1 \times 10^{-12}$ that of the atomic volume. As almost all the atomic mass is in the nucleus, it must be extremely dense.

Electron diffraction experiments allow more precise estimates of the size of nuclei to be made. The results of these experiments show that:

- all nuclei have approximately the same density, about $1 \times 10^{17}$ kg m$^{-3}$
- the greater the number of nucleons, the larger the radius of the nucleus
- the nucleon number, $A$, is proportional to the cube of the nuclear radius, $r$.

> *Unlike their nuclei, the atoms of different elements can have very different densities.*

An alternative way of writing this last point is:

- the nuclear radius, $r$, is proportional to the cube root of the nucleon number, $A$.

> The relationship between nucleon number, $A$, and nuclear radius, $r$, is:
> $$r = r_0 A^{1/3}$$
> where $r_0$ has the value $1.2 \times 10^{-15}$ m.
>
> **KEY POINT**

This relationship means that if the nuclei of $^{64}$Cu and $^{32}$S are compared, the radius of the copper nucleus is $2^{1/3} = 1.26$ times that of the sulphur, as the nucleon number of copper is twice that of sulphur. Similarly, the copper nucleus has a radius which is $4^{1/3} = 1.59$ times that of $^{16}$O.

## Stable and unstable nuclei

AQA A · A2 · OCR B · A2
EDEXCEL A · M1, M3C

> *Remember, isotopes of an element all have the same number of protons but different numbers of neutrons in the nucleus.*

Carbon-11, carbon-12 and carbon-14 are three isotopes of carbon. Of these, only carbon-12 is stable. It has equal numbers of protons and neutrons. The graph shows the relationship between the number of neutrons ($N$) and the number of protons ($Z$) for stable nuclei.

It can be seen from this graph that the condition for a nucleus to be stable depends on the number of protons:

- for values of $Z$ up to 20, a stable nucleus has equal numbers of protons and neutrons
- for values of $Z$ greater than 20, a stable nucleus has more neutrons than protons.

> *For stable nuclei with more than 20 protons, the neutron:proton ratio increases steadily to a value of around 1.5 for the most massive nuclei.*

Unstable nuclei above the stability line in the graph are **neutron-rich**; they can

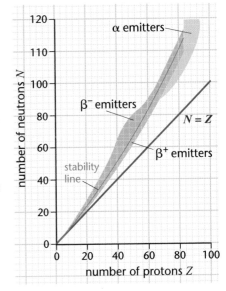

become more stable by decreasing the number of neutrons. They decay by β⁻ emission; this leads to one less neutron and one extra proton and brings the neutron–proton ratio closer to, or equal to, one. An example is:

$$^{24}_{11}\text{Na} \rightarrow {}^{24}_{12}\text{Mg} + {}^{0}_{-1}\text{e}$$

Unstable nuclei below the stability line decay by β⁺ emission; this increases the neutron number by one at the expense of the proton number. An example is:

$$^{11}_{6}\text{C} \rightarrow {}^{11}_{5}\text{B} + {}^{0}_{+1}\text{e}$$

> The N=Z line corresponds to a neutron–proton ratio of 1. As an alpha particle consists of two neutrons and two protons, its emission would hardly affect a neutron–proton ratio that is nearly 1.

The emission of an alpha particle has little effect on the neutron–proton ratio for isotopes that are close to the N=Z line and is confined to the more massive nuclei. For these nuclei, emission of an alpha particle changes the balance of the proton–neutron ratio in the favour of the neutrons. In the decay of thorium-228 shown below, the neutron–proton ratio increases from 1.53 to 1.55.

$$^{228}_{90}\text{Th} \rightarrow {}^{224}_{88}\text{Ra} + {}^{4}_{2}\text{He}$$

## Decay chains and dating

EDEXCEL A   M3C

> There are four decay chains, starting with actinium, neptunium, thorium and uranium. All except the neptunium chain are found naturally in rocks.

The decay of a nucleus may result in another unstable nucleus which in turn decays and may result in another unstable nucleus. This process is called a **decay chain** or **radioactive series** and it ends in the formation of a nucleus that is stable. The thorium decay chain starts with thorium-232 and ends with lead-208. The beginning of the chain is shown below:

$$^{232}_{90}\text{Th} \rightarrow {}^{228}_{88}\text{Ra} + {}^{4}_{2}\text{He} \rightarrow {}^{228}_{89}\text{Ac} + {}^{0}_{-1}\text{e} \rightarrow {}^{228}_{90}\text{Th} + {}^{0}_{-1}\text{e} \rightarrow {}^{224}_{88}\text{Ra} + {}^{4}_{2}\text{He} \rightarrow$$

A decay chain can be represented on a chart of nucleon number, A, against proton number, Z. The diagram below shows the complete decay chain for thorium-232. On this chart:

- alpha decay is shown by a movement of two units to the left and four units down
- beta-minus decay involves a movement of one unit to the right
- the chain branches at bismuth-212, which has two possible modes of decay.

> Bismuth-212 decays by both alpha and beta-minus emission.

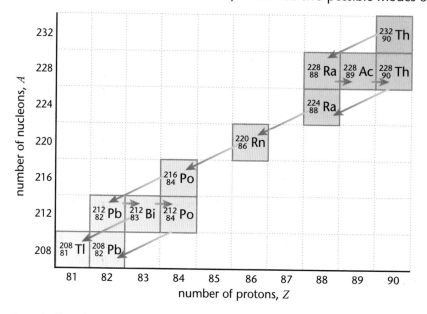

In a similar chain, the decay of uranium-238 ends with the stable isotope lead-206. Uranium-238 decays with a half-life of $4.5 \times 10^6$ years, the intermediate isotopes having relatively short half-lives. This allows the age of rocks to be estimated by comparing the numbers of uranium-238 atoms and lead-206 atoms present in the rock. A ratio of 1:1 indicates that the age of the rock is the same as the half-life of uranium-238. The oldest rocks found have this ratio of uranium-238 to lead-206 present, showing that the age of the Earth is approximately $4.5 \times 10^9$ years.

> This assumes that there was no lead-206 present when the rock was formed.

In **radiocarbon dating** a similar technique is used. Radiocarbon dating compares the concentrations of carbon-14 atoms and carbon-12 atoms in a sample of material. Living things maintain a constant level of carbon-14 which then decays after death. The lower the concentration of carbon-14, the longer the time since death. This method of dating can only be used for objects made from once-living material. The half-life of carbon-14 is 5730 years, so it is not suitable for dating objects that are relatively new (less than five hundred years since death) or very old (more than one hundred thousand years since death).

> Radiocarbon dating was used to show that the Turin shroud, once thought to have been used for the body of Christ, actually originated around 1300 AD.

## A conservation problem

| | | | |
|---|---|---|---|
| AQA A | A2 | NICCEA | A2 |
| AQA B | A2 | OCR A | A2 |
| EDEXCEL A | M3C | OCR B | A2 |
| EDEXCEL B | A2 | WJEC | A2 |

A carbon-12 nucleus consists of six protons and six neutrons. The mass of the atom is precisely 12u, by definition, so after taking into account the mass of the electrons, that of the nucleus is 11.9967 u. The mass of the constituent neutrons and protons is:

$$6\ m_p + 6\ m_n = 6(1.0073\ u + 1.0087\ u) = 12.0960\ u$$

The nucleus has less mass than the particles that make it up. This appears to contravene the principle of conservation of mass. Einstein established that *energy has mass*. The mass that you gain due to your increased energy when you walk upstairs is infinitesimally small, but at a nuclear level this mass cannot be ignored. Any change in energy is accompanied by a change in mass, and vice versa.

> Einstein's equation $E = mc^2$ gives the method for working out how much mass is associated with energy. Try using it to work out the increase in your mass when you walk upstairs.

When dealing with the nucleus and nuclear particles, energy and mass are so closely linked that their equivalence, and that of their units, has been established:

> **KEY POINT**
>
> $$1\ u = 930\ MeV$$
> where 1 eV (one electron volt) is the energy transfer when an electron moves through a potential difference of 1 volt.

> An electronvolt is a tiny amount of energy compared to the joule, which is too large a unit to use on an atomic scale. $1\ eV = 1.60 \times 10^{-19}$ J.

Using this relationship, the separate conservation rules regarding mass and energy can be combined into one so that (mass + energy) is always conserved in nuclear interactions.

To split a nucleus up into its constituent nucleons requires energy. It follows that a nucleus has less energy than the sum of the energies of the corresponding number of free neutrons and protons. So the fact that a nucleus has less energy than its nucleons would have in isolation, means that it also has less mass.

> **KEY POINT**
>
> The difference between the sum of the masses of the individual nucleons and the mass of the nucleus is called the **mass defect** or **nuclear binding energy**. It represents the energy required to separate a nucleus into its individual nucleons.

In the case of carbon-12 the mass defect, or nuclear binding energy, is equal to 0.093 u = 89.3 MeV.

As would be expected, the greater the number of nucleons, the greater the binding energy. The figure shows how the **binding energy per nucleon** varies with nucleon number. The most stable nuclei have the greatest binding energy per nucleon.

> The greater the binding energy per nucleon, the more energy (per nucleon) is required to split the nucleus up, giving it more stability.

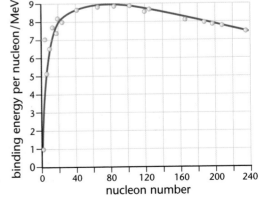

## Energy spectra

AQA A    A2    EDEXCEL A    M3C

A nucleus can only decay naturally to a state where it has less energy.

The kinetic energy of the decay products includes that of the recoiling nucleus. This is most significant in the case of alpha decay.

When a nucleus decays, the daughter nucleus has less mass and so more binding energy than the parent. The energy difference is the **decay energy**, the energy released by the nucleus as a result of its decay. This energy is transferred to:

- kinetic energy of the decay products
- energy of a gamma ray, if one is emitted
- energy of a neutrino or antineutrino, in the case of β emission.

There are different patterns in the energies of alpha, beta and gamma emissions.

**Energy being removed from a nucleus in alpha emission**

recoiling nucleus — alpha particle — gamma photon

- The energy of alpha particles depends on the source; some sources emit alpha particles that all have the same energy, other sources emit alpha particles with two or more possible energies.
- The energy of beta particles varies continuously over a range from zero up to a maximum value that depends on the source.
- In gamma emission, each source emits a line spectrum; the photons have only one or a small number of energies.

The energies of the emitted particles are different for different sources. The shorter the half-life, the greater the energy of the emissions.

These cloud chamber tracks are from a source that emits alpha particles of two separate energies.

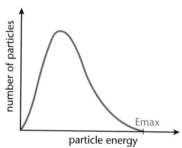

number of particles / particle energy / Emax

**The range of energies in beta emission**

Whilst alpha and gamma emissions correspond to certain energy values, in beta emission most of the particles are emitted with less than the maximum energy available. This led to the discovery of the neutrino, ν, and antineutrino, ν̄. The antineutrino is a particle emitted along with the electron in β⁻ decay and a neutrino is emitted along with a positron in β⁺ decay. The energies of these particles account for the difference between the energy released by the nucleus and the kinetic energies of the beta particle and the recoiling nucleus together with the energy of any gamma ray photon that is emitted.

Remember that gamma emission is usually due to a nucleus being left in an excited state following the emission of an alpha or beta particle.

## Progress check

1 Tritium, $^3_1$H, is a radioactive isotope of hydrogen. Use the $N$–$Z$ graph to explain how it decays and to write an equation for the decay.

2 The nuclear mass of oxygen-16 is 15.9905 u.
  a Use the data on page 111 to calculate the mass defect in u.
  b Calculate the energy equivalent in MeV.

3 a In what way is the energy spectrum of beta emissions different to that of alpha and gamma emissions?
  b Explain how this is accounted for.

b This is explained by the energy of the neutrino or antineutrino that accompanies a beta particle.
3 a The beta emission spectrum is continuous, whereas those of alpha and gamma only contain certain values of energy.
2 a 8(1.0073 u + 1.0087 u) = 16.1280 u − 15.9905 u = 0.1375 u.
  b 128 MeV
1 As tritium lies above the $N$–$Z$ line, it decays by beta-minus emission.
$^3_1$H→$^3_2$He+$^0_{-1}$e

## Sample question and model answer

An ice cube of mass 12.0 g at a temperature of –9°C is placed in a glass containing 100 g of water at 20°C. The mass of the glass is 250 g.
Calculate the equilibrium temperature when the ice has melted.     [8]

Specific heat capacities:   ice     2.1 J g$^{-1}$ K$^{-1}$
                          water   4.2 J g$^{-1}$ K$^{-1}$
                          glass   0.67 J g$^{-1}$ K$^{-1}$

Specific latent heat of ice = 340 J g$^{-1}$

At equilibrium, the ice has changed into water and all the water and the glass are at the same temperature. The glass and the original water have lost energy, and the melted ice has gained energy.
Assuming no energy transfer to or from the surroundings, these gains and losses must be equal. The equation opposite relates the energy gained by the ice to that lost by the glass and water.

energy gained by ice = energy lost by glass and water     1 mark

There are three separate components in working out the energy gained by the ice. These are:
- energy to warm the ice to 0°C
- energy to melt the ice
- energy to warm the melted ice to the final temperature.

energy gained by ice     = energy to warm the ice to 0°C + energy to melt the ice + energy to heat the melted ice to x°C     1 mark

= $m_{ice} \times c_{ice} \times \Delta\theta_{ice} + m_{ice} \times l_{ice} + m_{ice} \times c_{water} \times \Delta\theta_{water}$     1 mark

This total energy is balanced by that lost by both the glass and the water.

energy lost by glass and water

= $m_{water} \times c_{water} \times \Delta\theta_{water} + m_{glass} \times c_{glass} \times \Delta\theta_{glass}$     1 mark

Having established the physical principle, the numerical values are now substituted into the equation. It is necessary to have a symbol for the equilibrium temperature.

If the equilibrium temperature is x°C:

12.0 g × 2.1 J g$^{-1}$ K$^{-1}$ × 9 K + 12.0 g × 340 J g$^{-1}$ + 12.0 g × 4.2 J g$^{-1}$ K$^{-1}$ × x × K = 100 g × 4.2 J g$^{-1}$ K$^{-1}$ × (20 – x) K + 250 g × 0.67 J g$^{-1}$ K$^{-1}$ × (20 – x) K

2 marks

This equation now has to be solved to find the value of x.

226.8 J + 4080 J + 50.4 × x J = 420 × (20 – x) J + 167.5 × (20 – x) J
= 11 750 J – 587.5 × x J     1 mark

Now collect the terms involving numbers on one side of the equation, and those involving x on the other side to give:

637.9x = 7443 so x = 11.7

The final temperature of the mixture is 11.7°C.     1 mark

## Practice examination questions

1   A sample of an ideal gas has a volume of $1.50 \times 10^{-4}$ m³ at a temperature of 22°C and a pressure of $1.20 \times 10^5$ Pa.

(a)  Calculate the temperature of the gas in kelvin.                                          [1]

(b)  Calculate the number of moles in the sample.
$R = 8.3$ J mol⁻¹.                                                                              [3]

(c)  If the volume of the gas remains constant, calculate the temperature at which the pressure of the gas is $2.00 \times 10^5$ Pa.                                     [3]

(d)  (i)   Calculate the mean kinetic energy of the gas particles at a temperature of 22°C. The Boltzmann constant, $k = 1.38 \times 10^{-23}$ J K⁻¹.     [3]
     (ii)  At what temperature would this mean kinetic energy be doubled? Explain how you arrive at your answer.                                        [2]

2   Calculate the energy released when 1.8 kg of steam at 100°C is changed into water at 85°C.                                                                          [3]
specific latent heat of vaporisation of water = $2.25 \times 10^6$ J kg⁻¹
specific heat capacity of water                $= 4.20 \times 10^3$ J kg⁻¹ K⁻¹

3   The diagram shows how an electric immersion heater can be used to measure the specific heat capacity of a metal block.

(a)  (i)   What is the advantage of heating the block through a temperature difference of 10 K rather than 1 K?                                                  [1]

     (ii)  What is the advantage of heating the block through a temperature difference of 10 K rather than 100 K?                                                [1]

     (iii) Explain whether you would expect the experiment to give a higher value or a lower value than that given in a data book.                          [2]

     (iv)  Suggest one way of improving the reliability of the result.              [1]

(b)  Use the data to calculate a value for the specific heat capacity of the metal.                                                                                      [3]
heater power              = 24 W
time heater switched on = 350 s
mass of block             = 0.80 kg
initial temperature       = 22°C
final temperature         = 36°C.

4　The first law of thermodynamics can be written as:

$$\Delta U = Q + W$$

(a)　State the meaning of each symbol in the equation.　[3]

(b)　As a gas is being compressed, 4500 J of work are done on it and 4800 J of heat are removed from it. Explain what happens to the temperature of the gas as a result of this.　[2]

5　The diagram shows a heat engine.

(a)　Calculate the efficiency of the heat engine.　[2]

(b)　What is the maximum efficiency of the engine?　[2]

(c)　Suggest one reason why the efficiency of the heat engine is less than its theoretical maximum.　[1]

6　The diagram shows some possible outcomes when alpha particles are fired at gold foil.

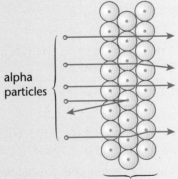

atoms in gold foil

(a)　Explain how these outcomes show that:
  (i)　the nucleus is positively charged　[2]
  (ii)　most of the atom is empty space.　[1]

(b)　In **deep inelastic scattering** electrons are used to penetrate the nucleus.
  (i)　What is meant by 'deep inelastic scattering'?　[2]
  (ii)　Explain why electrons are able to probe further into the nucleus than alpha particles can.　[2]
  (iii)　Explain why the proton and neutron are not fundamental particles.　[1]
  (iv)　How does deep inelastic scattering provide evidence for quarks?　[1]
  (v)　Describe the quark structure of a proton and a neutron.　[2]

## Practice examination questions *(continued)*

**7** 1.00 g of carbon is obtained from the wood of a living tree.
   (a) Calculate the number of carbon atoms present in 1.00 g, assuming that all the carbon atoms are carbon-12.
   The Avagadro constant, $N_A = 6.02 \times 10^{23}$ mol$^{-1}$.
   The mass of 1 mole of carbon-12 is 12 g precisely. [2]

   (b) Natural carbon consists of approximately:
   99% carbon-12
   1% carbon-13
   $1 \times 10^{-10}$ % carbon-14.

   (i) Explain whether the actual number of carbon atoms in a 1.00 g sample is likely to be significantly different from the answer to (a). [2]
   (ii) Estimate the number of carbon-14 atoms present in the sample. [1]

   (c) The rate of decay of carbon-14 in the sample is 0.25 Bq.
   (i) Calculate the value of $\lambda$, the decay constant. [2]
   (ii) Use this value to estimate the half-life of carbon-14. [2]

**8** For nuclei with up to 20 protons, a stable nucleus has approximately equal numbers of protons and neutrons.
   (a) Explain how, for nuclei with less than 20 protons, the type of decay undergone by unstable nuclei depends on the neutron–proton ratio. [2]

   (b) Describe how the decay of $^{13}_{6}$C results in the formation of a more stable nucleus. [2]
   (c) How does the neutron–proton ratio change for more massive stable nuclei? [2]

   (a) How does the structure of a **meson** differ from that of a **baryon**? [2]

   (b) The Feynman diagram illustrates the interaction between two protons.

**9**

   The exchange particle is a pion.
   (i) Which pion is exchanged in this interaction? [1]
   (ii) Explain why it is this pion rather than a different one. [2]

# Electromagnetic radiation

*The following topics are covered in this chapter:*

- *What is a wave?*
- *Simple wave properties*

- *More complex wave behaviour*
- *Wave properties across the spectrum*

## 4.1 What is a wave?

*After studying this section you should be able to:*

- *describe the difference between a longitudinal and a transverse wave*
- *use the wave equation to calculate speed, frequency and wavelength*
- *explain the meaning of phase and phase difference*

**LEARNING SUMMARY**

### Wave motion

| | | | |
|---|---|---|---|
| AQA A | A2 | NICCEA | M2 |
| AQA B | M2 | OCR A | M3 |
| EDEXCEL A | A2 | OCR B | M1 |
| EDEXCEL B | M1 | WJEC | M1 |

> The waves on a water surface are almost transverse, the particles move in an elliptical path.

Waves are used to transfer a signal or energy. They do this without an accompanying flow of material, although some waves, such as sound, can only be transmitted by the particles of a substance.

All waves consist of **vibrations** or **oscillations**. Sound and other compression waves are classified as **longitudinal** because the vibrations of the particles carrying the wave are along, or parallel to, the direction of wave travel. All electromagnetic waves are **transverse**; the vibrations of the electric and magnetic fields in these waves are at right angles to the direction of wave travel.

Waves transmitted through the body of water, or any other liquid, are longitudinal.

The diagram shows a longitudinal wave being transmitted by a spring and a transverse wave on a rope.

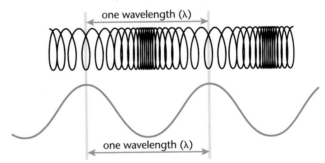

one wavelength ($\lambda$)

one wavelength ($\lambda$)

### Wave measurements

| | | | |
|---|---|---|---|
| AQA A | A2 | NICCEA | M2 |
| AQA B | M1, M2 | OCR A | M3 |
| EDEXCEL A | A2 | OCR B | M1 |
| EDEXCEL B | M1 | WJEC | M1 |

> Most waves that you come across everyday are progressive.
> Waves that are not progressive are discussed in the section 'more complex wave behaviour'.

These measurements apply to all **progressive** waves. A progressive wave is one that has a profile that moves through space.

- **Wavelength** (symbol $\lambda$) is the length of one complete cycle; a compression (squash) and rarefaction (stretch) in the case of a longitudinal wave, a peak and a trough in the case of a transverse wave – see diagram above.
- **Amplitude** (symbol $a$) is the maximum displacement from the mean position, see diagram on page 118.
- **Frequency** (symbol $f$) is the number of vibrations per second, measured in hertz (Hz).
- **Speed** (symbol $v$) is the speed at which the profile moves through space.
- **Period** (symbol $T$) is the time taken for one vibration to occur. It is related to frequency by the equation $T = 1/f$.

> For all waves, the wavelength, speed and frequency are related by the equation:
>
> $$speed = frequency \times wavelength$$
> $$v = f \times \lambda$$

## Wave graphs

| AQA A | A2 | NICCEA | M2 |
| AQA B | M2 | OCR A | M3 |
| EDEXCEL A | A2 | OCR B | M1 |
| EDEXCEL B | M1 | WJEC | M1 |

Both transverse and longitudinal waves can be represented by graphs of displacement against position. The position along a wavefront can be measured in two ways:

- as a distance from a point in space
- as an angle.

In angular measure, one complete cycle of the wave is represented by an angle of 360°. This is shown in the diagram.

> A displacement–position or displacement–time graph does not distinguish between a transverse wave and a longitudinal one.

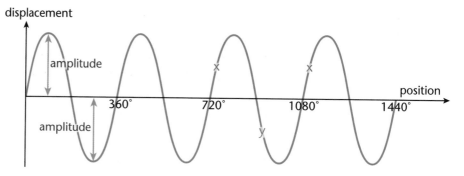

Using angular measure is a convenient way of comparing the **phase** of different parts of a wave. Two points on a wave are **in phase** if they have the same displacement and velocity. For progressive waves, this is only true for points exactly $n$ wavelengths or $360n°$ apart, where $n$ is an integer. Points marked x on the diagram are in phase.

> Try substituting values of $n$, for $n = 0, 1, 2$, etc. into the expression for 'an odd number of half wavelengths'. This will help you to understand the meaning of the phrase.

The point marked **y** is exactly **out of phase** with the points **x** as it has the opposite displacement and its velocity has the same value but in the opposite direction. This applies to any two points separated by an odd number of half wavelengths (i.e. 1/2, 3/2, 5/2 wavelengths, etc.). This can be expressed algebraically as $(2n+1)\lambda/2$, where $n$ is an integer.

The **phase difference** between two points on a wave describes their relative displacement and velocity and is normally expressed in degrees or wavelengths. A phase difference of $\lambda/2$ or 180° describes two points separated by an odd number of half wavelengths (i.e. exactly out of phase), while a phase difference of $\lambda/4$ or 45° compares two points that are separated by one-quarter, five-quarters, nine-quarters of a wavelength, etc.

> A common error is to misinterpret a CRO display as a graph of displacement against distance.
> This leads to a horizontal measurement, which represents time, being interpreted as a distance.

A **cathode ray oscilloscope** (CRO) can be used to display a different type of graph that represents wave motion. Rather than showing the displacement along a wave, the CRO plots the displacement of one point on the wave against time. The display on the CRO looks the same as that in the diagram, but the horizontal axis represents time rather than position on the wave.

To measure the **period** and **frequency** of a wave from a CRO display, the time–base control, which controls the rate at which the dot sweeps the screen horizontally, needs to be in the 'cal' (calibration) position. The diagram opposite shows a CRO display of a sound wave, with the time–base set to 2 ms cm⁻¹.

One cycle of the wave occupies a distance of 8 cm on the screen, so the period, $T = 8 \times 2$ ms = 16 ms = $1.6 \times 10^{-2}$ s.

The frequency of the wave, $f = 1/T = 1 \div 1.6 \times 10^{-2}$ s = 62.5 Hz.

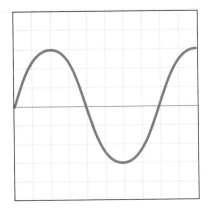

## Progress check

1 Describe the difference between a longitudinal and a transverse wave and give one example of each.

2 Calculate the frequency of a VHF radio broadcast that has a wavelength of 2.75 m. The speed of radio waves, $c = 3.00 \times 10^8$ m s$^{-1}$.

3 Two points on a wavefront have a phase difference of 180°. Describe their relative displacement and velocity.

1 In a longitudinal wave the vibrations are parallel to the direction of travel, e.g. sound or any other compression wave.
In a transverse wave the vibrations are at right angles to the direction of travel, e.g. light or any other electromagnetic wave.

2 $1.09 \times 10^8$ Hz.

3 The points have the same amount of displacement but in opposite directions, i.e. the displacement of each one is the negative of the displacement of the other one.
Similarly, the velocities are equal in size but opposite in direction.

# 4.2 Simple wave properties

*After studying this section you should be able to:*

- *explain the meaning of the terms* polarisation *and* intensity
- *use the relationships between refractive index, speeds of light in media and angles of incidence and refraction*
- *describe how optical fibres are used in communications*

**LEARNING SUMMARY**

## Polarisation

| | | | |
|---|---|---|---|
| AQA A | A2 | NICCEA | M2 |
| AQA B | M2 | OCR A | M3 |
| EDEXCEL A | A2 | OCR B | M1 |
| EDEXCEL B | M2 | WJEC | M1 |

Polarising material is used in some types of sunglasses and in filters for camera lenses. It cuts out reflected light from a flat surface such as an expanse of water.
Can you work out how it does this?

In order to receive a television or radio broadcast, the receiving aerial has to be lined up with that of the transmitter. This is because radio waves broadcast from aerials are polarised, the vibrations are only in one plane. For radio and television broadcasts the plane of polarisation is usually either vertical or horizontal.

Light waves are not normally polarised, and nor are radio waves from stars. In an unpolarised wave the vibrations are in all planes at right angles to the direction of travel. Polarisation does not apply to longitudinal waves.

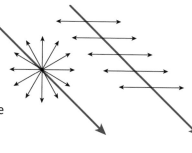

The diagram shows the vibrations in polarised and unpolarised waves. Light waves can become polarised as they pass through some materials and they are partially polarised when reflected.

## Intensity

| | | | |
|---|---|---|---|
| AQA B | M2 | OCR A | M3 |
| EDEXCEL A | M1 | OCR B | M2 |
| EDEXCEL B | M1 | WJEC | M1 |

A point source such as a star radiates energy in all directions. In this case the variation of intensity with distance follows an inverse square pattern, with:

$$I \propto 1/r^2$$

The signal strength of a radio broadcast decreases the further away the receiver is from the transmitting aerial. In a similar way, a sound appears fainter and a light seems dimmer further away from their origins. This is due to energy becoming spread over a wider area as the wave travels away from its source.

The intensity with which a wave is received depends on the power incident on the area of the detector.

> **KEY POINT**
>
> Intensity = power per unit area measured at right angles to the direction of travel.
>
> $$I = P/A$$
>
> Intensity is measured in W m$^{-2}$.

The diagram opposite shows the sound spreading out from a loudspeaker. The ear that is further away detects a quieter sound because the power is spread over a wider area, so less enters the ear.

## Reflecting light

| | | | |
|---|---|---|---|
| AQA B | M2 | OCR B | M1 |
| OCR A | M3 | | |

The image in a mirror is virtual, upright, the same size as the object and the same distance behind the mirror as the object is in front.

Mirrors form images because they reflect light in a regular way. When light is reflected at a mirror:

- the angles of incidence and reflection are equal
- the incident light, reflected light and the normal all lie in the same plane.

These are illustrated in the diagram opposite, which serves as a reminder that the angles of incidence and reflection are measured relative to the normal line, a line drawn at right angles to the surface.

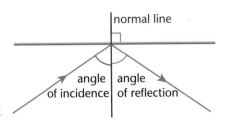

The second rule merely states that the event can be captured on a flat sheet of paper.

## Changing speed and direction

| | | | |
|---|---|---|---|
| AQA A | M1 | OCR A | M3 |
| AQA B | M2 | OCR B | M1 |
| EDEXCEL B | M1, M2 | WJEC | M1 |
| NICCEA | M2 | | |

When waves cross a boundary between two materials there is a change of speed, called **refraction**. If the direction of wave travel is at any angle other than along the normal line, this change in speed causes a change in direction. The diagram opposite shows what happens to light passing through glass. Note that there is always some reflection at a boundary where refraction takes place.

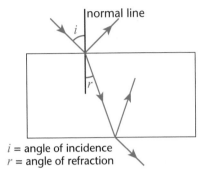

$i$ = angle of incidence
$r$ = angle of refraction

Refraction also causes virtual images. Images due to refraction are similar to the object, the only difference being that they are closer to the observer.

The change in speed is described by the **refractive index**.

> The refractive index of a material, $n$, is the ratio of the speed of light in a vacuum to the speed of light in the material.
> $$n = c_v \div c_m$$
> where $c_v$ is the speed of light in a vacuum and $c_m$ is the speed of light in the material.
> Refractive index is a ratio, it does not have a unit.
>
> **KEY POINT**

The refractive index is the factor by which the speed of light is reduced when it passes from a vacuum into the material. This is sometimes called the **absolute** refractive index to distinguish it from the **relative** refractive index between two materials.

> The refractive index between two materials, $_1n_2$ is the ratio of the speed of light in material 1 to the speed of light in material 2.
> $$_1n_2 = c_1 \div c_2 = n_1 \div n_2$$
>
> **KEY POINT**

The refractive index of air is 1.0003, so there is no difference in the refractive index of a material relative to air and relative to a vacuum when working to three significant figures.

The greater the change in speed when light is refracted, the greater the change in direction. At a given angle of incidence, there is a greater change in direction when light passes into glass, $n_g = 1.50$, than when light passes into water, $n_w = 1.33$. The values quoted here are absolute values, but in practice there is little difference between the refractive index of a material relative to a vacuum and that relative to air.

When light is refracted:

- the incident light, refracted light and the normal all lie in the same plane
- **Snell's law** relates the change in direction to the change in speed that takes place.

> Snell's law states that:
> $$\frac{\sin i}{\sin r} = \frac{c_1}{c_2} = {}_1n_2$$
>
> **KEY POINT**

So the sines of the angles are in the same ratio as the speeds in the media when refraction takes place.

## Critical angle

| AQA A | M1 | OCR A | M3 |
| AQA B | M2 | OCR B | M1 |
| EDEXCEL B | M1 | WJEC | M1 |
| NICCEA | M2 | | |

When light speeds up as it crosses a boundary, the change in direction is away from the normal line. For some angle of incidence, called the **critical angle**, the angle of refraction is 90°.

The diagram below shows what happens to light meeting a glass–air boundary at angles of incidence that are **a** less than, **b** equal to and **c** greater than the critical angle.

> The intensity of the internal reflection increases as the angle of incidence increases.

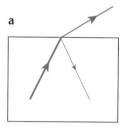

there is a weak reflection          there is a stronger reflection          all light is reflected

These diagrams show that:

> The abbreviation TIR is often used for total internal reflection.

- at angles of incidence less than the critical angle, both reflection and refraction take place
- at the critical angle, the angle of refraction is 90°
- at angles of incidence greater than the critical angle, the light is **totally internally reflected**.

Snell's law can be used to show that the relationship between the critical angle and refractive index is:

> Diamond has a very high refractive index and a low critical angle. Multiple internal reflections make diamonds sparkle.

$$n = 1/\sin C$$
where $C$ is the critical angle and $n$ is the absolute refractive index.

**KEY POINT**

## Fibre optics

| AQA A | M1 | OCR A | M3 |
| AQA B | M2 | OCR B | M1 |
| EDEXCEL B | M1 | WJEC | M1 |

Total internal reflection is used in prismatic binoculars, car and cycle reflectors and cats' eyes, but the greatest impact has been in the fields of medicine and communications.

Optical fibres enable light to travel round curves by repeated total internal reflection at the boundaries of the fibre, which is made from glass or plastic. Provided that light hits the boundary at an angle greater than the critical angle, none passes out of the fibre. The diagram opposite shows how light can be made to travel 'round the bend' of a fibre.

In medicine, optical fibres are used in **endoscopy** to look inside a patient. This can be done using a natural body opening or by 'keyhole surgery', which requires a small incision just wide enough to take the fibre. Two bundles of fibres are used in endoscopy; one to transmit light to inside the patient and the other to carry the reflected light to a small television camera.

Although they are commonly referred to as light, the pulses used are usually in the infra-red region of the electromagnetic spectrum.

In communications, optical fibres are used to transmit telephone conversations, television and radio programmes in the form of **digital signals**. The signals used can only have the values 1 or 0, where 1 is represented by a pulse of light from a laser diode and 0 is the absence of light. The diagram opposite shows how optical fibres are used in telephone communications.

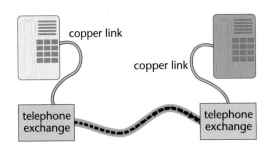

The advantages of using optical fibres rather than copper cables in communications are:

The range of a signal in a copper cable is only a few km, compared to up to 100 km in a glass fibre.

- the range of a signal in an optical fibre is much greater than that in a copper cable, so amplification is needed less frequently
- optical fibres do not pick up **noise** from changing magnetic fields.

However, as the diagram above shows, the signal in an optical fibre is still subject to distortion as it travels along the fibre. This is due to **multipath dispersion**, a process which causes the pulse to become elongated as some parts of the pulse travel further than others, for example when the cable goes round corners. In modern fibres the effect of this is reduced by concentrating the light pulses along the core of the fibre.

A similar effect occurs if 'light' containing more than one wavelength is used, as the different wavelengths travel at different speeds in the fibre.

**Multipath dispersion and its effect on a digital signal**

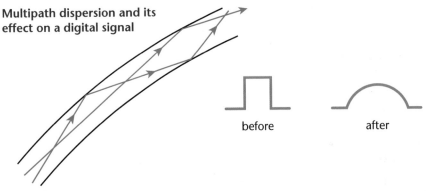

before      after

Transmission of information using digital, rather than analogue signals has two main advantages:

- any noise or distortion can be easily removed in a process called regeneration
- digital signals can carry more information than analogue signals.

Noise affects the amplitude of a signal. In AM (amplitude modulated) transmissions, the information is carried in the form of variations in amplitude of a wave. Any noise cannot be distinguished from the signal and so cannot be removed. When the signal is amplified, the noise is also amplified.

This is not the case with a digital signal, which only has certain allowed values. If the amplitude varies around these values, it can be restored. This is shown in the diagram below.

The 'allowed values' used in communications are normally 1 and 0, so regeneration is a straightforward task.

an analogue signal — becomes noisy and distorted — which is amplified

a digital signal — becomes noisy and distorted — and can be regenerated

## Progress check

The table gives the speed of light in different media.

| medium | speed of light/m s⁻¹ |
|--------|---------------------|
| vacuum | $3.00 \times 10^8$ |
| ice | $2.29 \times 10^8$ |
| ethanol | $2.21 \times 10^8$ |
| quartz | $1.94 \times 10^8$ |

**1 a** Calculate the values of the absolute refractive indexes of ice, ethanol and quartz.

**b** Calculate the refractive index for light travelling from ice into ethanol.

**2** A piece of ice is placed in ethanol.

**a** In which direction is light travelling for total internal reflection to take place?

**b** Calculate the value of the critical angle.

**3** The critical angle for a certain glass is 39°. Calculate the refractive index of the glass.

1 a ice 1.31  ethanol 1.36  quartz 1.55
b 1.04
2 a from ethanol to ice
b 75°
3 1.59

# 4.3 More complex wave behaviour

## Diffraction

| | | | |
|---|---|---|---|
| AQA A | A2 | NICCEA | M2 |
| AQA B | M2 | OCR A | M2, M3 |
| EDEXCEL A | A2 | OCR B | M2 |
| EDEXCEL B | M2 | WJEC | M1 |

Our everyday experience is that sound can travel round corners, but light travels in straight lines. The spreading out of waves as they pass through openings or the edges of obstacles is called **diffraction**. All waves can be diffracted, but the effect is more noticeable with long wavelength waves such as sound and radio waves than with short wavelength waves such as light and X-rays.

The diagrams show what happens when water waves pass through gaps of different sizes.

The amount of spreading when the wave has passed through the gap depends on the relative sizes of the gap and the wavelength.

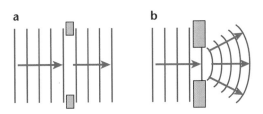

- For a gap that is many wavelengths wide, see diagram **a**, no detectable spreading takes place.
- Some spreading occurs when waves pass through a gap that is several wavelengths wide, see diagram **b**.
- The maximum amount of spreading corresponds to a gap that is the same size as the wavelength as in diagram **c**.

> When answering questions about diffraction, always emphasise the size of the gap compared to the wavelength.

Diagram **a** models what happens when light (wavelength approximately $5 \times 10^{-7}$ m) passes through a doorway, and diagram **c** models sound (wavelength approximately 1 m) passing through the same doorway.

> Diffraction explains why you can 'hear' round corners but you cannot 'see' round corners.

Diffraction also explains why long wavelength radio broadcasts can be detected in the shadows of hills and buildings, but short wavelength broadcasts cannot. Diffraction is also important in the design of loudspeakers to maximise the spreading of sounds of all wavelengths and in satellite transmissions to minimise the diffraction that occurs.

## Superposition and interference of sound

| | | | |
|---|---|---|---|
| AQA A | A2 | NICCEA | M2 |
| AQA B | M2 | OCR A | M3 |
| EDEXCEL A | A2 | OCR B | M2 |
| EDEXCEL B | M1 | WJEC | M1 |

The **superposition** of waves occurs when two or more waves cross. This is happening all the time, but the effects are so short-lived that we seldom notice them. One of the easiest effects of superposition to observe is **two-source interference**. This can be demonstrated with sound, light, surface water waves and 3 cm electromagnetic waves. The diagram overleaf shows the variation in the loudness of the sound in front of two loudspeakers vibrating in phase.

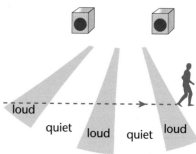

An observer walking along a line parallel to the loudspeakers notices that:

- there are regions where the sound is much louder than from one loudspeaker alone
- there are regions where the sound is barely audible
- in between these, the intensity of the sound varies from loud to quiet.

This is explained using the principle of superposition, which describes how two waves can reinforce each other or cancel each other out.

**Superposition of sound waves**

> The term 'vector sum' here means that the directions are taken into account, so that a positive displacement can be cancelled by a negative one.

> **KEY POINT**
> When two or more waves cross, the resultant displacement is equal to the vector sum of the individual displacements.

This means that displacements in the same direction cause reinforcement, while displacements in opposite directions cause cancellation. The diagram opposite shows how a loud sound and no sound can be produced when two waves of equal amplitude meet at a point.

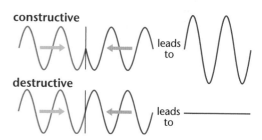

- If the waves are exactly in step (in phase) then constructive interference takes place, resulting in a loud sound.
- If the waves are exactly out of step (out of phase) then destructive interference takes place, resulting in no sound.

## Conditions for interference to be observed

| | | | |
|---|---|---|---|
| AQA A | A2 | NICCEA | M2 |
| AQA B | M2 | OCR A | M3 |
| EDEXCEL A | A2 | OCR B | M2 |
| EDEXCEL B | M1 | WJEC | M1 |

> The amplitudes do not need to be the same, but they should be comparable for the interference to be observed. If one wave has a much greater amplitude than the other then the effects of cancellation and reinforcement will not be noticeable.

The waves do not need to have the same amplitude for interference to take place. The effect of one wave having a greater amplitude than the other is that the destructive interference is not total. This happens when the observer is closer to one loudspeaker than to the other.

The diagram below represents the waves from two sources vibrating in phase. The solid and broken lines could represent peaks and troughs in the case of water waves or compressions and rarefactions in the case of sound waves.

> Try drawing similar patterns and investigating the effect of changing the wavelength and the separation of the sources.

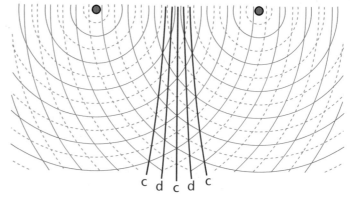

**Constructive and destructive interference from two wave sources**

Some lines of constructive interference (c) and destructive interference (d) have been marked in.

If the sources in the diagram on page 126 are in phase, then at any point equidistant from both sources the waves arrive in phase, giving constructive interference. This is the case for points along the central c line in this diagram. At all other points in the interference pattern, there is a path difference, i.e. the wave from one source has travelled further than that from the other source.

The conditions can be written as $n\lambda$ path difference for constructive interference and $(2n+1)\lambda/2$ path difference for destructive interference, where $n$ is an integer.

- **Constructive interference** takes place when the path difference is a whole number of wavelengths.
- **Destructive interference** takes place when the path difference is one and a half wavelengths, two and a half wavelengths, etc.

This is shown in the diagram.

the path difference here is half a wavelength, giving destructive interference

the path difference here is one wavelength, giving constructive interference

Clearly, for an interference pattern to be observed, the sources must not only be coherent but also of the same type.

These conditions for constructive and destructive interference only apply to two sources that are in phase. If the phase of one of the oscillators in the diagram on page 126 is reversed so that the sources are in antiphase, then the conditions for constructive and destructive interference are also reversed. The effect is to shift the interference pattern so that lines of constructive interference (c) become destructive and vice versa. Any value of phase difference between the two sources gives an interference pattern that is stationary, provided that the phase difference is not changing. Two sources with a **fixed phase difference** are said to be **coherent**. Coherent sources must have the same wavelength and frequency.

## Interference of light

| | | | |
|---|---|---|---|
| AQA A | A2 | NICCEA | M2 |
| AQA B | M2 | OCR A | M3 |
| EDEXCEL A | A2 | OCR B | M2 |
| EDEXCEL B | M1 | WJEC | M1 |

Interference patterns are easy to set up with water waves and sound because two sources can be driven from the same oscillator, giving coherence. Because of the way in which light is emitted in random bursts of energy from a source, it is not possible to have two separate sources that are coherent. Instead, diffraction is used to obtain two identical copies of the light from a single source. This is done by illuminating two narrow slits with a lamp that is parallel to the slits so that the same wavefronts arrive at each slit. A suitable arrangement is shown in the diagram below.

A colour filter can be used between the lamp and the slits to reduce the range of wavelengths interfering, but this also reduces the intensity of the interference pattern.

Bright and dark fringes on the screen are due to constructive and destructive interference between the two overlapping beams of light. The separation of the fringes depends on:

- the separation of the slits
- the wavelength of the light
- the distance between the slits and the screen.

This formula enables the wavelength of light to be measured from a simple experiment. The fringe spacing, $x$, should be obtained by measuring the separation of as many fringes as are visible and dividing by the number of fringes.

The separation of the fringes is related to the other variables by the formula:

$$\text{fringe spacing} = \frac{\text{wavelength} \times \text{distance from slits to screen}}{\text{slit separation}}$$
$$x = \lambda D/a$$

where $x$ is the distance between adjacent bright (or dark) fringes
$\lambda$ is the wavelength of the light
$D$ is the distance between the slits and the screen
$a$ is the distance between the slits.

As the wavelength of light is very small, the slits need to be close together and separated from the screen by a large distance for the interference fringes to be seen.

## Stationary (standing) waves

| AQA A | A2 | NICCEA | M2 |
| AQA B | M2 | OCR A | M3 |
| EDEXCEL A | A2 | OCR B | M2 |
| EDEXCEL B | M1 | WJEC | M1 |

A progressive wave is one whose profile moves through space, whereas a **stationary** wave has a static profile. The waves on the vibrating strings and in the vibrating air columns of musical instruments are stationary waves.

Stationary waves are caused by the superposition of two waves of the same wavelength travelling in opposite directions. They often arise when a wave is reflected at a boundary, but the waves can come from two separate coherent sources. The diagram shows the vibrations of a stationary wave on a string.

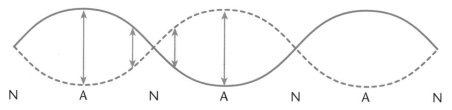

N    A    N    A    N    A    N

All points on a progressive wave vibrate with the same amplitude, but each point is only in phase with points precisely $n$ wavelengths away.

Stationary waves differ from progressive waves in a number of respects:
- there is no flow of energy along a stationary wave, although stationary waves often radiate energy
- within each loop of a stationary wave, all particles vibrate in phase and exactly out of phase (180° phase difference) with the particles in adjacent loops
- the amplitude of vibration varies with position in the loop
- there are **nodes** (points marked N in the diagram), where the displacement is always zero and **antinodes** (marked A in the diagram) which vibrate with the maximum amplitude.

A convenient way of measuring the wavelength of a progressive wave is to use it to set up a stationary wave, for example by superimposing the wave on its reflection from a barrier, and measuring the distance between a number of successive nodes or antinodes.

The wavelength of a stationary wave is twice the distance between two adjacent nodes or antinodes. The diagram opposite shows how a stationary wave (shown with a solid line) is formed from two progressive waves (shown with broken lines) travelling in opposite directions.

In (a) the waves interfere constructively, so each point on the wave has its maximum displacement.

formation of a stationary wave

In (b) each progressive wave has moved one quarter of a wavelength; the interference is now destructive, resulting in no displacement at all points on the wave.

Meters used to measure sound levels in dBA filter out the high and low frequencies so that more weighting is given to those that cause the loudest sounds.

## Attenuation and impedance

AQA A    M6

The decrease in intensity of a wave as it travels through a material is known as attenuation. Attenuation is due to a number of factors, including:

- the energy of the wave becoming spread over a wider area as it travels from its source
- interactions with the material including absorption, diffraction and scattering.

The opposition of a material to sound passing through it is called acoustic impedance. The acoustic impedance depends on the density of the material and the speed at which sound travels in it. Maximising the transfer of sound from one material to another depends on matching the impedances as closely as possible.

Sound travels faster in solids than in liquids and faster in liquids than in gases.

## Progress check

1 State two factors that reduce the upper limit of the hearing range.
2 A sound has an intensity of $5.0 \times 10^{-7}$ W m$^{-2}$. Calculate the intensity in dB.

2  57 db.
1  Age and exposure to noise.

# 5.4 Medical imaging and treatment

**After studying this section you should be able to:**

- *explain how the heart is controlled by electrical signals*
- *describe and compare the techniques used for imaging the body*
- *recall the methods used to calculate absorption of harmful radiation*

**LEARNING SUMMARY**

## Biological measurement

AQA A    M6

The heart is a pump controlled by electrical signals. One side takes in de-oxygenated blood from the body and pumps it to the lungs; the other side takes in oxygenated blood from the lungs and pumps it to the body. The diagram shows the heart and the one-way valves that only allow blood to flow in the directions shown by the arrows, from the atrium to the ventricle.

> The valves are necessary to ensure that blood does not flow from the ventricle to the atrium when the ventricle contracts.

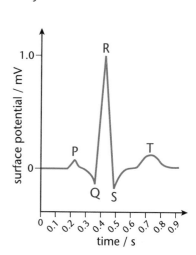

Electrical signals that control the action of the heart:

- are produced in the sino-atrial node
- are carried by nerve fibres
- cause contraction of the atria, followed by contraction of the ventricles to pump blood through the heart and round the body.

Body fluids also conduct the electrical signals, enabling them to be detected at the surface of the skin by electrodes. A display, such as that produced on a CRO of the change in voltage with time, is called an electrocardiagram (ECG).

In the ECG shown in the diagram:

> The pulse that causes relaxation of the atria occurs at the same time as, and is masked by, the QRS pulse.

- the pulse at P causes the atria to contract
- the QRS pulses causes the ventricles to contract
- the T pulse causes the ventricles to relax.

## Using ultrasound

AQA A    M6      OCR A    M5.2
EDEXCEL A    AS3      NICCEA    AS

Ultrasound consists of compression waves with a frequency above that of human hearing. It is useful for examining the internal structure of the body because it passes through flesh, but some reflection takes place at tissue boundaries. These reflections are used to generate images.

Pulses of ultrasound are produced and detected by a piezoelectric crystal:

> The alternating voltages due to the reflected waves are interpreted by a computer to produce an image.

- an alternating voltage applied to the crystal causes it to vibrate
- the voltage is removed to allow the crystal to detect any reflections
- when the crystal is made to vibrate by the reflected pulses it generates an alternating voltage.

Reflections at tissue boundaries are due to the tissues having different acoustic impedances (see section 5.3). The greater the difference in acoustic impedance, the more intense the reflection. For this reason a jelly-like substance is spread over the patient's skin during scanning, to improve the impedance matching and maximise the amount of energy transmitted into the body.

Types of ultrasound images include:

- an **A-scan** (amplitude modulated) which uses the echoes to show how the tissue boundaries are related to depth below the surface of the skin, A-scans are used to make precise determinations of the position of an object such as a brain tumour

- a **B-scan** (brightness modulated) which gives a two-dimensional picture, such as that of a fetus.

> The scanning of a fetus is now a routine procedure carried out on all pregnant women to check for abnormalities and determine the precise date when birth is due to take place.

Ultrasound scans have the advantage in that ultrasound is non-invasive and does not cause ionisation and damage to body tissue; their disadvantage is that the high frequencies needed for good resolution are attenuated more than lower frequencies, so they can only produce high-resolution scans close to the surface of the body.

## Producing X-rays

| AQA A | M6 | OCR A | M5.2 |
|-------|----|-------|------|
| EDEXCEL A | AS3 | NICCEA | AS |

X-rays are produced when high-speed electrons hit a target. In the X-ray tube shown in the diagram:

- electrons are emitted from the cathode by thermionic emission

- they are accelerated through the vacuum by the high-voltage anode

- when they strike the anode about 1% of the energy is transferred to X-rays, the rest heats the anode

**A rotating anode X-ray tube**

> X-rays are emitted when electrons strike the screen of a television or cathode ray tube.

- the anode rotates to avoid overheating of one small area.

The **intensity** of the beam is determined by the filament current which causes thermionic emission. The energy of the X-ray photons is determined by the anode voltage; increasing this voltage produces higher-energy, shorter-wavelength X-rays which are more penetrative.

## Using X-rays in diagnosis

| AQA A | M6 | OCR A | M5.2 |
|-------|----|-------|------|
| EDEXCEL A | AS3 | NICCEA | AS |

Whereas ultrasound images are produced from reflections, X-ray images rely on some absorption and some transmission by the body. As an X-ray beam passes through tissue it becomes attenuated (see page 129). The intensity varies exponentially with thickness.

> In exponential attenuation, the intensity of the beam is reduced by the same proportion in passing through equal thicknesses of material. This means that the half-thickness is independent of the initial intensity of the beam.

> **KEY POINT**
>
> The intensity of an X-ray beam, $I$, after passing through a distance $x$ of absorbing material, is given by:
>
> $$I = I_0 e^{-\mu x}$$
>
> Where $I_0$ is the initial intensity and $\mu$ is a constant, called the **linear attenuation coefficient**. The value of $\mu$ depends on the material and the energy of the X-rays.
>
> The **half-value thickness** of a material, $x_{1/2}$, is the thickness that halves the original intensity of the beam. This relationship between $\mu$ and $x_{1/2}$, is:
>
> $$\mu x_{1/2}, = \ln 2$$

The linear attenuation coefficient changes with changing density, $\rho$, of the material. The mass attenuation coefficient, $\mu_m$, defined as $\frac{\mu}{\rho}$, is independent of density.

An X-ray photograph is a shadow picture. A detector placed behind the part of the body being X-rayed shows areas of varying greyness, with the dark areas representing the least attenuation of the beam. When bones such as teeth are X-rayed, they show up as white. Suitable detectors are:

- black and white photographic film, but this requires a beam of high intensity

- an intensifying screen which contains a material that absorbs energy from the X-rays and re-emits it as light, a process known as fluorescence. The light then produces a photographic image. Film is more sensitive to light than to X-rays, so less exposure is required.

- a fluoroscopic image intensifier; the X-rays cause electrons to be ejected from a photocathode. These electrons are then accelerated in the same way as in a cathode ray tube, where they produce a bright image on a fluorescent screen. This technique is used when a 'moving' picture is required.

> Studying a moving picture enables medical staff to diagnose faults in the way that organs act or bones move.

When intestines are X-rayed, there is very little absorption by the soft tissue and the result is a low-contrast image. The contrast between the intestines and surrounding tissue is increased by giving the patient a 'barium meal'. This lines the intestines with a material that is opaque to X-rays, so they show up as white.

A CT or CAT (computerised axial topography) scan uses a rotating X-ray beam to produce an image of a slice through the body. The beam is rotated around the patient as the patient moves through the beam. The scattered X-rays are detected and used by a computer to build up a picture. Three-dimensional images can be created from images of the individual slices. CAT scans detect slight changes in attenuation and are often used for detecting tumours and cancers.

## Fibre optics

AQA A    M6    NICCEA    AS
OCR A    M5.2

> Fibres are normally clad with a glass of lower refractive index than the core in order to minimise the value of the critical angle.

An optical fibre can transmit light round bends and corners using total internal reflection. When light strikes the interface between the fibre core and its cladding it is reflected back inside the fibre provided that the angle of incidence is greater than the critical angle. This is shown in the diagram.

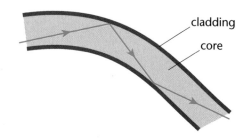

An endoscope is a bundle of fibres that can be used for viewing inside the body or performing surgery.

For viewing inside the body two sets of fibres are used:

- the fibres that take light into the body are non-coherent, they do not need to stay in the same position relative to other fibres

- the fibres that transmit the image out of the body are coherent, they are in the same positions relative to each other at each end of the bundle. The narrower the fibres used, the greater the resolution of the image.

> Tissue becomes heated by absorption of the energy from laser radiation.

Lasers are sources of high-intensity electromagnetic radiation that can be passed into the body along an optical fibre. This allows heating to be applied to small areas of tissue by absorption of the laser beam. Used in this way optical fibres can:

- act as a scalpel by cutting through tissue but without causing bleeding

- repair tears and stop bleeding by coagulating body fluids.

## Magnetic resonance imaging

OCR A ▷ M5.2 NICCEA ▷ AS

Magnetic resonance imaging (MRI) detects the presence of hydrogen nuclei. Since body tissue has a high water content, it can be used to produce an image of the body. It works by applying pulses of a high-intensity and high-frequency magnetic field which results in the emission of electromagnetic radiation from the nuclei by the following process.

Very powerful electromagnets are needed to produce the magnetic fields required by an MRI scanner.

- Hydrogen nuclei spin, giving them a magnetic field.
- When an external magnetic field is applied, they rotate around the direction of the field. This rotation is called precession, it is like that of a spinning top when it 'wobbles'.
- The frequency of precession is called the Larmor frequency.
- Application of a pulse of radiation at the Larmor frequency causes resonance, the nuclei absorb the energy into the precession.
- When the pulse is removed the nuclei lose the energy, emitting it as electromagnetic radiation. This takes place over a short period of time known as the relaxation time.

The emissions from the hydrogen nuclei are detected by a radio aerial and processed by a computer to give three-dimensional body imaging. MRI scanning is non-invasive and does not cause ionisation, but it is very expensive in terms of the capital equipment and running costs.

## Using ionising radiation

OCR A ▷ M5.2

Hydrogen peroxide is extremely corrosive and can cause extensive damage when formed within a cell.

Ionising radiation such as X-rays and gamma rays:

- can damage enzymes and DNA by ionisation
- can cause the formation of hydrogen peroxide from water, which in turn causes cell damage.

The effects of this include skin burns, cancer and sterility. For this reason, exposure to ionising radiation should always be kept to a minimum.

> **KEY POINT**
>
> Exposure, $X$, is defined in terms of the ionisation that the radiation causes:
>
> $$X = \frac{Q}{m}$$
>
> where $Q$ is the total charge of one sign of ion (either positive or negative) produced in a mass $m$ of air. Exposure is measured in C kg$^{-1}$.

This definition of exposure only applies to X-rays and gamma radiation. The absorbed dose of radiation measures the energy absorbed by a material.

> **KEY POINT**
>
> Absorbed dose, $D$, is defined as
>
> $$D = \frac{E}{m}$$
>
> where $E$ is the energy absorbed by mass $m$. Its unit is the gray (Gy) where 1 Gy = 1 J kg$^{-1}$.

The effect of absorbing radiation depends on the absorbed dose and the type of radiation involved. Alpha particles and neutrons are absorbed over a short distance whereas X-rays and gamma rays cause less damage for the same absorbed dose as their energy is distributed over a greater mass of material. The quality factor takes into account the effects of different radiations and is used to calculate equivalent effects of absorbing different radiations.

Effectively the dose equivalent is used to compare the energies absorbed by doses of different radiations.

> **KEY POINT**
>
> The **dose equivalent**, H, of a radiation is defined by:
>
> $$H = Q \times D$$
>
> where $Q$ is the quality factor of the radiation.
> Dose equivalent is measured in sievart (Sv).

X-rays and gamma rays have a quality factor of 1, while that for radiation which is absorbed more readily is higher, that for alpha particles being 20.

The effects of a dose of ionising radiation can be classified as stochastic (random) or non-stochastic:

- stochastic effects such as cancer are due to cell mutations; no minimum dose is required but the greater the dose the more likely it is that the effect will occur

- non-stochastic effects such as skin burns need a minimum dose to occur; the greater the dose above this minimum value, the worse the effect.

A similar effect is achieved by rotating a single beam around the patient so that the beam is always directed towards the tumour.

When X-rays and gamma rays are used to kill a malignant tumour several beams are used, aimed at the patient in different directions. The diagram shows how the beams intersect at the tumour. This ensures that the tumour receives a high dose while that in the surrounding tissue is relatively low.

this area receives the greatest
intensity of radiation

---

## Progress check

1 In an X-ray tube:
  **a** what determines the intensity of the X-ray beam
  **b** what determines the wavelength of the X-rays produced?

2 Suggest TWO reasons why non-invasive examination is preferred to invasive examination.

3 Why is ultrasound preferred to X-rays for producing an image of a fetus?

3 Ultrasound does not cause ionisation or damage body tissue.
2 There are no wounds, so infections cannot occur.
  The patient recovers more rapidly.
1 **a** The current in the filament.
  **b** The anode voltage.

## Sample question and model answer

Ultrasound is used for examining delicate organs and tissue. The diagram shows how an ultrasound probe is used to examine the retina.

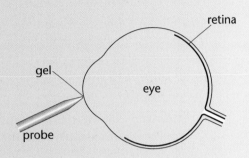

The depth of the eye is 2.5 cm.
The speed of sound in the eye tissue is $1.50 \times 10^3$ m s$^{-1}$.

(a) For the best results, the wavelength used for an ultrasound scan should be 1/200 of the depth of the tissue being scanned.
Calculate the optimum frequency for scanning a retina. [3]

> This is an example of where a question on an extension topic requires recall of material from the core, in this case the wave equations.

optimum wavelength = $1/200 \times 2.5 \times 10^{-2}$ m = $1.25 \times 10^{-4}$ m   1 mark
optimum frequency, $f = v/\lambda = 1.50 \times 10^3$ m s$^{-1} \div 1.25 \times 10^{-4}$ m   1 mark
= $1.20 \times 10^7$ Hz   1 mark

(b) Explain why the gel is used between the probe and the eyeball. [2]

> This is an example of impedance matching to maximise the energy transferred between two materials.

This maximises the energy transmitted into the eye (1 mark) by reducing the amount reflected (1 mark).

(c) In an ultrasound scan a piezoelectric crystal acts as both transmitter and detector of the ultrasound. It emits short pulses of waves, with a time gap between each pulse.

(i) Explain why there is a time gap between each pulse of waves. [2]

To allow the reflected waves to return to the crystal and be detected (1 mark) without interfering with the waves emitted by the crystal.   1 mark

(ii) Estimate the minimum time gap that should be left between pulses in this example. [3]

The minimum time should be that taken for the pulse to travel to the retina and for the reflection to be detected.   1 mark

> The factor of two is needed here because the pulse has to travel to the back of the eye and return.

time = distance ÷ speed = $2 \times 2.50 \times 10^{-2}$ m $\div 1.50 \times 10^3$ m s$^{-1}$   1 mark
= $3.33 \times 10^{-5}$ s   1 mark

## Practice examination questions

**1**

Sound is attenuated as it travels in the air.

(a) Explain the meaning of *attenuation*. [1]

(b) Sound from a loudspeaker has an intensity of $6.0 \times 10^{-5}$ W m$^{-2}$, measured at a distance of 1 m from the loudspeaker.
The threshold of hearing is $1.0 \times 10^{-12}$ W m$^{-2}$.

   (i) Calculate the intensity of the sound at a distance of 1 m from the loudspeaker in decibels. [3]

   (ii) Assuming that the sound is radiated in all directions, calculate the distance from the loudspeaker where the sound becomes inaudible. [3]

   (iii) What is the intensity, in dB, of a sound at the threshold of hearing? [1]

(c) The ear is most sensitive to sounds with a frequency around 3000 Hz.
Explain how the perceived loudness of sounds of equal intensity depends on their frequency. [3]

**2**

For ultra-sound scanning, a piezoelectric crystal produces a series of pulses of ultrasound and then detects the reflections.

An ultrasound scan is used to detect a tumour at a depth of 6.0 cm below the surface of the skin. The speed of sound in tissue is 1500 m s$^{-1}$.

(a) Explain whether an A-scan or B-scan is appropriate for this application. [2]

(b) Calculate the time that elapses between the pulse being sent and the echo being detected. [2]

(c) Explain why it is necessary to use a short pulse of ultrasound. [2]

(d) Explain why it is necessary to use pulses of ultrasound rather than a continuous wave. [2]

(e) The optimum frequency is one that produces a wave with a wavelength that is 1/200 the depth of the organ.
Calculate the optimum frequency for this application. [3]

(f) Before a scan is carried out, the skin of the patient is coated with a gel.
Explain the purpose of this gel. [2]

(g) CAT scans are also used to detect tumours.

   (i) Outline the principle of operation of a CAT-scanner. [3]

   (ii) Describe the advantages and disadvantages of using a CAT scan rather than an ultrasound scan to detect a tumour. [3]

# Astronomy and cosmology

## The following topics are covered in this chapter:

- *Looking up*
- *The model unfolds*
- *The expanding Universe*

- *Classifying stars*
- *Relativity*

## 6.1 Looking up

### After studying this section you should be able to:

- *describe the structure of reflecting, refracting and radio telescopes*
- *explain how the resolution of a telescope depends on its aperture*
- *recall that a charge-coupled device is used to give a brighter image*

**LEARNING SUMMARY**

### Optical telescopes

AQA A   MS

A **refracting telescope** uses two converging lenses. The objective lens produces a real image at its principle focus. The weaker this lens, the greater the size of the image produced and the longer the telescope. The eyepiece lens acts as a magnifying glass; by placing it so that its principal focus is at the position of the real image being magnified, the final image is formed 'at infinity' and the telescope is said to be in **normal adjustment**. This is shown in the diagram.

> Refracting telescopes are limited in the diameter of the objective lens that can be used, since the weight of a large lens can cause the glass to flow and change its shape.

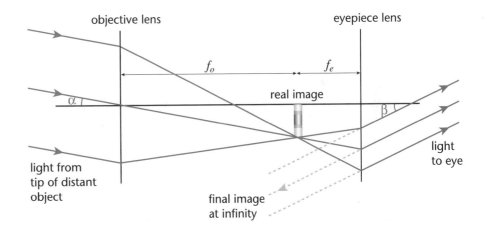

Since the size of the image on the retina depends on the angle subtended by the light reaching the eye, **angular magnification** is used to measure the factor by which the telescope increases this angle.

> **KEY POINT**
>
> Angular magnification, $M$, $= \dfrac{\text{angle subtended at eye by image}}{\text{angle subtended at eye by object}} = \dfrac{\beta}{\alpha}$
>
> For a telescope in normal adjustment, $M = \dfrac{f_o}{f_e}$

> As with a converging lens, the principal focus is the point where light parallel to the principal axis converges.

In the Cassegrain **reflecting telescope** a concave mirror is used to converge parallel light from a distant object. This would produce a real image at its principal focus, but the light is intercepted by a small convex mirror which produces an enlarged image for magnification by the eyepiece lens. This is shown in the diagram.

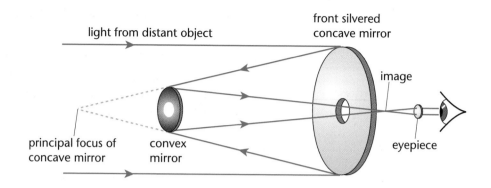

light from distant object — front silvered concave mirror

A front-silvered mirror is used to prevent loss and possible distortion of light as it passes through the glass of a rear-silvered mirror.

principal focus of concave mirror — convex mirror — image — eyepiece

Both types of telescope can produce blurring of the image called aberration.

- Chromatic aberration is due to light of different wavelengths being focused at different points; it is reduced by using combination lenses made with two different types of glass.

- Spherical aberration is caused by light that strikes the outer edges of a lens or mirror being brought to a focus slightly closer than those from the centre; it is reduced by using plano-convex lenses (only one face is curved) in refracting telescopes and parabolic mirrors in reflecting telescopes.

## Resolution

AQA A    M5

The resolving power of a telescope describes its ability to distinguish two stars that are close together, rather than identify them as a single star. When light passes through a circular aperture it is diffracted (see section 2.4). The bright central ring (known as the Airy disc) is surrounded by a number of low-intensity rings. The diagram shows the variation of intensity with angle from the centre of the Airy disc.

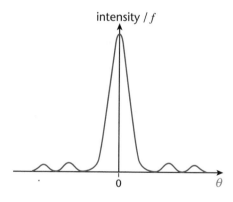

intensity / $f$

$0$          $\theta$

This is the separation required for both Airey discs to be discernible.

To be able to distinguish between two stars:

- the minimum separation of the images is when the central maximum of each is at the same place as the first minimum of the other

The symbol ≈ means 'of the order of' rather than precisely equal to.

- the Rayleigh criterion for this to happen is that the angle between the stars

$$\theta \approx \frac{\lambda}{D}$$ ,, where $D$ is the diameter of the telescope and $\theta$ is in radians

- $\theta$ is known as the resolving power of the telescope.

## Improving the image

AQA A    M5    EDEXCEL A    AS3

Charge-coupled devices are cheap to manufacture but they are limited in size.

Images that appear dim when viewed through a telescope can be viewed as much brighter images by storing them on a charge-coupled device:

- A charge-coupled device is a thin slice of silicon whose upper surface is divided into a large number of pixels

- When a photon of light is absorbed by a pixel, the energy releases an electron from the silicon onto the pixel.
- The charge on any pixel is proportional to the number of photons absorbed.

Information from each pixel is read and interpreted by a computer. The charge-coupled device has a much higher quantum efficiency than photographic film, at more than 70% compared to 4%.

This means that the probability of an individual photon liberating an electron is 0.7.

## Radio astronomy

AQA A    M5

Using wire mesh reduces the weight of the dish and the size of the structures needed to support it.

A **radio telescope** has a similar structure to that of a reflecting optical telescope, but they need to be much larger for resolution of the longer wavelengths involved. In a radio telescope:

- the reflector is a parabolic dish which can be made from mesh provided that the mesh size is less than $\lambda/20$, where $\lambda$ is the wavelength being detected
- the radio waves are detected by an aerial placed at the principal focus of the reflector
- the reflector should be manufactured to a precision of $\lambda/20$ to avoid distortion
- the resolving power is the same as that for an optical telescope, $\theta \approx \dfrac{\lambda}{D}$,
- doubling the diameter of a dish quadruples its area and the power received, so $P \propto d^2$
- the reflector is movable so that an individual object can be tracked or a specific area of the sky can be scanned.

## Progress check

1  What is the advantage of a reflecting optical telescope over a refracting one?
2  A radio telescope has a dish diameter of 75 m. What is its resolving power when detecting 0.50 m wavelength radio waves?

2  $6.7 \times 10^{-3}$ radians

1  It can be made of much larger diameter, allowing increased resolution and brighter images.

# 6.2 The model unfolds

After studying this section you should be able to:

- describe how the work of Copernicus, Kepler and Galileo enabled greater understanding of the Solar System and the Universe
- apply the three methods of measuring astronomical distances in appropriate situations
- explain how the evolution of a star depends on its mass

## Early models

OCR A    M5.1

**Copernicus'** model of the Universe (circa 1520), in which he proposed that the Sun and not the Earth, is at the centre, was the first model that explained the apparent backwards motion and the change in brightness of the planets. The stars were pictured as being fixed on a sphere beyond the planets.

**Kepler** later took the model one stage further (circa 1600) and established that the planets move in elliptical orbits. Kepler's three laws of planetary motion are:

1  planets follow elliptical orbits around the Sun, with the Sun at one focus

2  a line joining a planet to the Sun sweeps out equal areas in equal times

3  $T^2 \propto R^3$, where $T$ is the orbital period of a planet and $R$ is its mean distance from the Sun.

> Unlike Copernicus, Galileo did publish his findings. His support for the Copernican model lead to his imprisonment. The last years of his life were spent under house arrest.

In order to avoid conflict with the established teachings of the Church, Copernicus did not publish his ideas until just before his death in 1543. When **Galileo** (circa 1610) observed moons orbiting Jupiter, he realised that this confirmed the Copernican model and that the Church's teaching that 'everything in the Universe orbits the Earth' had to be wrong.

Having established his laws of motion on Earth, **Newton** applied these to the movement of the planets (circa 1690). This lead to the realisation that for Kepler's second law to be true there had to be a force directed inwards. He described that force in his law of universal gravitation (see section 3.1).

> **KEY POINT**
>
> Newton's law of gravitation can be used to show the relationship between the orbital period, $T$, and radius of orbit, $R$:
>
> $$T^2 = \left(\frac{4\pi^2}{GM_s}\right)R^3$$
>
> which is in agreement with Kepler's third law.

> Kepler's law was derived by analysing data, so it provides experimental support for Newton's law.

Although Newton's law and Kepler's laws supported each other, the orbit of Uranus did not. Uranus was seen to change orbital speed in a way that Newton's theory did not explain. Not, that is, until the theory was used to predict the existence of Neptune as being responsible for the irregularities in Uranus' orbit.

## The modern picture

EDEXCEL B    M5       OCR A    M5.1

The **Solar System** includes:

- the **Sun**, the star at its centre and its source of energy
- the **planets**, which can be seen by reflected sunlight
- planetary **satellites**, the natural satellites are the planets' moons
- **comets**, which unlike the planets can orbit the Sun in any plane and any direction.

The **Universe** includes:

- galaxies; a **galaxy** is a collection of stars, between a few million and hundreds of billions in number, held together by gravitational forces
- an uncountable number of stars
- **microwave radiation**, thought to be left over from the creation of the Universe, and corresponding to a temperature of 3K.

## Astronomical distances

AQA A ▸ M5    OCR A ▸ M5.1
EDEXCEL B ▸ M5

> The AU is used for measuring distances within the Solar System, the light year for close stars and the parsec for more distant stars.

Different units are used to measure distances in our Solar System, our galaxy and beyond our galaxy. The table shows these units and gives their equivalent in m.

| unit | definition | equivalent distance in m |
|------|-----------|--------------------------|
| astronomical unit (AU) | mean distance between Earth and Sun | $1.5 \times 10^{11}$ |
| light year | distance travelled by light in 1 year | $9.5 \times 10^{15}$ |
| parsec (pc) | distance between observations in AU / angle subtended (in arc seconds, where 1 arc second = 1/3600 degrees) | $3.1 \times 10^{16}$ |

The parsec is used to measure distances by observing the position of a star from opposite extremes of the Earth's orbit. At these extremes the distance between observations is 2 AU.

The next table shows typical sizes and masses of the Universe and its contents.

| object | distance in light years | mass in kg |
|--------|------------------------|-----------|
| nearest star | 4.3 | $10^{30}$ |
| Milky Way galaxy | $10^5$ measured across from opposite extremes | $10^{40}$ |
| Universe | $10^{10}$ measured across from opposite extremes | unknown, but thought to be around $10^{55}$ |

## The life of a star

AQA A ▸ M5    EDEXCEL B ▸ M5
EDEXCEL A ▸ AS3    OCR A ▸ M5.1

> When protons fuse together the resulting nucleus has less mass than its constituent particles. Einstein's equation $E = mc^2$ is used to calculate the energy released in fusion (see section 4.3).

When a star is formed:

- clouds containing dust, hydrogen and helium collapse due to gravitational forces
- the contraction causes heating
- the temperature in the core becomes hot enough for fusion reactions to occur
- energy is released as hydrogen nuclei fuse, eventually forming the nuclei of helium.

The star is now in its **main sequence**. How long it remains like this depends on its mass; the more massive the star, the shorter the main sequence.

When there is no longer sufficient hydrogen in the core of a star to maintain its energy production and brightness the star cools:

> Fusion to form nuclei more massive than iron requires energy to be supplied, so energy cannot be released by further fusion reactions when the core consists of iron.

- the core contracts but the outer layers expand, forming a **red giant** or **red supergiant**, depending on its size
- increased heating of the core creates temperatures hot enough for the helium nuclei to fuse, forming carbon and oxygen.

In a small star like our Sun, once the helium has fused the core contracts to form a **white dwarf**, and the outer layers are ejected. The star then cools to an invisible **black dwarf**.

| star forms in gas cloud | main sequence | red giant | outer layers flung off | white dwarf | black dwarf |

**The life cycle of a small star**

In a larger star:

- further fusion reactions occur and the star becomes a blue supergiant

- the nuclei of more massive elements are formed until the core is mainly iron, the most stable element

- the core now collapses and the heating causes the star to increase in brightness, it is a supernova, an exploding star

- in the explosion the outer layers are flung off to form a new dust cloud and the dense core remains as a neutron star or, in the case of a very massive star, a black hole, whose gravitational field is so strong that light cannot escape from it.

The boundary around a black hole, beyond which nothing can be seen, is called the event horizon. Its size is related to the mass of the black hole.

## The Schwarzschild radius

AQA A     M5

> **KEY POINT**
>
> The **Schwarzschild radius**, $R_s$ is the radius of the event horizon around a black hole:
>
> $$R_s = \frac{2GM}{c^2}$$
>
> where $M$ is the mass of the black hole, $G$ is the universal gravitational constant and $c$ is the speed of light.

## Stars and galaxies

EDEXCEL B     M5     OCR A     M5.1

The Sun is a small star situated on one arm close to the outer edge of a spiral galaxy called the Milky Way. The diagram shows the three shapes of galaxy: spiral, elliptical and irregular.

The Milky Way galaxy whirls around at enormous speeds but, like the planets in the Solar System, the orbital time of stars increases with increasing distance from the centre. When Newton's law of gravitation is applied to the Solar System, the mass of the planets is ignored as this has only a small effect on the movement of an individual planet. The mass of all the other stars in a galaxy cannot be ignored when applying the law to an individual star, but only those within the orbit of a star are taken into account as it is the gravitational attraction of these stars that provides the centripetal force.

**KEY POINT**

Newton's law applied to the orbit of a star gives the relationship between the orbital time, $T$, and the orbital radius, $r$:

$$T^2 = \frac{4\pi^2 r^3}{GM}$$

where $M$ is the total mass contained within the orbit of the star.

## Olber's paradox

OCR A ▶ M5.1

A paradox is an apparent contradiction.

Newton pictured a static Universe that stretched to infinity, with stars being uniformly distributed throughout. Otherwise, he argued, the Universe would collapse due to the gravitational attraction of its mass. Olber's paradox is that if this were the case the sky should be bright at night. To understand why:

- imagine that the Universe is split into shells of equal thickness, like the structure of an onion
- the intensity of the light received from a star in a shell of radius $r$ is four times that received from a star in a shell of radius $2r$

The volume of a shell of surface area $A$ and radius $r$ is approximately equal to $Ar$.

- because it has four times the volume, a shell of radius $2r$ contains four times as many stars as one of radius $r$
- it follows that the light from each shell is equally bright, and if there are an infinite number of shells the sky should be bright at night.

The contradiction arises because of the incorrect assumptions that:

- the Universe is infinite

The Universe is finite and expanding.

- the Universe is static.

## The cosmological principle

EDEXCEL B ▶ M5    OCR A ▶ M5.1

The cosmological principle is a statement that there is nothing special about the Earth. It states that:

- on a large scale, mass is distributed evenly throughout the Universe, although on a small scale there are regions of high and low concentration
- on a large scale, the Universe does not differ in any one direction compared to any other direction
- physical laws and principles apply equally throughout the Universe.

## Progress check

1   The second nearest star to Earth subtends an angle of 1.52 arc seconds when measured from opposite extremes of the Earth's orbit. Calculate its distance from the Earth in pc.

2   What causes the release of energy from a star in its main sequence?

3   What is meant by an *event horizon*?

3   The boundary around a black hole, beyond which nothing can be seen.
2   The fusion of protons into more massive nuclei, eventually forming helium.
1   1.32 pc.

# 6.3 The expanding Universe

*After studying this section you should be able to:*

- *explain how line spectra give evidence for the composition of stars*
- *describe how Doppler shift can be used to measure the speed of a star relative to the Earth*
- *explain how Hubble's law can be used to estimate the age of the Universe*

**LEARNING SUMMARY**

## Emission and absorption spectra

| AQA A | M5 | EDEXCEL B | M5 |
|---|---|---|---|
| EDEXCEL A | M4 | OCR A | M5.1 |

In section 2.5 it was explained how line spectra are emitted when an electric current passes in an ionised gas. Light is emitted when an electron loses energy either by being captured by a gas ion or by a transition to a lower energy level.

A similar effect causes a gas that is hot (typically in excess of 1000°C) to emit light. When this happens:

- electrons gain energy in the collisions due to thermal motion of the particles
- electrons can only absorb the amounts of energy associated with movement to a higher energy level
- electromagnetic radiation is emitted when an electron loses energy and moves to a lower energy level.

> The temperature at which a gas starts to emit light depends on the smallest energy level transition that results in the emission of a visible photon.

As each element has its own characteristic **emission spectrum**, the composition of a gaseous object that emits its own light can be determined by analysing this spectrum.

This is how helium was discovered in solar flares, but the spectrum obtained from the body of the Sun and other stars is not a line emission spectrum but a line **absorption spectrum**. It contains the full range of visible wavelengths with a large number of 'black' lines, where wavelengths are missing.

> Hot solids and liquids can emit all wavelengths, and appear white hot, because there are no restrictions on the allowed energy level transitions.

Inside a star the particles are very close together, and their energy levels overlap. So unlike a gas at normal temperatures and pressures on Earth, there are no limits on the energy level transitions that are possible. This explains why a star emits the full frequency range of the visible spectrum. The explanation of the black lines is:

- a gas can only absorb the wavelengths of electromagnetic radiation that it emits, since these correspond to the same energy level transitions in the opposite directions
- when white light from the body of the Sun passes through the cooler, gaseous surface layers, the elements present absorb their own characteristic wavelengths
- this energy is re-emitted in all directions, much of it back into the body of the Sun, leaving a dark line in the spectrum that reaches the Earth.

The composition of a star is determined from the absorption spectrum and the wavelengths missing from that spectrum, rather than the wavelengths it contains.

## Doppler shift

| | | | |
|---|---|---|---|
| AQA A | M5 | EDEXCEL B | M5 |
| AQA B | AS | OCR A | M5.1 |
| EDEXCEL A | M4 | OCR B | M4 |

Radar guns, used to measure the speed of a moving car, also use the Doppler effect.

You can tell whether an unseen aircraft is moving towards or away from you by the pitch of the engines. You receive higher frequency sounds when it is moving towards you than when it is moving away. This is due to the **Doppler effect**, the difference between the frequency of a wave emitted and that of the wave received when there is relative movement of the wave source and an observer.

> **KEY POINT**
>
> For light, the change in frequency, $\Delta f$, or wavelength, $\Delta\lambda$, due to relative motion is given by:
> $$\Delta f/f = \Delta\lambda/\lambda = v/c$$
> where $v$ is the speed of the source relative to the observer and $c$ is the speed of light.

The spectral lines in the absorption spectrum of a star observed from the Earth are subject to Doppler shift:

- a shift towards the blue end of the spectrum corresponds to a reduction in wavelength and an increase in frequency; this happens when a star is moving towards the Earth

- a shift towards the red end of the spectrum corresponds to an increase in wavelength and a reduction in frequency.

## Hubble and red-shift

| | | | |
|---|---|---|---|
| AQA A | M5 | EDEXCEL B | M5 |
| AQA B | AS | OCR A | M5.1 |
| EDEXCEL A | M4 | OCR B | M4 |

We live in an expanding Universe. Evidence for this comes from the shift in frequency and wavelength of the light received from other galaxies. Light from the nearby Andromeda galaxy is shifted towards the blue end of the spectrum, but for every other galaxy the shift is towards the red. The amount of shift depends on the speed of the galaxy relative to the Earth, known as the speed of recession.

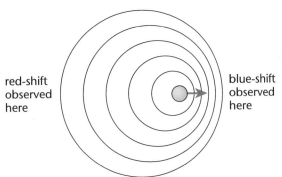

red-shift observed here

blue-shift observed here

**How movement of a star causes a difference between the wavelengths emitted and those received**

Hubble discovered that:

the speed of recession of a galaxy is proportional to its distance from the Earth.

This is known as **Hubble's law**, and it gives rise to Hubble's constant.

Hubble's law seems to state that the Earth is the centre of the expansion. This is not the case. Measurements taken from any point in an expanding Universe would give the same pattern.

> **KEY POINT**
>
> Hubble's law can be written as
> $$v \propto d \text{ or } v = Hd$$
> where $v$ is the speed of a galaxy relative to the Earth, $d$ is its distance from the Earth and $H$ is Hubble's constant.
> $H$ has units $s^{-1}$ but its value is not known precisely.
> An alternative unit for $H$ is km $s^{-1}$ Mpc$^{-1}$.

The expansion of the Universe gives evidence to support the **Big Bang** theory. According to this theory:

- the Universe started with an enormous explosion about 15 thousand million years ago
- initially the Universe was very small and very hot
- the Universe has been cooling as the galaxies have been moving away from each other ever since.

> The energy of the background radiation is equivalent to that radiated from an object at a temperature of about 3 K.

The theory is also supported by the existence of microwave background radiation that fills the whole Universe. This is thought to be electromagnetic radiation left over from the initial explosion. Due to the expansion of the Universe its wavelength has increased and its energy has decreased.

## The age of the Universe

| | | | |
|---|---|---|---|
| AQA A | M5 | EDEXCEL B | M5 |
| EDEXCEL A | M4 | OCR A | M5.1 |

> The age of the Universe is simply $\frac{1}{H}$.

Measurements of the value of the Hubble constant enable the age of the Universe to be estimated. This involves measuring both the relative speed and the distance of a star from the Earth. To estimate the distance to a star:

- for close stars (up to 300 light years) the apparent movement of the star across the fixed background of more distant stars is used. This involves measuring the position of the star when the Earth is at opposite points in its orbit.

- for more distant stars observations of the spectrum, brightness and periodic variation in brightness are needed.

Both methods only produce an estimate. In the case of close stars, movement of the star itself and the Sun also affect the apparent change in position due to the Earth's movement. For more distant stars, scattering of light from the star as it passes through the Earth's atmosphere leads to uncertainty in the measurements.

> The most recent data from the Hubble telescope puts the value of the Hubble constant at between $1.95 \times 10^{-18}$ s$^{-1}$ and $2.28 \times 10^{-18}$ s$^{-1}$.

Since the distance of a star, $d$, cannot be measured very precisely it follows that there is uncertainty about the value of the Hubble constant, $H$. Probably the only certain thing about it is that it is not constant! Its value is probably decreasing due to galaxies slowing down because of the effect of their gravitational pulls on each other.

The diagram shows how the speed of a galaxy relative to the Earth varies with distance.

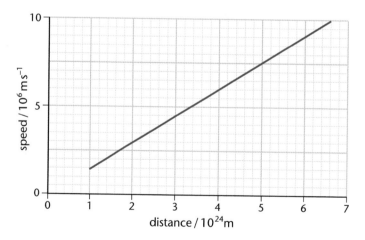

The Hubble constant can be calculated from any pair of values obtained from the graph. An estimate of the age of the Universe is then obtained by calculating the reciprocal of this constant.

## In the beginning

| AQA A | M5 | EDEXCEL B | M5 |
| EDEXCEL A | M4 | OCR A | M5.1 |

There is little direct evidence to support any theory about what happened in the first fraction of a second after the 'Big Bang'. The energies of the particles that existed would be so vast that it is impossible at the present time to accelerate particles to these energies in order to study their behaviour.

Before the first 0.01 s:

- electrons and positrons (anti-electrons) and quarks existed
- quarks combined to form nucleons and their anti-particles.

An anti-particle has the same mass as its corresponding particle, but the opposite charge (if any). When they collide they cease to exist, forming gamma rays or other particles.

After the first 0.01 s.

- annihilation of nucleons and anti-nucleons occurred, leaving an excess of nucleons
- after about 15 s, electron-positron annihilation left an excess of electrons
- after about 100 s, fusion of protons formed light nuclei such as helium
- after about $5 \times 10^5$ years, atoms formed from nuclei and electrons
- stars and galaxies began to form as there was less electrostatic repulsion between particles to oppose the gravitational attraction

The formation of neutral atoms resulted in less electrostatic repulsion.

- after about $5 \times 10^6$ years the formation of clusters of galaxies caused local variations in the density of the Universe.

## What about the future?

| EDEXCEL A | M4 | OCR A | M5 1 |
| EDEXCEL B | M5 | | |

There are three possibilities for the future of the Universe:

- it will continue to expand
- it will reach a steady size
- it will contract, ending in a 'Big Crunch'.

Whatever the answer, it depends on the gravitational attraction between the galaxies, and this in turn depends on the average density, known as the mass-energy density to emphasise that the energy in the Universe has considerable mass and contributes to its density.

The future of the Universe depends on its density. The critical density is that required to hold the Universe in a steady state; if the density is less than this then the Universe will continue to expand. If the density is greater than the critical density, the Universe will stop expanding and contract, resulting in the 'Big Crunch'.

Newton's law of gravitation, together with Hubble's law, can be used to show that:

A strict derivation of the relationship is only possible using relativity theory (see section 6.5).

> **KEY POINT**
>
> The critical density, $\rho_0$, is given by the expression
> $$\rho_0 = 3H^2/8\pi G$$

This gives a value of approximately $2.0 \times 10^{-26}$ kg m$^{-3}$, depending on the value assumed for Hubble's constant.

The following graphs show the variation in the size of the Universe for the open Universe (infinite expansion) and closed Universe (final contraction) models.

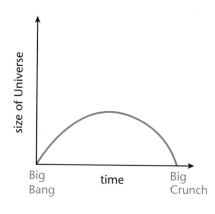

---

## Progress check

1 Explain why dark lines in an absorption spectrum correspond to bright lines in an emission spectrum of the same element.

2 A star is moving away from the Earth at a speed of $1.0 \times 10^6$ m s$^{-1}$.

Calculate the fractional difference in wavelength, $\Delta\lambda/\lambda$ between the waves received and those emitted.

3 Use data from the graph on page 146 to calculate a value for the Hubble constant.

1 A gaseous element can only emit and absorb certain frequencies or wavelengths. These frequencies or wavelengths are the same for emission and absorption. When white light passes through the gas it absorbs the frequencies in its emission spectrum, leaving dark lines.

2 0.0033

3 $1.5 \times 10^{-18}$ s$^{-1}$

# 6.4 Classifying stars

*After studying this section you should be able to:*

- *compare the apparent magnitudes and calculate absolute magnitudes of stars*
- *describe how stars are classified in terms of their brightness, temperature and composition*
- *explain how an absorption spectrum gives information about the elements present in a star*

**LEARNING SUMMARY**

## Star magnitudes

| AQA A | M5 | EDEXCEL B | M5 |
|-------|----|-----------|----|
| EDEXCEL A | AS3 | OCR A | M5.1 |

> The magnitude scale is a logarithmic scale, equal changes on this scale correspond to the brightness changing by the same factor.

The brightness of the light received from a star, known as its apparent magnitude, depends on both its actual brightness and its distance from the Earth. A star of magnitude 1 is defined as appearing 100 times as bright as a star of magnitude 6, so one unit on the magnitude scale corresponds to a change in brightness by a factor of 2.5.

> **KEY POINT**
>
> The apparent magnitude, $m$, of a star is related to its intensity, $I$, by the expression:
> $$m = -2.5 \log I + \text{constant}$$

In using this relationship:

- the value of the constant depends on both the distance of the star from the Earth and its surface area

- since the equation is only used to compare apparent magnitudes, the constant has no significance as it does not appear in the relationship $m_1 - m_2$, the difference in the apparent magnitudes of two stars

- the ratio of intensities of two stars of apparent magnitudes $m_1$ and $m_2$ is equal to $m_1 - m_2 = 2.5^{m_1 - m_2}$.

> Apparent magnitudes are compared by the intensities of the images on photographic film or other detector exposed for a fixed time.

The smaller the apparent magnitude of a star, the brighter it appears to be. Stars can have negative values of apparent magnitude; the brightest star seen from the Earth is the Sun, with an apparent magnitude of −27.

To compare actual stellar magnitudes the effect of distance on the detected brightness needs to be taken into account. This is done by calculating the magnitudes that stars would have if they were the same distance from the Earth.

> **KEY POINT**
>
> The **absolute magnitude**, $M$, of a star is the apparent magnitude it would have if it were a distance of 10 pc from the Earth.
> It is calculated from the apparent magnitude using the expression
> $$m - M = 5 \log (d/10)$$
> where $d$ is the distance of the star from the Earth in pc.

When the magnitudes of stars are plotted against their surface temperatures, the result is a Hertzsprung-Russell diagram.

The H-R diagram shows that:

- different types of star occupy specific regions on the diagram

- most stars are in the main sequence band, with the more massive stars in the upper left corner

- the white dwarfs are hot but dim because of their small size.

Note that, on the diagram, movement towards the right corresponds to decreasing temperature.

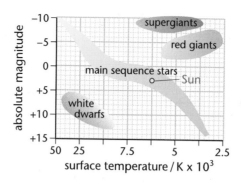

## Stars are black bodies

| AQA A | M5 | EDEXCEL B | M5 |
|---|---|---|---|
| EDEXCEL A | AS3 | OCR A | M5.1 |

All objects emit electromagnetic radiation over a range of wavelengths that depends on the temperature of the object. Increasing the temperature of an object results in:

- an increase in the power radiated
- the wavelength at which the maximum power is radiated becoming shorter.

A black body absorbs all wavelengths of electromagnetic radiation and reflects none. However, it need not appear black, as solid objects at a temperature greater than about 600°C emit visible radiation. The rate of emission of radiation from a black body at different temperatures is shown in the graph opposite.

The area between the graph line and the wavelength axis represents the total power radiated. The graph shows that this increases with increasing temperature.

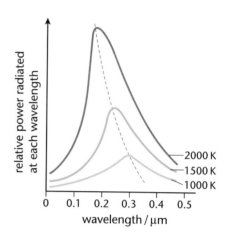

The dotted line shows how the wavelength at which the maximum power is emitted depends on the temperature of the black body. Knowledge of this wavelength enables the surface temperature of a star to be estimated using Wien's law.

Note that the unit 'm K' is 'metre × Kelvin', not milliKelvin, which would be written as mK.

> **KEY POINT**
>
> Wien's law relates the wavelength at which the maximum power is emitted by a black body, $\lambda_{max}$, to its temperature:
> $$\lambda_{max}T = 2.90 \times 10^{-3} \text{ m K}$$
> Where $T$ is the temperature of the object in K.

In practice it may be difficult to determine $\lambda$ precisely because:

- the instrument used to detect the radiation may not have the same sensitivity for all wavelengths
- the atmosphere absorbs some wavelengths of light more than others, so the intensity of these wavelengths may be reduced by more than that of others as the light travels through the atmosphere.

The power radiated by a black body depends on both its surface area and its temperature.

When applying Stefan's law to stars, it is assumed that stars are black bodies in that they are unselective in the wavelengths of radiation that they emit and absorb.

> **KEY POINT**
>
> **Stefan's law** states that:
>
> the power radiated per m² of a black body is proportional to its (absolute temperature)⁴
> $$P = \sigma AT^4$$
> where $\sigma$ is Stefan's constant and has the value $5.67 \times 10^{-8}$ W m⁻² K⁻⁴.

When applying Stefan's law:

- the power radiated is calculated from measurements of the intensity of the light received and the distance of the star from the Earth, using the inverse square law

- the inverse square law states that the intensity decreases as the inverse of the square of the distance, $I \propto \dfrac{1}{d^2}$

- the inverse square law applies to light or other electromagnetic radiation travelling in a vacuum, but absorption and scattering of light by the atmosphere reduces the reliability of calculations based on this law.

All the radiation from a star passes through a spherical surface of radius *d* with uniform intensity if the centre of the surface is at the star. Doubling the radius of the sphere causes its surface area to quadruple, so the power incident on unit area, the intensity, is quartered.

If the temperature of a star is calculated from Wien's law, then Stefan's law can be used to estimate the area needed for the star to have the same power output as the Sun ($3.8 \times 10^{26}$ W). The surface area, and hence the radius of the star, can be calculated by comparing the absolute magnitudes.

## Spectral classes

AQA A — M5    EDEXCEL B — M5
EDEXCEL A — AS3    OCR A — M5.1

Analysis of the emission and absorption spectra of a star enables it to be classified according to its temperature and composition. The table shows the characteristics of the seven classes of star in order of decreasing surface temperature.

| type | colour | surface temp/K $\times 10^3$ | absorption of hydrogen Balmer lines | elements detected in absorption spectrum |
|---|---|---|---|---|
| O | blue | 40 | weak | helium ions |
| B | pale blue | 20 | fairly strong | helium atoms |
| A | white | 10 | strong | hydrogen |
| F | pale yellow | 7 | medium | ionised metals such as calcium and iron |
| G | yellow | 6 | weak | calcium ions and iron atoms |
| K | orange | 5 | weak | calcium and iron atoms, together with some molecules |
| M | red | 3 | weak | molecules of metal oxides |

This table shows that:

- in the surface layers, ionised helium is only present in the hottest stars where the temperature is high enough to ionise helium atoms

- stars in groups B and A show strong absorption by hydrogen in the Balmer series; this is the part of the hydrogen emission spectrum due to exciting electrons from the first excited energy level above the ground state (see section 2.5). This energy level is given the value $n = 2$, where $n = 1$ corresponds to the ground state. Absorption of these spectral lines shows that the temperature is cool enough for hydrogen atoms to exist but hot enough for them to be in an excited state

- the relatively cool stars in classes F and G show absorption lines due to ionised metals

- stars in classes K and M are cool enough for molecules such as those of metal oxides to exist without being separated into atoms or ionised.

Analysing an absorption spectrum obtained at the Earth's surface can be misleading, as the following diagram shows.

This graph shows the reduction in intensity of all wavelengths as the radiation passes through the atmosphere, with particular wavelengths being absorbed more than others. Information from spectra obtained from satellites gives a fuller picture than that from spectra obtained at ground level.

## Quasars

AQA A      MS

Quasars are the most distant objects that can be detected. They are very bright objects which were first discovered by detection of the radio signals that some of them emit. Analysis of the light from a quasar reveals that:

- the spectrum is an emission spectrum, with very bright lines on a continuous spectrum
- these lines show enormous red-shift, indicating that quasars are moving with very high speeds.

Analysis of the red-shift indicates that some quasars are moving at speeds greater than 90% of the speed of light at distances of more than ten billion ($1 \times 10^{10}$) light years. They are relatively small and yet they appear to emit thousands of times as much energy as a typical galaxy.

However, this may be misleading. If some other mechanism is responsible for the red-shift, they could turn out to be much closer and less bright than was at first believed.

## Progress check

1 The apparent magnitude of Venus is −4, and that of the Sun is −27. By what factor does the Sun appear brighter than Venus?

2 Suggest why the magnitudes of red giants are greater than some main sequence stars that have a higher surface temperature.

3 Explain why dark lines in an absorption spectrum correspond to bright lines in the emission spectrum of the same element.

3 A gaseous element can only emit and absorb certain frequencies or wavelengths. These frequencies or wavelengths are the same for emission and absorption. When white light passes through the gas it absorbs the frequencies in its emission spectrum, leaving dark lines.

2 The red giants have a much greater surface area.

1 $1.4 \times 10^9$

# 6.5 Relativity

**After studying this section you should be able to:**

- recall the postulates of special relativity
- calculate the amount of time dilation and length contraction when a moving object is observed
- explain why there is a maximum speed to which an object can be accelerated

**LEARNING SUMMARY**

## Does time pass at a constant rate?

AQA A ▸ M8    OCR A ▸ M5.1

In everyday life, we sense and judge speed relative to the surface of the Earth or vehicles that we travel in. If you are in a car travelling at 25 m s$^{-1}$ and another car travelling at the same speed is approaching you, then its speed relative to your vehicle is 50 m s$^{-1}$. If, on the other hand, it is in front of you and travelling in the same direction, its speed relative to you is zero.

This led Einstein to use the idea of a frame of reference as something that speeds are measured relative to. In special relativity frames of reference are not accelerating.

> Since there is no such thing as absolute motion, the laws of physics have to be the same for all observers in relative motion, provided they are not accelerating.

> **KEY POINT**
>
> The laws of physics apply equally in all inertial frames of reference.
> This is the first postulate (assumption) of Einstein's theory of special relativity.

In the diagram, the driver of the red car sees the blue car approaching at 50 m s$^{-1}$ but, from the frame of reference of the stationary observer, the blue car is approaching at 30 m s$^{-1}$.

What about the light that passes between the two vehicles? Is its speed affected by their motion? The answer is no.

> **KEY POINT**
>
> The speed of light in a vacuum always has the same value.
> This is the second postulate of Einstein's theory.

> Far more muons reach the surface of the Earth than is predicted by a half-life of 2 microseconds.

The invariance in the speed of light leads to the conclusion that neither time nor distance have absolute values. Each of the drivers of the cars in the diagram would notice that time passes more slowly in the other vehicle. This effect is known as time dilation. It increases the apparent half-life of muons created in the upper atmosphere from 2 microseconds to 20 microseconds, measured by an observer on Earth. We do not normally notice the effects of time dilation because we move at speeds very much less than the speed of light. The muons move through the atmosphere at a speed which is within 1% of that of light.

As well as having different measurements of time, the car drivers in the diagram have different measurements of distance. Each driver sees the other car as being shorter than its driver sees it; this is known as length contraction. Length contraction only occurs in the direction of motion.

> **KEY POINT**
>
> The factor by which time is dilated and length is contracted is $\sqrt{1-\dfrac{v^2}{c^2}}$, where $v$ is the speed of the object relative to the observer and $c$ is the speed of light.

It follows that time $t_0$ and length $l_0$ measured within a frame of reference are observed by someone moving relative to that frame to have values:

- $t = t_0 / \sqrt{(1 - v^2/c^2)}$
- $l = l_0 \sqrt{(1 - v^2/c^2)}$

## How is mass affected?

AQA A    M8    OCR A    M5.1

A consequence of length contraction is that as an object speeds up under the action of a constant force, its apparent acceleration decreases. This causes an apparent increase in mass.

> For everyday objects moving at everyday speeds, there is no significant increase in mass.

> **KEY POINT**
>
> The apparent mass, m, is given by the expression
> $$m = \frac{m_0}{\sqrt{(1 - v^2/c^2)}}$$
> where $m_0$ is the **rest mass**, the mass of the object measured within its frame of reference.

This expression shows that the mass of an object tends towards infinity as its speed approaches the speed of light. It is therefore impossible to accelerate an object beyond the speed of light – the maximum possible speed.

## General relativity

OCR A    M5.1

> This is known as the principle of equivalence of inertial and gravitational forces.

Einstein stated that gravitational forces are equivalent to other forces that cause acceleration. Imagine an astronaut in a spacecraft, away from the effects of any gravitational fields. If the craft accelerates at a rate of 9.8 m s⁻², the astronaut experiences the same force that she would if she were in the capsule and it was stationary on the Earth.

> The path of the light appears to be curved because of the movement of the spacecraft since the light was emitted by the torch.

$R = mg$

astronaut
stationary
on Earth

$F = ma$

astronaut
accelerating
in space

Furthermore, light from her torch would appear to follow a curved path. So if light follows a curved path within an accelerating frame of reference, it should also follow a curved path within a gravitational field.

This was confirmed in the solar eclipse of 1991, where two stars appeared to move further apart when the Sun and Moon were between the paths of light coming from the stars. The apparent change in position was due to the light changing direction as it passed through the Sun's gravitational field.

Gravitational fields also cause time dilation, with clocks running at different speeds in gravitational fields of different strengths.

Einstein's theory of general relativity was used to explain the changes observed in Mercury's orbit, which Newton's laws could not explain. Mercury's perihelion, the point of closest approach to the Sun, rotates around the Sun. Calculations based on the general theory of relativity predicted the amount of rotation of the perihelion to be the same as that observed.

## Progress check

1 A muon has a half-life of 2 microseconds, measured within its frame of reference. Calculate the observed half-life when the muon moves at $0.99c$.

2 Explain why a moving spacecraft appears shorter but not narrower to an external observer.

3 Explain whether light follows a curved path as it travels past the Earth, above the height of the atmosphere.

3 Yes, due to the effect of the Earth's gravitational field.
2 Length contraction only occurs in the direction of motion.
1 14 microseconds

## *Sample question and model answer*

(a) Explain the meaning of the *Doppler Effect* and describe, in outline, how it is used to measure the speeds of stars relative to the Earth. [4]

The Doppler Effect refers to the change in wavelength and frequency when a source of waves is moving relative to an observer (1 mark). The wavelength is increased when the source is moving away from the observer and there is a decrease in wavelength when it moves towards the observer (1 mark).

An outline description means that you should describe the physical principles without going into great detail of how the measurements are made.

The faster a star is moving, the greater the change in wavelength (1 mark). To measure the speed of a star, the wavelength received is compared to that emitted by a stationary source (1 mark).

b) Measurements of *red shift* often show different values for different stars in the same galaxy. Suggest a reason for this.

Stars within a galaxy are rotating around the centre of the galaxy (1 mark). Depending on the direction of movement of a particular star, this can cause an increase or a decrease in its velocity relative to the Earth and consequently an increase or decrease in the amount of red shift (1 mark).

The effect of the rotation of stars within a galaxy is to broaden the spectral lines received from the galaxy as a whole, showing that stars are moving with a range of speeds relative to the Earth.

(c) In the light from a star, the wavelength of a hydrogen line in its absorption spectrum is measured to be $5.153 \times 10^{-7}$ m. A laboratory measurement of the same spectral line gives its wavelength as $4.861 \times 10^{-7}$ m.

(i) Calculate the speed of movement of the star relative to the Earth.
$c = 3.00 \times 10^8$ m s$^{-1}$ [2]

$\Delta\lambda/\lambda = v/c$

$v = c \times \Delta\lambda/\lambda = 3.00 \times 10^8$ m s$^{-1}$ $\times$ $(5.153 \times 10^{-7} - 4.861 \times 10^{-7}$ m$)/$ $4.861 \times 10^{-7}$ m (1 mark) $= 1.80 \times 10^7$ m s$^{-1}$ (1 mark)

There is no mark for the recall of this relationship as it will be given in the question or on a separate data sheet.

(ii) Explain whether the star is moving towards or away from the Earth. [2]

The star is moving away from the Earth (1 mark) as the wavelength of light received is greater than that emitted (1 mark).

The cue word explain here means that you must give a reason to justify your answer to gain full marks.

## Practice examination questions

**1**

(a) (i) Explain the meaning of the terms *apparent magnitude* and *absolute magnitude*. [2]

   (ii) Explain why the absolute magnitude is preferred to the apparent magnitude when comparing the magnitudes of stars. [2]

(b) Sirius is at a distance of 2.7 pc from the Earth. Its apparent magnitude is −1.5. Calculate the absolute magnitude of Sirius. [2]

(c) The absolute magnitude of the Pole star is −4.9. Its apparent magnitude is 2.0. Calculate the distance of the Pole Star from the Earth. [2]

(d) White dwarfs are hotter than red giants, but they are less bright. Suggest a reason for this. [2]

**2**

The hydrogen absorption spectrum contains a dark line corresponding to a wavelength of $6.563 \times 10^{-7}$ m.

(a) Explain why the lines in an absorption spectrum are dark. [3]

(b) In the spectrum detected from a star, this line has a wavelength of $6.840 \times 10^{-7}$ m.

   (i) Explain this apparent increase in wavelength. [2]

   (ii) Calculate the speed of the star relative to the Earth.
$c = 3.00 \times 10^{8}$ m s$^{-1}$. [3]

   (iii) The value of the Hubble constant, $H_0 = (2.12 \times 10^{-18} \pm 0.07)$ s$^{-1}$. Calculate the range of distances from the Earth within which the star lies. [3]

**3**

The diagram shows three possible ways in which the size of the Universe could change in the future.

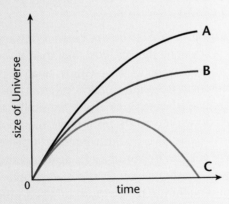

With reference to the graph, explain how the future of the Universe depends on the amount of mass that it contains. [6]

# Chapter 7
# Synoptic assessment

## What is synoptic assessment?

Unit tests only examine the material in that unit, in isolation from any other content of the AS and A2 courses.

Synoptic assessment:

- takes place at the end of the course
- counts for 15–20% of the marks for A Level, and 30 – 40% of the marks for A2
- tests understanding, knowledge and skills in contexts that bring together different areas of the subject
- emphasises skills and principles.

## What skills do I need?

All the skills that you have developed throughout your A Level course. In particular, you need to be able to:

- make links between different areas of physics – for example, how the inverse square law applies to radial fields and the spreading of light from a point source and the similarities and differences between radioactive decay and capacitor discharge
- apply knowledge and understanding in unfamiliar contexts – for example, knowledge and understanding of force and motion to vehicle safety features
- handle data – for example, use a graph to test a relationship between two quantities
- interpret and evaluate information – for example, explain the meaning of scientific terms used in a passage.

## What types of question are asked?

There are a number of different types of synoptic question. The awarding bodies use a combination of these.

- Passage analysis You are given a passage which may be adapted from a book or a scientific journal. You may then be asked questions that test your comprehension of the passage, analyse data, perform calculations and draw inferences from information given in the passage.
- Structured questions These are similar to the questions in unit tests, but they will draw on content from more than one unit.
- Unstructured questions Instead of being split up into separate parts, these give you the opportunity to develop arguments and ideas. When answering these questions, you should always refer to physical principles such as conservation of momentum or electromagnetic induction.

## How should I prepare for synoptic assessment?

Synoptic assessment can cover any area of the full A Level specification. It is important therefore that you:

- know all the formulae and relationships for your specification, both AS and A2, that are not provided on a data sheet
- can interpret graphs, including the significance of gradients and areas
- revise and review all the key concepts in the full AS and A2 specification

# 7.1 Drawing it all together

**After studying this section you should be able to:**

- describe how similar physical models can be applied to a variety of situations
- make comparisons between the action of a spring and a capacitor and also between electric and gravitational fields
- explain the similarities between radioactive decay and capacitor discharge

**LEARNING SUMMARY**

## Comparing a spring and a capacitor

EDEXCEL A    M6

In each case the outcome variable, extension of the spring or charge stored on the capacitor, is proportional to the input variable, force applied to the spring or voltage across the capacitor. This is shown in the table.

|  | *input variable* | *outcome variable* | *relationship* |
|---|---|---|---|
| *spring* | force, $F$ | extension, $x$ | $F = kx$ |
| *capacitor* | voltage, $V$ | charge stored, $Q$ | $V = Q/C$ |

The stiffness, $k$, of a spring is equivalent to $1/C$ for a capacitor as:

- the greater the value of $k$ the smaller the extension for a given force
- the greater the value of $C$ the greater the charge stored for a given voltage.

The expressions for the energy stored in a stretched spring or a charged capacitor are also similar.

In each case the energy stored can be related to the area between the graph line and the force or voltage axis.

> **KEY POINT**
>
> In each case the energy stored is equal to:
>
> $$\tfrac{1}{2} \times \text{input variable} \times \text{outcome variable}$$
> $$\tfrac{1}{2}Fx \text{ and } \tfrac{1}{2}VQ$$
>
> for the spring and the capacitor respectively.

## Electric and gravitational fields

| AQA B | M5 | NICCEA | M4 |
|---|---|---|---|
| EDEXCEL A | M6 | WJEC | M5 |
| EDEXCEL B | M5 | | |

This statement is true for point masses and point charges. In the case of large masses and charges, it only applies to the space beyond the mass or charge.

All radial fields show an inverse square law in the relationship between the field strength and distance from the centre of the field.

The difference between gravitational and electric fields is:

- gravitational fields can only exert attractive forces on masses
- the force exerted by an electric field on a charge can be either attractive or repulsive.

The similarities between the forces exerted and the field strengths at a distance $r$ from the centre of the field are shown in the table.

Remember that $k = 1/4\pi\varepsilon_0$

| *type of field* | *size of force on point mass/charge* | *field strength* |
|---|---|---|
| *electric* | $F = kQq/r^2$ | $E = kQ/r^2$ |
| *gravitational* | $F = GMm/r^2$ | $g = GM/r^2$ |

For these fields:

- the constant $k$ is for an electric field is equivalent to the constant $G$ for a gravitational field
- the field strength is defined as the force per unit mass/charge.

## Capacitor discharge and radioactive decay

EDEXCEL A    M6

> The rate of decay or discharge falls as the number of undecayed nuclei or charge on the capacitor falls.

Unlike radioactive decay, the discharge of a charged capacitor is not a random process, but they are both examples of exponential change. This means that:

- the rate at which the charge on the capacitor changes is proportional to the charge on the capacitor
- the rate at which the number of undecayed nuclei changes is proportional to the number present
- in equal time intervals the charge on the capacitor or the number of undecayed nuclei changes in the same ratio.

In radioactive decay the time interval used is the half-life, $t_{1/2}$. In capacitor discharge the time interval used is the time constant, $\tau = RC$.

The similarities and differences between the treatment of capacitor discharge and radioactive decay are summarised in the table.

| | amount or number present | rate of change | time interval |
|---|---|---|---|
| *capacitor discharge* | $Q = Q_0 e^{-t/RC}$ | current, $I = Q/RC$ | $\tau$ = time for charge to fall to 1/e of its initial value |
| *radioactive decay* | $N = N_0 e^{-\lambda t}$ | activity, $A = \lambda N$ | $t_{1/2}$ = time for number of undecayed nuclei to fall to ½ of its initial value |

The decay constant $\lambda$ in radioactive decay is equivalent to $1/\tau$ in capacitor discharge. This is because:

- the larger the value of $\lambda$ the greater the rate of decay
- the larger the value of $\tau$ the smaller the rate of discharge.

> $t_{1/2}$ for a capacitor is the time it takes for the charge to fall to half its initial value.

However, there is no reason why different mathematical treatments should be applied to these situations. If the concept of half-life, or half-charge, is applied to capacitor discharge another similar relationship emerges:

$$t_{1/2}/\tau = \ln 2 \quad \text{and} \quad \lambda t_{1/2} = \ln 2$$

This again reveals the equivalence of $\lambda$ and $1/\tau$.

## Progress check

1 a If two identical springs of stiffness $k$ are joined in series, what is the stiffness of the combination?

   b If two identical capacitors of capacitance $C$ are joined in series, what is the capacitance of the combination?

2 Give one way in which the force between two point masses is different from that between two positive point charges.

3 What fraction of undecayed nuclei remain in a radioactive sample after time $t = 1/\lambda$ elapses?

3 1/e

2 Either one is gravitational and the other is electric or one is attractive and the other is repulsive.

b C/2

1 a k/2

# 7.2 Accelerators and detectors

*After studying this section you should be able to:*

- *describe how a Van de Graaff generator and a linac are used to accelerate particles for firing at fixed targets*
- *explain how a cyclotron is used for colliding beam experiments*
- *explain the principles of operation of particle detectors*

**LEARNING SUMMARY**

## Van de Graaff generator

EDEXCEL A  M6   OCR B  M5
EDEXCEL B  M4

Remember, 1 eV is the energy gained by an electron when accelerated through a potential difference of 1 V.

The **Van de Graaff generator** is used to accelerate particles in a straight line. It is capable of giving them energies up to several MeV. The principle of a high voltage Van de Graaff generator is shown in the diagram.

- Positive charge is sprayed onto a revolving belt at A by a high voltage source which ionises the air and repels positively-charged particles onto the belt.

Negative particles are sprayed when the surrounding gas is ionised by the strong electric field, and the negative ions are repelled by the negative electrode.

- A negative charge is induced on the inner surface of the dome, with a positive charge being induced on the outer surface.
- Negatively-charged particles are sprayed onto the belt at B.

This results in a very high voltage being created between the dome and earth. The voltage can be used to accelerate charged particles by passing them through a series of electrodes connected so that ions pass through them in stages. At each stage the ions are accelerated and focused.

The diagram below shows an arrangement suitable for accelerating negative ions such as electrons. To accelerate positive ions, the potential difference is reversed.

The resistors are used to create a 'step' in potential from each cylinder to the next.

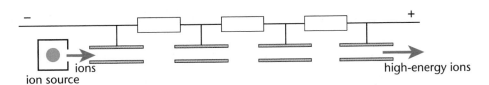

## The linac

AQA B  M5   OCR B  M5
EDEXCEL A  M6   NICCEA  M6
EDEXCEL B  M4   WJEC  M5

A **linac** (short for linear accelerator) provides much higher-energy particles than a Van de Graaff machine is capable of producing. Charged particles accelerated by a linac can achieve energies of several GeV, at which their speed is approaching that of the speed of light.

No particle can ever reach the speed of light because, according to Einstein's theory, the mass of a particle increases with increasing speed and approaches infinity as its speed approaches that of light.

The diagram below shows a linac that is used to accelerate protons.

In a proton-accelerating linac:

- the protons pass through a large number of cylindrical tubes
- the protons travel at a constant speed within each tube and are accelerated as they pass into the next one
- the acceleration is achieved by a high-frequency, high-voltage alternating potential difference applied to the tubes
- the polarity of the potential difference applied to each tube changes to positive as the proton leaves the previous tube
- each tube is longer than the previous one because the accelerated protons are travelling increasing distances in the time taken for half of one complete cycle of the applied voltage to occur.

> As the protons speed up, they travel increasing distances in time $1/2f$. This is the time it takes to travel through one tube and pass into the next, where $f$ is the frequency of the voltage supply.

The source of protons is normally a Van de Graaff accelerator, so that the protons entering the linac already have energies of MeV. Linacs that accelerate electrons use high-frequency electromagnetic waves that travel inside metal cavities.

The World's largest linear accelerator is at Stanford in California. It is nearly 4 km long and it can accelerate electrons and positrons to energies of 50 GeV. Linear accelerators:

- produce a narrow beam of high intensity
- suffer little energy loss from the beam as it travels along the accelerator.

> There are plans to build a new electron-positron linac that will be 30 km long and will accelerate these particles to energies of the order of $10^{13}$ eV.

## Ring accelerators

| | | | |
|---|---|---|---|
| AQA A | M4 | OCR B | M5 |
| AQA B | M5 | NICCEA | M6 |
| EDEXCEL A | M6 | WJEC | M5 |
| EDEXCEL B | M4 | | |

**Ring accelerators** can produce even higher energy particles than linacs currently in use. They can accelerate particles to energies of 1000 GeV, without requiring an enormously long tube. Protons can be accelerated to speeds approaching the speed of light in a 200 m-diameter ring. The principle of the cyclotron is explained in section 3.5. In a cyclotron:

- the particles follow a spiral path within two semi-circular dees
- they are accelerated each time they cross from one dee into the other
- a magnetic field perpendicular to the dees keeps the particles in circular motion.

> This is how a transmitting aerial emits radio waves; electrons emit radiation when their velocity changes.

One major disadvantage of ring accelerators is that the charged particles radiate energy due to the continual change in velocity as they change direction – a problem that does not arise in a linear accelerator.

## Targets

| | | | |
|---|---|---|---|
| AQA B | M5 | EDEXCEL A | M6 |

High-energy particles from accelerators are used in the quest for more knowledge and understanding of the behaviour of sub-atomic particles. To achieve this, high-energy particles are fired at targets and then the tracks left by the collision debris are examined.

The particles produced by an accelerator can collide with a fixed target or another beam of particles in a colliding beam experiment.

A fixed target could be liquid hydrogen or water or a metal such as iridium. In a fixed target experiment:

- conservation of momentum applies to these collisions, so the momentum of the particles after the collision is equal to that of the accelerated particle that collides with the target
- the kinetic energy of the particles formed as a result of the collision comes from that of the accelerated particle
- very little energy, typically only 5%, is available to create new particles.

When beams of particles moving in opposite directions collide:

- conservation of momentum applies, so if the particles have equal masses and speeds their momentum after the collision is zero
- no kinetic energy of the colliding particles remains after the collision, so all of the initial kinetic energy is available to create new particles
- these collisions can result in the creation of a new particle that has no kinetic energy but is high in mass.

## Tracks

**AQA B**   AS   **EDEXCEL A**   M6

There are a number of ways of detecting the particles created in high-energy collisions.

In a **cloud chamber**, particles pass through a super-saturated vapour which condenses on the ionised air particles created by collisions, leaving vapour trails.

A similar effect occurs in a **bubble chamber**. Here, the particles pass through a superheated liquid, and bubbles of vapour form in the path of the particle. Liquid hydrogen is often used as it acts both as the fixed target and the detector.

Because the liquid in a bubble chamber is much denser than the gas-vapour mixture in a cloud chamber, collisions are more frequent and high-energy particles are stopped in a shorter distance. A magnetic field applied perpendicular to the plane of the chamber causes charged particles to follow curved paths.

The diagram below shows cloud chamber and bubble chamber tracks.

**Alpha particle tracks in a cloud chamber**

**Bubble chamber tracks**

These tracks show that:

- the alpha particles in the cloud chamber have two different values of energy and momentum
- the bubble chamber shows the tracks left by particles of opposite signs, since there are deflections in opposite directions due to the magnetic field
- there is a range of particle energy and momentum.

In a **spark chamber**, ionising particles cause sparks as they pass between electrodes and ionise an inert gas such as neon. The passage of particles through the chamber can be detected in two ways:

- by photographing the sparks
- by detecting the electrical pulses that pass in the electrodes.

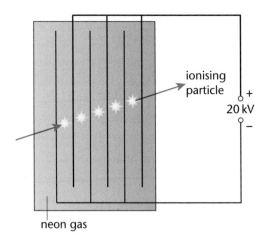

ionising particle

20 kV

neon gas

The principle of a spark chamber is shown in the diagram above.

The **drift chamber** is a development of the spark chamber and works on a similar principle. It consists of a mesh of fine wires, each of which is at the opposite potential to its neighbours. A drift chamber:

> The track of a particle through a drift chamber is produced by computer analysis of the pulses of current in the wires.

- takes into account the time taken for ionised particles to reach a wire and be detected
- can pinpoint the position of a particle to within 50 μm
- requires a very pure inert gas
- gives a three-dimensional picture of the track of a particle.

## Progress check

1 Explain why the tubes in a linac are of increasing length.

2 Why is there little energy loss from particles in a linac compared to that in a ring accelerator?

3 Suggest why X-rays do not leave tracks in a cloud chamber.

3 X-rays are only weakly ionising.
2 The particles do not radiate energy due to changes in direction.
1 The time taken for the particles to pass from one tube to the next is fixed. As the particles speed up, they travel increasing distances in this time.

# Sample synoptic questions and model answers

**1**

The gravitational field due to a point mass and the electric field due to a point positive charge show some similarities and differences.

One similarity is that both follow inverse square laws.

(a) (i) Write down expressions for the gravitational field strength due to a point mass, $m$, and the electric field strength due to a point positive charge, $q$. [2]

$g = Gm/r^2$      1 mark
$E = q/4\pi\varepsilon_0 r^2$      1 mark

> The definitions of gravitational and electric field strength are 'force per unit mass' and 'force per unit positive charge'.

(ii) What feature of these expressions shows that the fields follow an inverse square law? [1]

The $r^2$ term in the denominator shows that the field strengths vary as the inverse of the square of the distance from the mass or charge.      1 mark

> All radial fields follow inverse square laws.

(iii) State one other way in which the definitions of gravitational field strength and electric field strength are similar. [1]

They are both proportional to the size of the mass or charge.      1 mark

(iv) State one way in which the definitions of gravitational field strength and electric field strength are different. [2]

The definition of gravitational field strength does not specify the sign of the mass as mass can only have positive values.      1 mark
The definition of electric field strength specifies that it is the size of the force on a positive charge.      1 mark

> A negative charge experiences a force the same size as that experienced by a positive charge, but in the opposite direction.

(b) Gravitational potential and electric potential are properties of points within a field.
How is potential related to the potential energy of:

(i) a mass in a gravitational field? [1]
Potential energy = potential × mass.      1 mark

(ii) a charge within an electric field? [1]
Potential energy = potential × charge (including the sign of the charge).      1 mark

> Potential means 'the energy per unit mass/charge placed at that point'.

(c) Explain why the expressions for gravitational potential and electric potential have opposite signs. [4]

Gravitational forces can only be attractive (1 mark) so a point in a gravitational field always has a negative potential since energy would be needed to remove the mass to infinity (1 mark). The direction of an electric field is taken to be the direction of the force on a positive charge (1 mark). A positive charge in the field of another positive charge has a positive amount of energy since work would have to be done on the charge to move it from infinity to a point in the field (1 mark).

## Sample synoptic questions and model answers *(continued)*

**2**

A motorist sees a bright light in the sky and thinks that it is the planet Venus. Later, he notices that the movement across the sky is too rapid for a planet and realises that he is looking at a low-flying aircraft.

At its brightest, Venus reflects electromagnetic radiation from the Sun with a power of $2.8 \times 10^{17}$ W. On this occasion, the distance between Venus and the Earth was $8.0 \times 10^{10}$ m.

The power of the lights of a large aircraft is 1500 W.

Estimate the distance of the aircraft from the motorist for its lights to appear as bright as Venus. State the assumptions that it is necessary to make and explain whether your answer is likely to be too high or too low. [12]

The marks are awarded here for setting up a workable mathematical model – like the Earth, only half of Venus is receiving radiation from the Sun at any one time.

Assuming that Venus acts as a disc (1 mark) and reflects radiation over a hemisphere (1 mark):

You are not expected to know that the surface area of a sphere = $4\pi r^2$, this information is given in a data book.

Intensity of light from Venus at the Earth
$$= P/2\pi r^2 \qquad \text{1 mark}$$
$$= 2.8 \times 10^{17} \text{ W} \div (2 \times \pi \times (8.0 \times 10^{10} \text{ m})^2) \qquad \text{1 mark}$$
$$= 7.0 \times 10^{-6} \text{ W m}^{-2} \qquad \text{1 mark}$$

At a distance $d$, for the aircraft lights to have the same intensity, assuming that they radiate light equally in all directions over a hemisphere   1 mark
$$1500 \text{ W} \div (2 \times \pi \times d^2) = 7.0 \times 10^{-6} \text{ W m}^{-2} \qquad \text{1 mark}$$
$$d = \sqrt{(1500 \text{ W} \div (2 \times \pi \times 7.0 \times 10^{-6} \text{ W m}^{-2})} \qquad \text{1 mark}$$
$$= 5.8 \times 10^3 \text{ m or about 6 km} \qquad \text{1 mark}$$

It is important when making an estimate to round off the answer to a sensible number of significant figures.

This is assuming that the light from Venus and the aircraft contains the same proportion of visible radiation.   1 mark

The reasons for the answer are important here. No marks are awarded for a 50–50 guess.

If this is the case, then at a distance of 6 km the aircraft is likely to appear brighter than Venus, since its lights are focused into a beam, (1 mark) so I would expect my answer to be too low (1 mark).

## *Practice examination questions*

**1**

A mass spectrometer is a device for comparing atomic masses. Positive ions pass through slits A and B before entering a region where there is both an electric and a magnetic field.

(a) Suggest why the ions must first pass through slits A and B. [1]

(b) State the direction of:
   (i) the electric force on the ions
   (ii) the magnetic force on the ions
   (iii) the magnetic field. [3]

(c) Show that an ion can only pass through slit C if its velocity is equal to the ratio of the field strengths. [2]

(d) (i) Explain why an ion follows a circular path after leaving slit C. [2]
   (ii) An ion of carbon-12 has a mass of $2.0 \times 10^{-26}$ kg and carries a single positive charge of $1.6 \times 10^{-19}$ C. It emerges from slit C with a speed of $2.4 \times 10^{5}$ m s$^{-1}$.
   The magnetic field strength is 0.15 T.
   Calculate the radius of its circular path. [3]

(e) Radiocarbon dating is a method of estimating the age of objects made from materials that were once alive. In a living object the ratio of carbon-12 atoms to carbon-14 atoms is $10^{12}$ to one.
   Suggest how a mass spectrometer could be used to estimate the age of wood from an old table. [3]

**2**

In a Van de Graaff generator, charge from a moving belt is transferred onto a spherical metal dome which is insulated from the ground.

The graph below shows how the potential of a dome of radius 0.12 m depends on the charge on it.

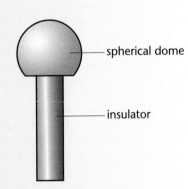

## Practice examination questions (continued)

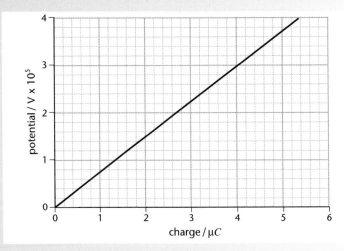

(a) Calculate the capacitance of the dome. [3]

(b) Calculate the amount of work that has to be done to place 4.0 µC of charge on the dome. [3]

The dome loses charge when the electric field strength at its surface reaches a value of $3.0 \times 10^6$ N C$^{-1}$.

(c) (i) Calculate the charge on the dome when the electric field strength at its surface has this value. [3]

$$k = 1/4\pi\varepsilon_0 = 9.0 \times 10^9 \text{ N m}^2 \text{ C}^{-2}$$

(ii) What is the potential of the dome when the electric field strength at its surface has this value? [1]

(d) (i) Calculate the force on an electron, charge = $-1.6 \times 10^{-19}$ C, near the surface of the dome when the electric field strength is $3.0 \times 10^6$ N C$^{-1}$. [3]

(ii) Suggest how the dome loses charge. [3]

**3**

Read the passage and answer the questions that follow.

Seismometers on the Moon's surface have detected very slight tremors in the body of the Moon. Some of these occur at irregular intervals and appear to originate at the Moon's surface. Others are regular and are repeated on a 14-day cycle. These regular Moonquakes originate deep below the Moon's surface and could be due to tidal stresses caused by the Moon's elliptical orbit.

It is not known whether the Earth undergoes similar tidal stresses, since the small magnitude of the tremors means that they would be indistinguishable from movement caused by other factors.

(a) Suggest one cause of the irregular tremors. [1]
(b) What 'other factors' could cause small Earth tremors? [2]
(c) Here is some data about the Moon and its orbit.

Mass of Moon = $7.4 \times 10^{22}$ kg
Orbital period = 27.3 days
Closest distance between Moon and Earth = $3.56 \times 10^8$ m
Greatest distance between Moon and Earth = $4.06 \times 10^8$ m.

Estimate the difference between the greatest and smallest forces that the Earth exerts on the Moon. State clearly any assumptions and approximations that you make and what effect these are likely to have on your answer. [12]

# Practice examination answers

## 1 Mechanics

**1** (a) $\Delta E_p = mg\Delta h$ [1] = 40 kg × 10 m s$^{-2}$ × 0.60 m [1] = 240 J [1].

(b) $\frac{1}{2}mv^2$ = 240 J [1]  $v = \sqrt{(2 \times 240 \text{ J} \div 40 \text{ kg})}$ [1] = 3.5 m s$^{-1}$ [1].

(c) The child is accelerating [1] upwards [1].

(d) $F = mv^2/r$ [1] = 40 kg × (3.5 m s$^{-1}$)$^2$ ÷ 3.2 m [1] = 150 N [1].

(e) The downward gravitational pull [1] of the Earth [1].

(f) (i) 400 N [1] (ii) 550 N [1].

**2** (a) Taking momentum to the right as positive, 0.30 kg m s$^{-1}$ [1] – 0.16 kg m s$^{-1}$ [1] = 0.14 kg m s$^{-1}$ [1].

(b) 0.14 kg m s$^{-1}$ ÷ 1.00 kg [1] = 0.14 m s$^{-1}$ [1] to the right [1].

(c) Total kinetic energy before collision = 0.107 J [1].
Kinetic energy after collision = 9.8 × 10$^{-3}$ J [1].
The collision is inelastic as there is less kinetic energy after the collision than before [1].

**3** (a) $v = 2\pi r/T$ [1] = 2 × π × 4.24 × 10$^7$ m × (24 h × 60 min h$^{-1}$ × 60 s min$^{-1}$) [1] = 3.08 × 10$^3$ m s$^{-1}$ [1].

(b) Acceleration = $v^2/r$ [1] = (3.05 × 10$^3$ m s$^{-1}$)$^2$ ÷ 4.24 × 10$^7$ m [1] = 0.22 m s$^{-2}$ [1] towards the centre of the Earth [1].

(c) The gravitational pull [1] of the Earth [1].

**4** (a) 0.06 kg × 15 m s$^{-1}$ [1] – – 0.06 kg × 24 m s$^{-1}$ [1] = 2.34 kg m s$^{-1}$ [1].

(b) Force = rate of change of momentum [1] = 2.34 kg m s$^{-1}$ [1] ÷ 0.012 s [1] = 195 N [1].

(c) 195 N

(d) From kinetic energy of the ball and racket [1] to potential energy of the squashed ball and stretched strings [1] and back to kinetic energy of the ball [1].

**5** (a) 600 N × cos 20° [1] = 564 N [1].

(b) Friction between the ground and the log [1]; 564 N [1].

(c) $W = F \times s$ [1] = 564 N × 250 m [1] = 1.41 × 10$^5$ J [1].

(d) It is transferred to the log and the ground [1] as heat [1].

**6** (a) $v^2/r$ [1] = (14 m s$^{-1}$)$^2$ ÷ 100 m [1] = 1.96 m s$^{-2}$ [1].

(b) $F = ma$ [1] = 800 kg × 1.96 m s$^{-2}$ [1] = 1.57 × 10$^3$ N [1].

(c) (i) The frictional push [1] of the road on the tyres [1].
(ii) The frictional push [1] of the tyres on the road [1].

(d) Friction between the tyres and the road is reduced [1] and may not be sufficient for the centripetal force required [1].

**7** (a) 5.0 × 10$^5$ kg × 60 m s$^{-1}$ [1] = 3.0 × 10$^7$ kg m s$^{-1}$ [1].

(b) 3.0 × 10$^7$ kg m s$^{-1}$ ÷ 20 s [1] = 1.5 × 10$^6$ N [1].

(c) Resistive forces oppose the motion [1] so a greater forwards force is needed to produce this unbalanced force [1].

(d) The aircraft gains momentum in the forwards direction [1]. The exhaust gases gain an equal amount of momentum in the opposite direction [1].

**8** (a) $\Delta E_p = mg\Delta h$ [1] = 3.2 × 10$^5$ kg × 10 m s$^{-2}$ × 180 m [1] = 5.76 × 10$^8$ J [1].

(b) $\frac{1}{2}mv^2$ = 5.76 × 10$^8$ J [1] $v = \sqrt{(2 \times 5.76 \times 10^8 \text{ J} \div 3.2 \times 10^5 \text{ kg})}$ [1] = 60 m s$^{-1}$ [1].

(c) $\frac{1}{2}$ × 3.2 × 10$^5$ kg × (60 m s$^{-1}$)$^2$ – $\frac{1}{2}$ × 3.2 × 10$^5$ kg × (3 m s$^{-1}$)$^2$ [1] = 5.7 × 10$^8$ J [1].

## 2 Waves

**1** (a) The frequency an object vibrates at, when it is displaced from its normal position [1] and released [1].
   (b) When the frequency of a forcing vibration [1] is equal to the natural frequency [1].
   (c) $f = v/\lambda$ [1] = 340 m s$^{-1}$ × 0.15 m [1] = 2.27 × 10$^3$ Hz [1].

**2** (a) $a$ represents acceleration [1], $k$ represents the spring constant [1], $x$ represents the displacement [1] and $m$ represents the value of the mass [1].
   (b) (i) 0.13 m [1].
       (ii) $f = 1/T$ [1] = 1 ÷ 1.6 s = 0.625 Hz [1].
   (c) $k = 4\pi^2 f^2 m$ [1]
       $= 4\pi^2 \times (0.625)^2 \times 0.5$ [1]
       $= 7.7$ N m$^{-1}$ [1].
   (d) (i) $v_{max} = 2\pi fA = 0.51$ m s$^{-1}$ [1], $E_k$ at this speed = ½ × 0.50 kg × (0.51 m s$^{-1}$)$^2$ [1] = 6.5 × 10$^{-2}$ J [1].
       (ii) At zero displacement [1].

**3** (a) (i) The minimum possible energy of an electron [1].
       (ii) An electron that is just free has zero energy [1]. An electron in the ground state needs to gain energy to become free [1].
   (b) (i) 12.75 eV × 1.60 × 10$^{-19}$ J eV$^{-1}$ [1] = 2.04 × 10$^{-18}$ J [1].
   (ii) $f = E/h$ [1] = 2.04 × 10$^{-18}$ J ÷ 6.63 × 10$^{-34}$ J s = 3.08 × 10$^{15}$ Hz [1].
   (iii) Ultraviolet [1].
   (iv) Danger, any one from skin cancer, blindness [1]. Use, any one from security marking, fluorescent lights [1].

**4** (a) (i) To produce two identical wave sources [1] by diffraction [1].
       (ii) This is close to the size of the wavelength [1] to ensure adequate spreading [1].
   (b) At a point equidistant from the gaps [1]. The waves are in phase [1] and have the greatest combined amplitude [1].
   (c) Separation of maxima, $x = \lambda D/a$ [1] = 2.8 × 10$^{-2}$ m × 1.20 m ÷ 0.10 m [1] = 0.336 m [1]. The distance between the strongest and weakest signals is half this, i.e. 0.168 m [1].
   (d) The slits are many wavelengths wide for light [1] so no diffraction occurs [1].

**5** (a) (i) $hf$ is the energy of a photon [1], $\Phi$ is the work function, i.e. minimum photon energy that causes emission [1] and $(\frac{1}{2}mv^2)_{max}$ is the maximum amount of kinetic energy of an emitted electron.
       (ii) Some electrons need more than the minimum energy to liberate them, so they have less than the maximum kinetic energy [1].
   (b) (i) $E = hf = 2.02 \times 10^{15}$ Hz × 6.63 × 10$^{-34}$ J s [1] = 1.34 × 10$^{-18}$ J [1].
       (ii) 4.5 × 10$^{-6}$ W m$^{-2}$ × 2.5 × 10$^{-6}$ m$^2$ ÷ 1.34 × 10$^{-18}$ J [1] = 8.40 × 10$^6$ [1].
       (iii) $hf - (\frac{1}{2}mv^2)_{max}$ [1] = 1.34 × 10$^{-18}$ J − 4.05 × 10$^{-19}$ J = 9.35 × 10$^{-19}$ J [1] ÷ 1.60 × 10$^{-19}$ J eV$^{-1}$ = 5.8 eV [1].

**6** (a) $A = 0.28$ m [1], $f = 0.67$ Hz [1].
   (b) $v_{max} = 2\pi fA = 1.18$ m s$^{-1}$ [1].
   (c) $v = 2\pi f\sqrt{(A^2 - x^2)}$ [1] = 0.82 m s$^{-1}$ [1]

# 3 Fields

**1** (a) $F = kQ_1Q_2/r^2$ [1] $= 9.0 \times 10^9$ N m$^2$ C$^{-2} \times (1.6 \times 10^{-19}$ C$)^2 \div (5.2 \times 10^{-11}$ m$)^2$ [1] $=$ $8.5 \times 10^{-8}$ N [1].

   (b) (i) $E = kQ/r^2$ [1] $= 9.0 \times 10^9$ N m$^2$ C$^{-2} \times 1.6 \times 10^{-19}$ C $\div (5.2 \times 10^{-11}$ m$)^2$ [1] $=$ $5.3 \times 10^{11}$ N C$^{-1}$ [1].

   (ii) From the proton towards the electron [1].

   (c) $v = \sqrt{(r \times F \div m_e)}$ [1] $= \sqrt{(5.2 \times 10^{-11}$ m $\times 8.5 \times 10^{-8}$ N $\div 9.1 \times 10^{-31}$ kg$)}$ [1] $=$ $2.2 \times 10^6$ m s$^{-1}$ [1].

**2** (a) (i) 1250 V [1]

   (ii) $E = qV$ [1] $= 10 \times 1.6 \times 10^{-19}$ C $\times 1250$ V [1] $= 2.0 \times 10^{-15}$ J [1].

   (b) $4.0 \times 10^{-15}$ J [1].

   (c) (i) Top plate positive and lower plate negative [1].

   (ii) $Vq/d = mg$ [1] $V = mgd/q = 5.0 \times 10^{-5}$ kg $\times 10.0$ m s$^{-2} \times 6.0 \times 10^{-3}$ m $\div 1.6 \times 10^{-18}$ C [1] $= 1.9 \times 10^{12}$ V [1].

**3** (a) (i) $g = GM_E/r^2$ [1] $= 6.7 \times 10^{-11}$ N m$^2$ kg$^{-2} \times 6.0 \times 10^{24}$ kg $\div (3.8 \times 10^8$ m$)^2$ [1] $=$ $2.8 \times 10^{-3}$ N kg$^{-1}$ [1].

   (ii) $2.8 \times 10^{-3}$ N kg$^{-1} \times 7.4 \times 10^{21}$ kg [1] $= 2.1 \times 10^{19}$ N [1].

   (iii) $T = 2\pi\sqrt{(rM_M/F)}$ [1] $= 2\pi\sqrt{(3.8 \times 10^8$ m $\times 7.4 \times 10^{21}$ kg $\div 2.1 \times 10^{19}$ N$)}$ [1] $=$ $2.3 \times 10^6$ s [1].

   (b) (i) $F = GMm/r^2$ [1] $= 6.7 \times 10^{-11}$ N m$^2$ kg$^{-2} \times 7.4 \times 10^{21}$ kg $\times 1.0$ kg $\div (3.8 \times 10^8$ m$)^2$ [1] $= 3.4 \times 10^{-6}$ N [1].

   (ii) The Sun has a greater mass than the Moon [1] but it is very much further away from the Earth [1] so its gravitational field strength at the surface of the Earth is less [1].

   (iii) In position A the gravitational pulls of the Moon and the Sun are in the same direction [1], in position B they are in opposite directions, so the resultant force on the water is smaller [1].

**4** (a) From N to S (radially outwards) [1].

   (b) Out of the paper [1].

   (c) The direction of the force on the electromagnet reverses [1] when the current reverses direction [1].

   (d) $F = BIl = BI2\pi rN$ [1] $= 0.85$ T $\times 0.055$ A $\times 2 \times \pi \times 2.5 \times 10^{-2}$ m $\times 150$ [1] $= 1.1$ N [1].

**5** (a) (i) $E_k = eV$ [1] $= 1.6 \times 10^{-19}$ C $\times 2500$ V [1] $= 4.0 \times 10^{-16}$ J [1].

   (ii) $v = \sqrt{(2E_k/m)}$ [1] $= \sqrt{(2 \times 4.0 \times 10^{-16}$ J $\div 9.1 \times 10^{-31}$ kg$)}$ [1] $= 3.0 \times 10^7$ m s$^{-1}$ [1].

   (b) (i) Down (from top to bottom of the paper) [1].

   (ii) There is an unbalanced force on the electrons [1] which is always at right angles to the direction of motion [1].

   (iii) $r = mv/Bq$ [1] $= 9.1 \times 10^{-31}$ kg $\times 3.0 \times 10^7$ m s$^{-1} \div (2.0 \times 10^{-3}$ T $\times 1.6 \times 10^{-19}$ C$)$ [1] $= 8.4 \times 10^{-2}$ m

**6** (a) (i) $\phi = BA$ [1] $= 0.15$ T $\times 8.0 \times 10^{-2}$ m $\times 5.0 \times 10^{-2}$ m [1] $= 6.0 \times 10^{-4}$ Wb [1].

   (ii) 0 [1].

   (b) (i) The reading is a maximum when the plane of the coil passes through the horizontal position [1] as the rate of change of flux is greatest [1]. The reading decreases to zero when the plane of the coil is horizontal [1] as the rate of change of flux decreases to zero [1].

   (ii) Generator/dynamo [1].

   (c) Any three from: increase the speed of rotation, increase the number of turns on the coil, increase the area of the coil, increase the strength of the magnetic field [1 mark each].

## 4 Particle physics

**1** (a) 298 K [1].
 (b) $n = pV/RT$ [1] $= 2.02 \times 10^5$ Pa $\times 2.20 \times 10^{-6}$ m$^3 \div (8.3$ J mol$^{-1}$ K$^{-1} \times 298$ K) [1] $= 1.80 \times 10^{-4}$ [1].
 (c) (i) The total energy of the particles [1].
   (ii) Kinetic [1].
   (iii) It doubles [1] as internal energy is directly proportional to absolute temperature [1].
   (iv) At constant pressure, the volume is directly proportional to absolute temperature [1].
   The new volume $= 2 \times 2.20 \times 10^{-6}$ m$^3 = 4.40 \times 10^{-6}$ m$^3$ [1].

**2** (a) (i) $^{239}_{92}$U [1]
   (ii) The nucleus splits [1] into two smaller nuclei [1] and spare neutrons [1].
   (iii) The neutrons can cause other nuclei to fission [1] which releases neutrons to cause more fissions [1].
 (b) (i) The total mass-energy of the products [1] is less than that of the original nucleus and neutron [1].
   (ii) Kinetic energy [1] of the fission products [1].
   (iii) The energy is removed by a coolant [1] and used to generate steam [1] which drives a steam turbine [1].
 (c) (i) The control rods absorb excess neutrons [1].
   (ii) The moderator slows down the neutrons [1].

**3** (a) $t = mc\Delta\theta/P$ [1] $= 1.50$ kg $\times 4.2 \times 10^3$ J kg$^{-1}$ K$^{-1} \times 90$ K $\div 2.40 \times 10^3$ W [1] $= 236$ s [1].
 (b) The element absorbs energy [1], the casing absorbs energy [1], energy is lost through conduction/convection/evaporation/radiation [1].

**4** (a) Gamma [1], gamma can penetrate tissue but beta is absorbed by tissue [1].
 (b) The beta radiation emitted by iodine-131 is absorbed and causes ionisation [1], resulting in the damage of cells/tissue [1].
 (c) Iodine-123 has a shorter half-life [1] so less is needed to produce the same rate of decay [1].
 (d) After 6 half-lives [1] = 78 hours [1].

**5** (a) (i) Electrons are not repelled by the nucleus [1]. They have wave-like behaviour so are diffracted by a nucleus [1].
   (ii) The higher the energy, the shorter the wavelength [1]. The wavelength needs to be short so that it is of the order of the size of a nucleus [1] for diffraction to occur [1].
 (b) (i) $R = 1.20 \times 10^{-15}$ m $\times 4^{1/3}$ [1] $= 1.90 \times 10^{-15}$ m [1].
   (ii) $3.81 \times 10^{-15}$ m [1].
 (c) The densities are the same [1]. The radius of the sulphur nucleus is twice that of the helium nucleus [1] so its mass should be $2^3 = 8$ times as great if it has the same density, which it is [1].

**6** (a) This shows that most alpha particles do not go close to a region of charge [1], so most of the volume of an atom is empty space [1].
 (b) There is a tiny concentration of charge in the atom [1] which must be the same charge as the alpha particles [1] to repel them [1]

## 5 Medical physics

**1** (a) The intensity decreases [1].

 (b) (i) Intensity = 10 log $(I/I_0)$ [1] = 10 log $(6.0 \times 10^{-5}$ W m$^{-2}$ ÷ $1.0 \times 10^{-12}$ W m$^{-2})$
 [1] = 78 dB [1].

 (ii) $r_0 = \sqrt{r_1^2(I/I_0)}$ [1] = $\sqrt{(6.0 \times 10^{-5}$ W m$^{-2}$ ÷ $1.0 \times 10^{-12}$ W m$^{-2})}$ [1] = $7.7 \times 10^3$ m [1].

 (iii) 0 [1].

 (c) The perceived loudness of a sound is greatest at a frequency of 3000 Hz [1]. It decreases [1] as the frequency becomes lower or higher [1].

**2** (a) An A-scan [1] as this measures depth below the surface of the skin [1].

 (b) $t = 2 \times s$ ÷ $v$ [1] = 0.12 m ÷ 1500 m s$^{-1}$ = $8.0 \times 10^{-5}$ s [1].

 (c) So that there is no interference [1] between the pulse and its reflection [1].

 (d) There needs to be a time lag when the crystal is not emitting ultrasound [1] so that it can detect the reflection [1].

 (e) $\lambda = 6.0 \times 10^{-2}$ m ÷ 200 = $3.0 \times 10^{-4}$ m [1]; $f = v/\lambda$ = 1500 m s$^{-1}$ ÷ $3.0 \times 10^{-4}$ m [1] = $5.0 \times 10^6$ Hz [1].

 (f) The gel maximises the energy transmitted into the body [1] by reducing the amount that is reflected [1].

 (g) (i) A CAT-scanner uses an X-ray beam [1] that rotates around the body [1] and produces a three-dimensional image of the body [1].

 (ii) Advantages: the CAT-scan produces a three-dimensional image [1], tumours are more easily detected [1]. Disadvantage: X-rays can be more damaging to body cells and tissue [1].

## 6 Astronomy and cosmology

**1** (a) (i) Apparent magnitude is a measure of the brightness when viewed from the Earth [1].
 Absolute magnitude is the apparent magnitude the star would have at a distance of 10 pc [1].

 (ii) The perceived brightness of a star decreases with increasing distance [1].
 Absolute magnitude takes account of this [1].

 (b) $M = m - 5$ log $(d/10)$ [1] = 1.3 [1].

 (c) $d = 10$ log$^{-1}$ $(m-M)/5$ [1] = 240 pc.

 (d) White dwarfs are smaller than red giants [1] so radiation is emitted from less surface area [1].

**2** (a) An absorption spectrum is what remains of the full spectrum after passing through an element [1]. Elements absorb the same wavelengths as they emit [1] so the absorption spectrum has these wavelengths removed from the full spectrum, leaving dark lines [1].

 (b) (i) The star is moving away from the Earth [1], causing the wavelengths received to be longer than those emitted [1].

 (ii) $v = c\Delta\lambda/\lambda$ [1] = $3.00 \times 10^8$ m s$^{-1}$ × (0.277 ÷ 6.563) [1] = $1.27 \times 10^7$ m s$^{-1}$ [1].

 (iii) $d = v/H$ [1] = $5.78 \times 10^{24}$ m [1] to $6.18 \times 10^{24}$ m [1].

**3** A shows an open Universe [1], the Universe will continue to expand if there is insufficient mass for gravitational forces to stop this [1].
B shows a steady-state [1] where the expansion stops due to gravitational attraction of the mass in the Universe, for this to happen the Universe needs to have the critical mass [1].
C shows the Universe contracting to a 'Big Crunch' [1], this will only happen if the mass of the Universe is greater than the critical mass [1]

## 7 Synoptic assessment

**1** (a) To produce a parallel beam [1].
   (b) (i) from right to left [1]
       (ii) from left to right [1]
       (iii) into the paper [1].
   (c) To move in a straight line, $Bqv = Eq$ [1] so $v = E/B$ [1].
   (d) (i) The only force acting is the magnetic force [1], this is always at right angles to the direction of motion [1].
       (ii) $r = mv/Bq$ [1] $= 2.0 \times 10^{-26}$ kg $\times 2.4 \times 10^5$ m s$^{-1}$ ÷ (0.15 T $\times 1.6 \times 10^{-19}$ C) [1] $= 0.20$ m [1].
   (e) The ratio of carbon-14 atoms to carbon-12 atoms decreases after a tree is felled. Use a sample of carbon from the wood in the table in the spectrometer; the carbon-14 atoms follow a path with a greater radius of curvature than that of the carbon-12 atoms [1]. The ratio of carbon-14 to carbon-12 can be determined by using counters [1].

**2** (a) $C = Q/V$ [1] $= 4.8 \times 10^{-6}$ C $\times 3.6 \times 10^5$ V [1] $= 1.33 \times 10^{-11}$ F [1].
   (b) $W = ½QV$ [1] $= ½ \times 4.0 \times 10^{-6}$ C $\times 3.0 \times 10^5$ V [1] $= 0.60$ J [1].
   (c) (i) $Q = Er^2/k$ [1] $= 3.0 \times 10^6$ N C$^{-1} \times (0.12$ m$)^2$ ÷ $9.0 \times 10^9$ N m$^2$ C$^{-2}$ [1] $= 4.8 \times 10^{-6}$ C [1].
       (ii) $V = kQ/r = Er$ [1] $= 3.0 \times 10^6$ N V$^{-1} \times 0.12$ m [1] $= 3.6 \times 10^5$ V [1].
   (d) (i) $F = Eq$ [1] $= 3.0 \times 10^6$ N C$^{-1} \times 1.6 \times 10^{-19}$ C [1] $= 4.8 \times 10^{-13}$ N [1].
       (ii) The force on the outer electrons ionises the air [1], movement of these ions causes further ionisation [1], the dome discharges as it attracts ions of the opposite charge to it [1].

**3** (a) Collisions with meteors [1].
   (b) Any two from traffic, wind, tides [2].
   (c) The mean speed of the Moon is calculated by assuming that the path is circular [1] with a mean radius of $3.81 \times 10^8$ m [1].
   mean speed = $(2 \times \pi \times 3.81 \times 10^8$ m$)$ ÷ $(27.3 \times 24 \times 60 \times 60$ s$)$ [1]
   $= 1.01 \times 10^3$ m s$^{-1}$ [1].
   The gravitational force between the Moon and the Sun is the centripetal force [1].
   Force at closest distance = $mv^2/r$ [1]
   $= 7.4 \times 10^{22}$ kg $\times (1.01 \times 10^3$ m s$^{-1})^2$ ÷ $(3.56 \times 10^8$ m$)$ [1]
   $= 2.1 \times 10^{20}$ N [1]
   Force at greatest distance = $1.9 \times 10^{20}$ N [1].
   The difference between these forces = $2 \times 10^{19}$ N.

   The actual difference is likely to be greater than this because of the assumption that the Moon travels at a constant speed [1]. In fact it speeds up as it gets closer to the Earth and slows down as it moves further away [1]. The effect of this would be to make the largest force larger and the smallest force smaller [1].

# Index

absolute magnitude 149–50
absorbed dose 133
absorption spectrum 63, 144, 151–2
accommodation 125
acoustic impedance 129, 131
activity 103–4
Airy disc 138
alpha 88, 102–3, 105–6, 133–4, 163
alpha particle scattering 107, 113
amplitude 43–5, 47–9, 50, 51, 54, 58, 127
angular magnification 137
angular velocity 36–7
antinode 58
apparent magnitude 149
aqueous humour 124
artificial transmutation 111
A-scan 131
astigmatism 125
astronomical unit 141
atomic mass unit 107
atomic number 103, 107
atrium 130
attenuation 129, 131–2
auditory nerve 127
background radiation 102
back-scattering 113
becquerel 103–4
bending 122
beta-minus 102–3, 105–6, 111, 112
beta-plus 102–3, 105–6
Big Bang 146–8
Big Crunch 147–8
black body 150–1
black dwarf 141–2
black hole 142
blue supergiant 142, 150
bone 120–2, 132
B-scan 131
capacitance 23, 78–80
capacitor 78–82, 159, 160
cathode ray oscilloscope 51, 76–7, 84, 126, 130
Celsius scale 99
centripetal acceleration 23, 37, 87, 88
centripetal force 23, 37–8
chain reaction 109–11
charge-coupled device 138–9
chromatic aberration 138
ciliary muscles 124
classes of star 141–2, 149–52
cochlea 127

coherent 55, 132
colour vision 125–6
comet 140
computerised axial topography 132
concave lens 123–5
cones 125–6
conservation of energy 34–5, 92, 108
conservation of momentum 28–9, 163
constructive interference 54–8
control rod 110–11
converging lens 123–5, 137
convex lens 123–5
coolant 110–11
Copernicus 140
cornea 124
cosmological principle 143
Coulomb's law 74–5, 113
critical density 147
cyclotron 88–9, 162
damping 45
de Broglie 61
decay constant 104–5, 160
decibel 127–8
depth of field 125
destructive interference 55–6, 58
diffraction 53, 55–7, 59, 61–2, 114, 115
diffraction grating 56–7
displacement 36, 43, 46–9, 51, 54, 58
diverging lens 123–5
Doppler effect 145
dose equivalent 134
driving force 26
ear drum 127
elastic collision 30
electric field 50, 59, 68, 74–7, 102, 159
electrocardiagram 130
electromagnetic spectrum 59, 144
electron beam 76–7, 114
electron capture 103
electron diffraction 61–2, 114
electronvolt 60, 108
emission spectrum 56, 62–3, 144, 151, 152
endoscope 132
energy level 62–3
energy transfer 33–5
equipotential surface 72, 76
event horizon 142
excited state 63
exponential change 81–2, 104–5, 160

exposure 133
eye lens 124–5
far point 125
farad 78–9
Faraday's law 90, 91–2
fission 109, 110–11
Fleming's left hand rule 83, 84, 92
fluorescence 91, 132
flux density 84, 90–1
focal length 123–5
forced vibration 43, 44
frame of reference 153
free vibration 43
free-fall acceleration 35, 70
frequency 23, 37, 46–9, 50, 51, 59, 60, 63, 89, 127–8, 145
fusion 108–9, 141, 147
galaxy 141, 142, 145–7
gamma 102–3, 112
gamma rays 59, 61, 62, 133–4
general relativity 154–5
geo-synchronous orbit 88
gravitational field 68–73, 77, 87, 142, 154–5, 159
gravitational field strength 23, 46, 69, 70–3, 77, 159
gravitational potential 71–3, 77
gravitational potential energy 23, 33, 34–5, 43, 71–2
ground force 121
ground state 63
half-life 105, 111–12, 160
half-value thickness 131
heart 130
henry 85
Hertzsprung-Russell diagram 149
Hubble's constant 145
Hubble's law 145, 147
hypermetropia 125
ideal gas 99, 100–1
ideal gas equation 100–1
induced fission 109
inelastic collision 30
infra-red 34, 59–60, 62–3
intensifying screen 132
intensity 52, 103, 126, 127–8, 131, 138, 149, 151–2
interference 53–8, 59, 61–2
internal energy 98–9
inverse square law 69–70, 75, 151, 159
ionisation 62, 102–3, 133, 151

isotope 104, 107, 111–12
joint 120–1
joule 60
kelvin 99
Kepler 140
kinetic energy 23, 30, 33, 34–5, 43, 60–1, 76, 98, 101, 109, 163
Larmor frequency 133
length contraction 154
lens formula 124–5
lens power 123
Lenz's law 90, 92
lever 120–1
lifting 122
light year 141
line spectrum 62–3, 144
linear attenuation coefficient 131–2
long sight 125
longitudinal 50, 51–2
loudness 53–4, 127–9
magnetic field 50, 59, 68, 83–5, 88–9, 90–3, 102, 115, 133
magnetic field strength 83–5, 88–9, 90
magnetic flux 90–3
magnetic resonance imaging 44, 133
main sequence 141–2
mass attenuation coefficient 132
mass defect 108
mechanical advantage 121
microwave radiation 141, 146
microwaves 44, 59–60
Milky Way 141, 142
moderator 110
molar gas constant 23, 100–1
momentum 23, 28–31, 61
motor 83
muscle 120–1, 122
myopia 125
natural frequency 43, 44
near point 125
neutral point 83
neutron 62, 102–3, 105–11, 133
neutron diffraction 115–16
neutron star 142
newton 30–1
Newton's first law 25–6, 27
Newton's law of gravitation 69–70, 140, 142, 147
Newton's third law 26–7
node 58
non stochastic 134
normal adjustment 137
nuclear binding energy 108
nuclear radius 114–15

nuclear reactor 110–11
nucleon emission 103, 147
nucleus 107, 113–14
Olber's paradox 143
optic nerve 124
optical centre 123
optical fibre 132
oscillation 43–5, 46–9, 50, 59
ossicles 127, 128
outer ear 127
oval window 127
parsec 141
path difference 54–5, 57
pendulum 46
period 50, 51
permeability 85
permittivity 75
persistence of vision 126
phase 51, 54–5, 100
phase difference 48, 51, 55
photoelectric effect 60–1, 62
photon 60, 62–3, 131, 139
piezoelectric crystal 130
Planck constant 60–1
plane of polarisation 52
planet 27, 87, 140
plutonium 111
polarisation 52
positron 102–3, 147, 162
potential 71–2, 75–7
potential energy 23, 33, 34, 43, 70–3
presbyopia 125
principal axis 123
principal focus 123, 137–9
principle of moments 121
principle of superposition 54
progressive wave 50–1, 57–8
proton 62, 88–9, 102–3, 105–8, 162
proton-neutron ratio 105–6
quantum of charge 60, 74
quasar 152
radian 36, 51
radiation 102–6, 111, 160
radio telescope 139
radio waves 59–60
radioactivity 102–6, 111, 160
Rayleigh criterion 138
real image 123–4, 137
rebound 29
recoil 30
red giant 141–2, 150
red supergiant 141, 150
red-shift 145, 152
refracting telescope 137
relaxation time 133

resistive force 25, 26, 33, 35, 45
resolving power 138, 139
resonance 44, 45
resonant frequency 44, 133
rest mass 154
retina 124, 126
rocket propulsion 31
rods 125–6
satellite 72, 87–8, 140
Schwarzschild radius 142
short sight 125
simple harmonic motion 46–9
sino-atrial node 130
skeleton 120, 122
Solar System 87, 140–2
solenoid 85
special relativity 153
specific heat capacity 99
spherical aberration 138
standing wave 57–8, 63
stationary wave 57–8
Stefan's law 150–1
stochastic 134
stopping potential 61
Sun 87, 140–1, 144, 149–51, 155
supernova 142
superposition 53–4
tendon 120
tesla 84
thermodynamic scale 99, 101
threshold frequency 60–1
threshold wavelength 60
time constant 81–2, 160
time dilation 153–5
time period 36–7, 47
total internal reflection 132
transformer 23, 91, 93
transducer 127
transverse 50, 51, 59
ultrasound 130–1
ultraviolet 59
uranium 109–11
ventricle 130
virtual image 123–4
vitreous humour 124
wavelength 23, 50–1, 53, 55–7, 59, 61, 114, 115, 138, 139, 144, 145, 150, 152
weber 90
white dwarf 141–2, 149–50
Wien's law 150–1
work 32–3, 34
work function 61
X-ray diffraction 114, 115–16
X-rays 59, 61, 103, 114, 115, 131–4

# Physics

## Graham Booth

# Contents

Specification lists   4

AS/A2 Level Physics courses   12

Different types of questions in A2 examinations   14

Exam technique   16

What grade do you want?   18

Essential mathematics   19

Four steps to successful revision   24

## Chapter 1 Mechanics

1.1   Newton I and II   25

1.2   Momentum and the second law   28

1.3   Energy to work   32

1.4   Motion in a circle   36

Sample question and model answer   39

Practice examination questions   40

## Chapter 2 Waves

2.1   Oscillations, resonance and damping   43

2.2   Simple harmonic motion   46

2.3   Wave properties   50

2.4   Complex wave behaviour   53

2.5   Quantum phenomena   59

Sample question and model answer   64

Practice examination questions   65

## Chapter 3 Fields

3.1   Gravitational fields   68

3.2   Electric fields   74

3.3   Capacitors   78

3.4   Magnetic fields   83

3.5   Circular orbits   87

3.6   Electromagnetic induction   90

Sample question and model answer   94

Practice examination questions   95

## Chapter 4    Particle physics

| | | |
|---|---|---|
| 4.1 | Thermal physics | 98 |
| 4.2 | Radioactive decay | 102 |
| 4.3 | Energy from the nucleus | 107 |
| 4.4 | Probing matter | 113 |
| | Sample question and model answer | 117 |
| | Practice examination questions | 118 |

## Chapter 5    Medical physics

| | | |
|---|---|---|
| 5.1 | Body mechanics | 120 |
| 5.2 | Using lenses | 123 |
| 5.3 | Hearing | 127 |
| 5.4 | Medical imaging and treatment | 130 |
| | Sample question and model answer | 135 |
| | Practice examination questions | 136 |

## Chapter 6    Astronomy and cosmology

| | | |
|---|---|---|
| 6.1 | Looking up | 137 |
| 6.2 | The model unfolds | 140 |
| 6.3 | The expanding Universe | 144 |
| 6.4 | Classifying stars | 149 |
| 6.5 | Relativity | 153 |
| | Sample question and model answer | 156 |
| | Practice examination questions | 157 |

## Chapter 7    Synthesis and synoptics

| | | |
|---|---|---|
| 7.1 | Drawing it all together | 159 |
| 7.2 | Accelerators and detectors | 161 |
| | Sample question and model answer | 165 |
| | Practice examination questions | 167 |

| | | |
|---|---|---|
| | Practice examination answers | 169 |
| | Index | 175 |

# Specification lists

## AQA A Physics

| MODULE | SPECIFICATION TOPIC | CHAPTER REFERENCE | STUDIED IN CLASS | REVISED | PRACTICE QUESTIONS |
|---|---|---|---|---|---|
| **Module 4 (M4)** *Waves, fields and nuclear energy* | Oscillations and waves | 2.1, 2.2, 2.3, 2.4 | | | |
| | Capacitance | 3.3 | | | |
| | Gravitational and electric fields | 1.4, 3.1, 3.2, 3.5, 7.1 | | | |
| | Magnetic effects of currents | 3.4, 3.5, 3.6, 7.2 | | | |
| | Nuclear applications | 4.3 | | | |
| **Module 5–9 (M5–M9)** *Common topic: Nuclear Instability* | Nuclear instability | 4.2, 4.4 | | | |
| **Module 5 (M5)** *Astrophysics* | Lenses and optical telescopes | 5.1, 6.1 | | | |
| | Radio astronomy | 6.1 | | | |
| | Classification of stars | 6.2, 6.4 | | | |
| | Cosmology | 6.3, 6.4 | | | |
| **Module 6 (M6)** *Medical physics* | Physics of the eye and ear | 5.2, 5.3 | | | |
| | Biological measurement and imaging | 5.4 | | | |
| **Module 7 (M7)** *Applied physics* | | | | | |
| **Module 8 (M8)** *Turning points in physics* | | | | | |
| **Module 9 (M9)** *Electronics* | | | | | |

## Examination analysis

The specification comprises two compulsory modules and one option module which has a section of questions on the common topic and a section of questions on the option. In module tests all the questions are compulsory.

| | | | |
|---|---|---|---|
| **Module 4** | 15 compulsory multiple choice questions and several short, structured questions | 1 hr 30 min test | 30% |
| **Module 5–9** | Structured questions on nuclear instability and one option topic **and** coursework **or** practical examination | 1 hr 15 min test  1 hr 45 min practical | 30% |
| **Module 10** | Synoptic test  Structured questions on modules 1–4 and nuclear instability | 2 hr test | 40% |

# AQA B Physics

| MODULE | SPECIFICATION TOPIC | CHAPTER REFERENCE | STUDIED IN CLASS | REVISED | PRACTICE QUESTIONS |
|---|---|---|---|---|---|
| **Module 4 (M4)** _Further Physics_ | Circular motion | 1.4, 3.5 | | | |
| | Oscillations | 2.1, 2.2 | | | |
| | Work and energy | 1.3 | | | |
| | Molecular kinetic theory | 4.1 | | | |
| | Heating and working | 4.1 | | | |
| | Capacitance and exponential decay | 3.3 | | | |
| | Momentum concepts | 1.2 | | | |
| | Quantum phenomena | 2.5 | | | |
| **Module 5 (M5)** _Fields and their applications_ | Electric and gravitational fields | 3.1, 3.2, 7.1 | | | |
| | Magnetic fields | 3.4, 3.5, 3.6 | | | |
| | Nuclear energy particle, accelerators and detectors | 4.3, 7.2 | | | |

# Examination analysis

The specification comprises three compulsory modules. In module tests all the questions are compulsory.

| | | | |
|---|---|---|---|
| **Module 4** | Structured questions | 1 hr 30 min test | 30% |
| **Module 5** | Structured questions on module 5 and synoptic questions on modules 1–5 including a comprehension question. | 2 hr test | 40% |
| **Module 6** | Practical examination | 3 hr test | 30% |

## Edexcel A  Physics

| MODULE | SPECIFICATION TOPIC | CHAPTER REFERENCE | STUDIED IN CLASS | REVISED | PRACTICE QUESTIONS |
|---|---|---|---|---|---|
| **Module 4 (M4)** *Waves and our Universe* | Circular motion and oscillations | 1.4, 2.1, 2.2 | | | |
| | Waves | 2.3, 2.5 | | | |
| | Superposition of waves | 2.4 | | | |
| | Quantum phenomena | 2.5 | | | |
| | The expanding Universe | 2.4, 2.5, 6.3 | | | |
| **Module 5 (M5)** *Fields and forces* | Gravitational fields | 3.1, 3.5 | | | |
| | Electric fields | 3.2 | | | |
| | Capacitance | 3.3 | | | |
| | Magnetic fields | 3.4 | | | |
| | Electromagnetic induction | 3.6 | | | |
| **Module 6 (M6)** *Synthesis* | Analogies in physics | 3.3, 4.2, 7.1 | | | |
| | Accelerators | 3.5, 7.2 | | | |

## Examination analysis

The specification comprises three compulsory modules. In module tests all the questions are compulsory.

| **Module 4** | Structured questions | 1 hr 20 min test | 30% |
|---|---|---|---|
| **Module 5** | Structured questions and | 1 hr 30 min test | 15% |
| | Practical examination | 1 hr test | 15% |
| **Module 6** | Synoptic test 1 comprehension question 1 question based on module 6 2 questions based on any area except the options topics | 2 hr test | 40% |

## Edexcel B Physics

| MODULE | SPECIFICATION TOPIC | CHAPTER REFERENCE | STUDIED IN CLASS | REVISED | PRACTICE QUESTIONS |
|---|---|---|---|---|---|
| **Module 4 (M4)** _Moving with physics_ | Transport on track | 1.2, 3.3, 3.4, 3.6 | | | |
| | The medium is the message | 3.2, 3.3, 3.4 | | | |
| | Probing the heart of matter | 1.2, 1.4, 2.5, 3.2, 3.5, 4.4, 7.2 | | | |
| **Module 5 (M5)** _Physics from creation to collapse_ | Reach for the stars | 3.1, 4.1, 4.2, 4.3, 6.3, 6.4, 7.1 | | | |
| | Build or bust? | 2.1, 2.2 | | | |

## Examination analysis

The specification comprises three compulsory modules. In module tests all the questions are compulsory.

| | | | |
|---|---|---|---|
| **Module 4** | Structured and free-response questions | _1 hr 30 min test_ | _30%_ |
| **Module 5** | Structured and free-response questions | _1 hr test_ | _20%_ |
| | Coursework | | _20%_ |
| **Module 6** | Synoptic test<br>1 comprehension question<br>about 3 structured and free-response questions | _1 hr 30 min test_ | _30%_ |

## OCR A Physics

| MODULE | SPECIFICATION TOPIC | CHAPTER REFERENCE | STUDIED IN CLASS | REVISED | PRACTICE QUESTIONS |
|---|---|---|---|---|---|
| **Module 4 (M4)** Forces, fields and energy | Dynamics | 1.1, 1.2 | | | |
| | Work and energy | 1.3 | | | |
| | Motion in a circle | 1.4 | | | |
| | Oscillations | 2.1, 2.2 | | | |
| | Gravitational fields | 3.1 | | | |
| | Electric fields | 3.2 | | | |
| | Capacitors | 3.3 | | | |
| | Electromagnetism | 3.4, 3.5 | | | |
| | Electromagnetic induction | 3.6 | | | |
| | Thermal physics | 4.1 | | | |
| | The nuclear atom | 4.3, 4.4 | | | |
| | Radioactivity | 4.2, 4.3 | | | |
| **Module 5.1 (M5.1)** Cosmology | Models of the known Universe | 6.2 | | | |
| | Stars and galaxies | 6.2 | | | |
| | Structure of the Universe | 6.2, 6.3 | | | |
| | Information from stellar observation | 6.3, 6.4 | | | |
| | How the Universe may evolve | 6.3 | | | |
| | Relativity | 6.5 | | | |
| **Module 5.2 (M5.2)** Health physics | Body mechanics | 5.1 | | | |
| | The eye and sight | 5.2 | | | |
| | The ear and hearing | 5.3 | | | |
| | Medical imaging | 5.4 | | | |
| | Medical treatment | 5.4 | | | |

## Examination analysis

The specification comprises three compulsory modules. In module tests all the questions are compulsory.

| | | | |
|---|---|---|---|
| **Module 4** | Structured and free-response questions | 1 hr 30 min test | 30% |
| **Module 5** | Test on one optional topic Structured and free-response questions | 1 hr 30 min test | 30% |
| **Module 6** | Synoptic test: questions on module 1–4 | 1 hr 15 min test | 20% |
| | Coursework or practical examination | 1 hr 30 min test | 20% |

## OCR B Physics

| MODULE | SPECIFICATION TOPIC | CHAPTER REFERENCE | STUDIED IN CLASS | REVISED | PRACTICE QUESTIONS |
|---|---|---|---|---|---|
| **Module 4 (M4)** *Rise and fall of the Clockwork Universe* | Creating models | 1.3, 2.1, 2.2, 3.3, 4.2 | | | |
| | Out into space | 1.2, 1.3, 1.4, 3.1 | | | |
| | Our place in the Universe | 6.3 | | | |
| | Matter: very simple | 4.1 | | | |
| | Matter: hot or cold | 4.1 | | | |
| **Module 5 (M5)** *Field and particle pictures* | Electromagnetic machines | 3.4, 3.6 | | | |
| | Charge and field | 3.2, 3.5, 7.2 | | | |
| | Probing deep into matter | 2.5, 3.4, 4.4 | | | |
| | Ionising radiation and risk | 4.2, 4.3 | | | |

## Examination analysis

The specification comprises three compulsory modules. In module tests all the questions are compulsory. They consist of short focused questions and longer, structured questions with some questions being open-ended.

| Module 4 | Structured and open-ended questions | 1 hr 10 min test | 18.4% |
| | Coursework: practical investigation | | 15% |
| Module 5 | Structured and open-ended questions | 1 hr 10 min test | 21.6% |
| | Coursework: research report | | 15% |
| Module 6 | Synoptic test: comprehension/data analysis and structured questions from any area of the specification | 1 hr 30 min test | 30% |

## WJEC Physics

| MODULE | SPECIFICATION TOPIC | CHAPTER REFERENCE | STUDIED IN CLASS | REVISED | PRACTICE QUESTIONS |
|---|---|---|---|---|---|
| **Module 4 (M4)** *Oscillations and energy* | Vibrations | 1.4, 2.1, 2.2 | | | |
| | Momentum concepts | 1.1, 1.2 | | | |
| | Energy concepts | 1.3, 4.1 | | | |
| | Molecular kinetic theory | 4.1 | | | |
| | Capacitance | 3.3 | | | |
| | Alternating currents | 2.3 | | | |
| **Module 5 (M5)** *Fields, forces and nuclei* | Uniform and radial fields of force | 3.1, 3.2 | | | |
| | B-fields | 3.4 | | | |
| | Electromagnetic induction | 3.6 | | | |
| | Radioactivity and radioisotopes | 4.2, 4.3 | | | |
| | Nuclear energy | 4.3 | | | |
| | Probing matter | 2.4, 2.5, 4.4 | | | |

## Examination analysis

The specification comprises three compulsory modules. In module tests all questions are compulsory.

| | | | |
|---|---|---|---|
| **Module 4** | Theory test, 5 short and 2 longer questions | *1 hr 30 min test* | *30%* |
| **Module 5** | Theory test, 5 short and 2 longer questions which include some synoptic assessment | *1 hr 30 min test* | *30%* |
| **Module 6** | Synoptic test: questions covering all areas of the specification coursework. 4 questions, 2 structured, 1 comprehension and 1 with data analysis | *2 hr test* | *25%* |
| | Coursework: practical investigatory task | | *15%* |

## NICCEA Physics

| MODULE | SPECIFICATION TOPIC | CHAPTER REFERENCE | STUDIED IN CLASS | REVISED | PRACTICE QUESTIONS |
|---|---|---|---|---|---|
| **Module 4 (M4)** *Energy, oscillations and fields* | Momentum and energy | 1.2, 1.3, 4.1 | | | |
| | Thermal physics | 4.1 | | | |
| | Uniform circular motion | 1.4 | | | |
| | Oscillations | 2.1, 2.2 | | | |
| | Fields | 3.1, 3.2, 3.5 | | | |
| **Module 5 (M5)** *Electromagnetism and nuclear physics* | Capacitors | 3.3 | | | |
| | Electromagnetism | 3.4, 3.6 | | | |
| | Electron physics | 2.3, 3.2, 3.4 | | | |
| | Atomic and nuclear physics | 2.5, 4.2, 4.3, 4.4 | | | |
| **Module 6 (M6)** *Particle physics* | Particle physics | 3.5, 4.3, 4.4, 7.2 | | | |

## Examination analysis

The specification comprises three compulsory modules. In module tests, all questions are compulsory.

| *Module 4* | Theory test: short-answer questions and a synoptic, data analysis question | *1 hr 30 min test* | *30%* |
|---|---|---|---|
| *Module 5* | Theory test: short-answer questions and a synoptic, comprehension | *1 hr 30 min test* | *30%* |
| *Module 6* | Theory test: short-answer questions and a synoptic, free-response question | *1 hr test* | *16.6%* |
| | Practical examination | *1 hr 30 min* | *23.4%* |

# AS/A2 Level Physics courses

## AS and A2

All Physics A Level courses are in two parts, with three separate units or modules in each part. Most students start by studying the AS (Advanced Subsidiary) course. Some then go on to study the second part of the A Level course, called the A2. It is also possible to study the full A Level course, both AS and A2, in any order.

## How will you be tested?

### Assessment units

For A2 Physics, you will be tested by three assessment units. These form 50% of the weighting for the full A Level, the remaining 50% being the assessment of AS.

Each unit can normally be taken in either January or June. Alternatively, you can study the whole course before taking any of the unit tests. There is a lot of flexibility about when exams can be taken and the diagram below shows just some of the ways that the assessment units may be taken for AS and A Level Physics.

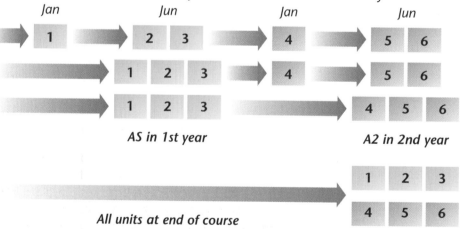

If you are disappointed with a module result, you can resit each module once. You will need to be very careful about when you take up a resit opportunity because you will have only one chance to improve your mark. The higher mark counts.

### A2 and Synoptic assessment

All full A Level specifications have to include an element of synoptic assessment, one that draws together ideas from all areas of physics studied at AS and A2 Level. Details of these synoptic assessments are given in the specification lists. The synoptic assessment can examine any area of the specification except any optional topics. Questions will assess the extent to which you can apply your knowledge and understanding of physics from more than one course unit, for example a question set in the context of renewable energy could examine understanding of the motion of satellites (the Moon), simple harmonic motion (the movement of tides) and electromagnetic induction (electricity generation).

Synoptic assessment will often include comprehension and data handling exercises and require you to apply fundamental physical principles in unfamiliar situations. In some specifications the synoptic assessment is by a separate test, and in others it is built into all the A2 module tests. However it is done, it carries 30–40% of the marks for A2 and 15–20% of the total marks for the full A Level award.

## Coursework

Coursework may form part of your A2 Level Physics course, depending on which specification you study. Where students have to undertake coursework, it may be a written project or for the assessment of practical skills or both. If in doubt, check with your teacher or the specification that you are studying.

## Key Skills

To gain the key skills qualification, which is equivalent to an AS Level, you will need to demonstrate that you have attained level 3 in the areas of Communication, Application of number and Information technology. Part of the assessment can be done as normal class activity and part is by formal test.

# What skills will I need?

For A2 Physics, you will be tested by assessment objectives: these are the skills and abilities that you should have acquired by studying the course. The assessment objectives for A2 Physics are shown below.

## Knowledge with understanding

- recall of facts, terminology and relationships
- understanding of principles and concepts
- drawing on existing knowledge to show understanding of the responsible use of physics in society
- selecting, organising and presenting information clearly and logically.

## Application of knowledge and understanding, analysis and evaluation

- explaining and interpreting principles and concepts
- interpreting and translating, from one form into another, data presented as continuous prose or in tables, diagrams and graphs
- carrying out relevant calculations
- applying knowledge and understanding to familiar and unfamiliar situations
- assessing the validity of physical information, experiments, inferences and statements

You must also present arguments and ideas clearly and logically, using specialist vocabulary where appropriate.

## Experimental and investigative skills

Physics is a practical subject and part of the assessment of AS Physics will test your practical skills. This may be done during your lessons or you may be tested in a more formal practical examination. You will be assessed on four main skills:

- planning
- implementing
- analysing evidence and drawing conclusions
- evaluating evidence and procedures

# Different types of questions in A2 examinations

In order to assess these abilities and skills, a number of different types of question are used.

In A2 Level Physics, unit tests include short-answer questions, structured questions requiring both short answers and more extended answers, together with free-response and open-ended questions. Multiple choice question papers are not used, although it is possible that some short-answer questions will use a multiple choice format, where you have to choose the correct response from a number of given alternatives.

## Short-answer questions

A short-answer question may test recall or it may test understanding by requiring you to undertake a short, one–stage calculation. Short-answer questions normally have space for the answers printed on the question paper. Here are some examples (the answers are shown in blue):

What is the relationship between electric current and charge flow?

Current = rate of flow of charge.

The current passing in a heater is 6 A when it operates from 240 V mains. Calculate the power of the heating element.

$P = I \times V = 6\,A \times 240\,V = 1440\,W$

Which of the following is the correct unit of acceleration?

A  $m\,s^{-1}$
B  $s\,m^{-1}$
C  $m\,s^{-2}$
D  $m^2\,s^{-1}$

C  $m\,s^{-2}$

## Structured questions

Structured questions are in several parts. The parts are usually about a common context and they often become progressively more difficult and more demanding as you work your way through the question. They may start with simple recall, then test understanding of a familiar or an unfamiliar situation. The most difficult part of a structured question is usually at the end, where the candidate is sometimes asked to suggest a reason for a particular phenomenon or social implication. Most of the practice questions in this book are structured questions, as this is the main type of question used in the assessment of A2 Level Physics.

When answering structured questions, do not feel that you have to complete one question before starting the next. The further you are into a question, the more difficult the marks are to obtain. If you run out of ideas, go on to the next question. Five minutes spent on the beginning of that question are likely to be much more fruitful than the same time spent wracking your brains trying to think of an explanation for an unfamiliar phenomenon.

Here is an example of a structured question that becomes progressively more demanding.

(a)  A car speeds up from $20\,m\,s^{-1}$ to $50\,m\,s^{-1}$ in $15\,s$.

Calculate the acceleration of the car.

acceleration = increase in velocity ÷ time taken
$$= 30 \, \text{m s}^{-1} \div 15 \, \text{s} = 2 \, \text{m s}^{-2}$$

(b)  The total mass of the car and contents is 950 kg.

Calculate the size of the unbalanced force required to cause this acceleration.

force = mass × acceleration
$$= 950 \, \text{kg} \times 2 \, \text{m s}^{-2} = 1900 \, \text{N}$$

(c)  Suggest why the size of the driving force acting on the car needs to be greater than the answer to (b)

The driving force also has to do work to overcome the resistive forces, e.g. air resistance and rolling resistance.

## Extended answers

In A2 Level Physics, questions requiring more extended answers will usually form part of structured questions. They will normally appear at the end of structured questions and be characterised by having at least three marks (and often more, typically five) allocated to the answers as well as several lines (up to ten) of answer space. These questions are also used to assess your abilities to communicate ideas and put together a logical argument.

The correct answers to extended questions are less well-defined than to those requiring short answers. Examiners may have a list of points for which credit is awarded up to the maximum for the question, or they may first of all judge the quality of your response as poor, satisfactory or good before allocating it a mark within a range that corresponds to that quality.

As an example of a question that requires an extended answer, a structured question on the use of solar energy could end with the following:

Suggest why very few buildings make use of solar energy in this country compared to countries in southern Europe.          [5]

Points that the examiners might look for include:

- the energy from the Sun is unreliable due to cloud cover
- the intensity of the Sun's radiation is less in this country than in southern Europe due to the Earth's curvature
- more energy is absorbed by the atmosphere as the radiation has a greater depth of atmosphere to travel through
- fossil fuels are in abundant supply and relatively cheap
- the capital cost is high, giving a long payback time
- photo-voltaic cells have a low efficiency
- the energy is difficult to store for the times when it is needed the most

Full marks would be awarded for an argument that put forward three or four of these points in a clear and logical way.

## Free-response questions

Little use is made of free-response and open-ended questions in A2 Level Physics. These types of question allow you to choose the context and to develop your own ideas. Examples could include 'Describe a laboratory method of determining *g*, the value of free-fall acceleration' and 'Outline the evidence that suggests that light has a wave-like behaviour'. When answering this type of question it is important to plan your response and present your answer in a logical order.

# Exam technique

## Links from AS

A2 Physics builds from AS Level Physics. This Study Guide has been written so that you will be able to tackle A2 Physics from an AS Physics background.

You should not need to search for important Physics from AS because this has been included where needed in each chapter. If you have not studied Physics for some time, you should still be able to learn Advanced Level Physics using this text and the AS Study Guide.

## What are examiners looking for?

Examiners use instructions to help you to decide the length and depth of your answer.

If a question does not seem to make sense, you may have misread it – read it again!

### State, define or list

This requires a short, concise answer, often recall of material that can be learnt by rote.

### Explain, describe or discuss

Some reasoning or some reference to theory is required, depending on the context.

### Outline

This implies a short response, almost a list of sentences or bullet points.

### Predict or deduce

You are not expected to answer by recall but by making a connection between pieces of information.

### Suggest

You are expected to apply your general knowledge to a 'novel' situation, one which you have not directly studied during the A2 Physics course.

### Calculate

This is used when a numerical answer is required. You should always use units in quantities and significant figures should be used with care.

Look to see how many significant figures have been used for quantities in the question and give your answer to this degree of precision.

If the question uses 3 (sig figs), then give your answer to 3 (sig figs) also.

## Some dos and don'ts

### Dos

**Do answer the question**

- No credit can be given for good Physics that is irrelevant to the question.

**Do use the mark allocation to guide how much you write**

- Two marks are awarded for two valid points – writing more will rarely gain more credit and could mean wasted time or even contradicting earlier valid points.

**Do use diagrams, equations and tables in your responses**

- Even in 'essay-type' questions, these offer an excellent way of communicating physics.

**Do write legibly**

- An examiner cannot give marks if the answer cannot be read.

**Do write using correct spelling and grammar. Structure longer essays carefully**

- Marks are now awarded for the quality of your language in exams.

### Don'ts

**Don't fill up any blank space on a paper**

- In structured questions, the number of dotted lines should guide the length of your answer.
- If you write too much, you waste time and may not finish the exam paper. You also risk contradicting yourself.

**Don't write out the question again**

- This wastes time. The marks are for the answer!

**Don't contradict yourself**

- The examiner cannot be expected to choose which answer is intended. You could lose a hard-earned mark.

**Don't spend too much time on a part that you find difficult**

- You may not have enough time to complete the exam. You can always return to a difficult calculation if you have time at the end of the exam.

# What grade do you want?

Everyone would like to improve their grades but you will only manage this with a lot of hard work and determination. You should have a fair idea of your natural ability and likely grade in Physics and the hints below offer advice on improving that grade.

## For a Grade A

You will need to be a very good all-rounder.

- You must go into every exam knowing the work extremely well.
- You must be able to apply your knowledge to new, unfamiliar situations.
- You need to have practised many, many exam questions so that you are ready for the type of question that will appear.

The exams test all areas of the syllabus and any weaknesses in your Physics will be found out. There must be no holes in your knowledge and understanding. For a Grade A, you must be competent in all areas.

## For a Grade C

You must have a reasonable grasp of Physics but you may have weaknesses in several areas and you will be unsure of some of the reasons for the Physics.

- Many Grade C candidates are just as good at answering questions as the Grade A students but holes and weaknesses often show up in just some topics.
- To improve, you will need to master your weaknesses and you must prepare thoroughly for the exam. You must become a better all-rounder.

## For a Grade E

You cannot afford to miss the easy marks. Even if you find Physics difficult to understand and would be happy with a Grade E, there are plenty of questions in which you can gain marks.

- You must memorise all definitions.
- You must practise exam questions to give yourself confidence that you do know some Physics. In exams, answer the parts of questions that you know first. You must not waste time on the difficult parts. You can always go back to these later.
- The areas of Physics that you find most difficult are going to be hard to score on in exams. Even in the difficult questions, there are still marks to be gained. Show your working in calculations because credit is given for a sound method. You can always gain some marks if you get part of the way towards the solution.

## What marks do you need?

As a rough guide, you will need to score an average of 40% for a Grade E, 60% for a Grade C and 80% for a Grade A:

| average | 80% | 70% | 60% | 50% | 40% |
|---------|-----|-----|-----|-----|-----|
| grade | A | B | C | D | E |

# Essential mathematics

This section describes some of the mathematical techniques that are needed in studying A2 Level Physics.

## Quantities and units

Physical quantities are described by the appropriate words or symbols, for example the symbol $R$ is used as shorthand for the value of a *resistance*. The quantity that the word or symbol represents has both a numerical value and a unit e.g. 10.5 $\Omega$. When writing data in a table or plotting a graph, only the numerical values are entered or plotted. For this reason headings used in tables and labels on graph axes are always written as (physical quantity)/(unit), where the slash represents division. When a physical quantity is divided by its unit, the result is the numerical value of the quantity.

Resistance is an example of a **derived** quantity and the ohm is a derived unit. This means that they are defined in terms of other quantities and units. All derived quantities and units can be expressed in terms of the seven **base** quantities and units that make up the SI, or International System of Units. The quantities, their units and symbols are shown in the table. The candela is not used in A Level Physics.

| quantity | unit | symbol |
|----------|------|--------|
| length | metre | m |
| mass | kilogram | kg |
| time | second | s |
| electric current | ampere | A |
| temperature difference | kelvin | K |
| amount of substance | mole | mol |
| luminous intensity | candela | cd |

## Equations

The equations that you use in Physics are relationships between physical quantities. Since it is not possible for one quantity to equate to a different quantity, an equation must be homogenous, i.e. the units on each side of the equation must be the same. This is useful for:

- finding the units of a constant such as resistivity
- checking the possible correctness of an equation; if the units on each side are the same the equation may be correct but if they are different it is definitely wrong.

It is important to distinguish between a physical quantity and its unit. The value of a physical quantity includes both the numerical value and the unit it is measured in.

When working out the units of an equation, make it clear that you are dealing with units only by writing units of (quantity) = (unit).

Often equations need to be used in a different form to that in which they are given or memorised. The equation needs to be **rearranged**. When rearranging an equation it is important to remember that:

- the same mathematical operation must be applied to each side of the equation
- addition or subtraction should be done first, followed by multiplication or division and finally roots and powers.

For example, to rearrange the equation $v^2 = u^2 + 2as$ to enable $u$ to be calculated:

1  subtract $2as$ from each side of the equation to give $v^2 - 2as = u^2$

2  square root each side of the equation to give $\sqrt{(v^2 - 2as)} = u$

An equation that relates the quantity to the product or quotient of two others can be rearranged using the 'magic triangle', a triangle divided into three.

To use the triangle, start with the quantities that are multiplied or divided in the relationship. For example, in $V = IR$, you would write $I$ and $R$ into the triangle first. Two quantities that are multiplied together are written side-by-side at the bottom of the triangle. Where one quantity is divided by the other these are written into the triangle as they would be in a division, with one quantity 'over the other'.

The remaining quantity is now written into the vacant slot. This is illustrated below.

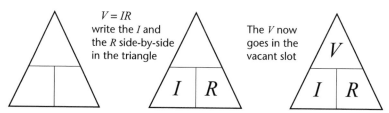

To use the triangle to make $I$ the subject of the equation, cover up the $I$ and you are left with '$V$ over $R$', so $I = V/R$.

## Drawing graphs

Graphs have a number of uses in Physics:

- they give an immediate, visual display of the relationship between physical quantities
- they enable the values of quantities to be determined
- they can be used to confirm or disprove a hypothesis about the relationship between variables.

Accuracy is the property of a measuring instrument. Precision refers to the reading taken; the precision depends on the user as well as the instrument being used.

When plotting a graph, it is important to remember that values determined by experiment are not exact. They are subject to both the accuracy and precision of any measuring instrument and the person using it, as well as any changes that the measuring instrument itself might cause. Consider measuring the temperature of a liquid with a simple thermometer.

- The manufacturer specifies the accuracy, the extent to which the thermometer is reliable. Typically, on a standard laboratory thermometer, this is ±1°C.
- When taking a reading, the user may read to the nearest half degree or nearest degree. This describes the precision.
- Inserting the thermometer into a substance may change the temperature of the substance as the two reach thermal equilibrium.

For these reasons, having plotted experimental values on a grid, the graph line is drawn as the best straight line or smooth curve that represents the points. Where there are 'anomalous' results, i.e. points that do not fit the straight line or curve, these should always be checked. If in doubt, ignore them but do add a note in your experimental work to explain why you have ignored them and suggest how any anomalous results could have arisen.

## Measurements and graphs

Quantities such as velocity, acceleration and power are defined in terms of a **rate of change** of another quantity with time. This **rate of change** can be determined by calculating the gradient of an appropriate graph. For example, *velocity* is the *rate of change of displacement with time*. Its value is represented by the gradient of a displacement-time graph. Different techniques are used to determine the gradient of a straight line and a smooth curve. For a straight line:

- determine the value of $\Delta y$, the change in the value of the quantity plotted on the $y$-axis, using the whole of the straight line part of the graph
- determine the corresponding value of $\Delta x$
- calculate the gradient as $\Delta y \div \Delta x$

For a smooth curve, the gradient is calculated by first drawing a tangent to the curve and then using the above method to determine the gradient of the tangent. To draw a tangent to a curve:

- mark the point on the curve where the gradient is to be determined
- use a compass to mark in two points on the curve, close to and equidistant from the point where the gradient is to be determined
- join these points with a ruler and extend the line beyond each point.

These techniques are illustrated in the diagrams below.

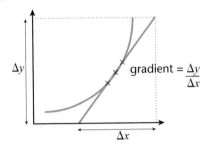

The area between a graph line and the horizontal axis, often referred to as the 'area under the graph' can also yield useful information. This is the case when the product of the quantities plotted on the axes represents another physical quantity, for example, on a *speed-time* graph this area represents the distance travelled. In the case of a straight line, the area can be calculated as that of the appropriate geometric figure. Where the graph line is curved, then the method of 'counting squares' is used.

- Count the number of complete squares between the graph line and the horizontal axis.
- Fractions of squares are counted as '1' if half the square or more is under the line, otherwise '0'.
- To work out the physical significance of each square, multiply together the quantities represented by one grid division on each axis.

These techniques are illustrated in the diagrams below

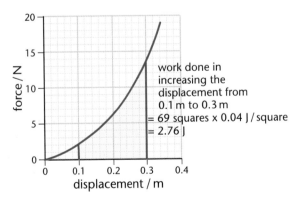

## Equations from graphs

By plotting the values of two variable quantities on a suitable graph, it may be possible to determine the relationship between the variables. This is straightforward when the graph is a straight line, since all straight line graphs have an equation of the form $y = mx + c$, where $m$ is the gradient of the graph and $c$ is the value of $y$ when $x$ is zero, i.e. the intercept on the $y$-axis. The relationship between the variables is determined by finding the values of $m$ and $c$.

The straight line graph below shows how the displacement, *s*, of an object varies with time, *t*.

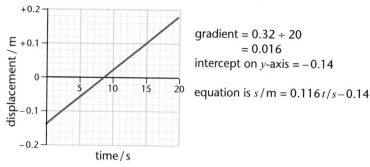

gradient = 0.32 ÷ 20
= 0.016
intercept on *y*-axis = −0.14

equation is $s/m = 0.116\,t/s - 0.14$

If the relationship between two variables is known to be of the form $y = kx^n$, the values of *k* and *n* can be determined by drawing a logarithmic graph. Since log *y* = log *k* + *n*log *x*, a graph of log *y* against log *x* has gradient equal to *n* and the intercept on the *y* axis is equal to log *k*.

In the case of an exponential change such as in capacitor discharge or radioactive decay, where a relationship is of the form $y = e^{-kx}$, the natural log function, ln, should be used. Applying this function to each side of the equation gives (since natural log and exponential are inverse functions) ln *y* = −*kx*. The gradient of a plot of ln *y* against *x* represents the constant *k*.

## Trigonometry and Pythagoras

In the right-angled triangle shown below, the sides are labelled o (opposite), a (adjacent) and h (hypotenuse). The relationships between the size of the angle θ and the lengths of these sides are:

- sin θ = o/h
- cos θ = a/h
- tan θ = o/a

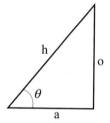

Pythagoras' theorem gives the relationship between the sides of a right-angled triangle: $h^2 = a^2 + o^2$

## Multiples and submultiples

For A2 Physics, you are expected to be familiar with the following multiples of units:

| name | multiple | symbol |
|------|----------|--------|
| micro- | $10^{-6}$ | μ |
| milli- | $10^{-3}$ | m |
| kilo- | $10^3$ | k |
| mega- | $10^6$ | M |
| giga- | $10^9$ | G |

For example, the symbol MHz means $1 \times 10^6$ Hz and mN means $1 \times 10^{-3}$ N.

## Equations that you need to know

This is a list of equations that may be needed but are not provided in topic tests or end-of-course examinations. Not all of these equations are needed for every specification as some are in the AS not the A2 specification. The table shows the equations needed for each A2 specification. Where an equation is shown as being required in the AS specification, it could be required in the synoptic assessment at A2.

All other equations that are needed will be provided on a data sheet, along with any data such as the values of physical constants.

| equation | symbol form | AQA A | AQA B | ED A | ED B | OCR A | OCR B | WJEC | NICCEA |
|---|---|---|---|---|---|---|---|---|---|
| speed = $\dfrac{\text{distance}}{\text{time taken}}$ | $v = \dfrac{\Delta d}{t}$ | AS | AS | AS | AS | AS | AS | AS | AS |
| force = mass × acceleration | $F = ma$ | AS | AS | AS | AS | AS | AS | AS | AS |
| acceleration = $\dfrac{\text{change in velocity}}{\text{time taken}}$ | $a = \dfrac{\Delta v}{t}$ | AS | AS | AS | AS | AS | AS | AS | AS |
| density = $\dfrac{\text{mass}}{\text{volume}}$ | $\rho = \dfrac{m}{V}$ | AS | | AS | AS | AS | | | |
| momentum = mass × velocity | $p = mv$ | AS | • | AS | • | • | • | • | AS |
| work done = force × distance moved in direction of force | $W = Fs$ | AS | • | AS | AS | AS | AS | A2 | A2 |
| power = $\dfrac{\text{energy transferred}}{\text{time taken}}$ = $\dfrac{\text{work done}}{\text{time taken}}$ | $P = \dfrac{\Delta W}{t}$ | AS | • | AS | AS | AS | • | • | • |
| weight = mass × gravitational field strength | $W = mg$ | AS | AS | AS | AS | AS | AS | AS | AS |
| kinetic energy = ¹/₂ × mass × speed² | $E_k = \frac{1}{2}mv^2$ | AS | AS | AS | AS | AS | AS | A2 | AS |
| change in gravitational potential energy = mass × gravitational field strength × change in height | $E_p = mgh$ | AS | AS | AS | AS | AS | AS | A2 | AS |
| pressure = $\dfrac{\text{force}}{\text{area}}$ | $P = \dfrac{F}{A}$ | | | AS | | AS | AS | | |
| pressure × volume = number of moles × molar gas constant × absolute temperature | $pV = nRT$ | AS | • | AS | • | • | • | • | • |
| charge = current × time | $q = It$ | AS | AS | AS | AS | AS | AS | AS | AS |
| potential difference = current × resistance | $V = IR$ | AS | AS | AS | AS | AS | AS | AS | AS |
| electrical power = potential difference × current | $P = IV$ | AS | AS | AS | AS | AS | AS | AS | AS |
| potential difference = $\dfrac{\text{energy transferred}}{\text{charge}}$ | $V = \dfrac{W}{q}$ | AS | AS | AS | AS | AS | AS | AS | AS |
| resistance = $\dfrac{\text{resistivity} \times \text{length}}{\text{cross-sectional area}}$ | $R = \dfrac{\rho l}{A}$ | AS | AS | AS | AS | AS | AS | AS | AS |
| energy = potential difference × current × time | $E = VIt$ | AS | AS | AS | AS | AS | AS | AS | AS |
| wave speed = frequency × wavelength | $v = f\lambda$ | • | AS | • | AS | AS | AS | AS | AS |
| centripetal force = $\dfrac{\text{mass} \times \text{speed}^2}{\text{radius}}$ | $F = \dfrac{mv^2}{r}$ | • | • | • | • | • | • | • | • |
| force between two charges | $F = \dfrac{kQ_1 Q_2}{r^2}$ | • | • | • | • | • | • | • | • |
| force between two masses | $F = \dfrac{Gm_1 m_2}{r^2}$ | • | • | • | • | • | • | • | • |
| capacitance = $\dfrac{\text{charge stored}}{\text{voltage}}$ | $C = \dfrac{Q}{V}$ | • | • | • | • | • | • | • | • |
| the transformer equation | $\dfrac{V_p}{V_s} = \dfrac{N_p}{N_s}$ | | • | • | • | | • | | • |

# Four steps to successful revision

## Step 1: Understand

- Study the topic to be learned slowly. Make sure you understand the logic or important concepts.
- Mark up the text if necessary – underline, highlight and make notes.
- Re-read each paragraph slowly.

**GO TO STEP 2**

## Step 2: Summarise

- Now make your own revision note summary:
  What is the main idea, theme or concept to be learned?
  What are the main points? How does the logic develop?
  Ask questions: Why? How? What next?
- Use bullet points, mind maps, patterned notes.
- Link ideas with mnemonics, mind maps, crazy stories.
- Note the title and date of the revision notes
  (e.g. Physics: Mechanics, 3rd March).
- Organise your notes carefully and keep them in a file.

This is now in **short term memory**. You will forget 80% of it if you do not go to Step 3. **GO TO STEP 3**, but first take a 10 minute break.

## Step 3: Memorise

- Take 25 minute learning 'bites' with 5 minute breaks.
- After each 5 minute break test yourself:
  Cover the original revision note summary
  Write down the main points
  Speak out loud (record on tape)
  Tell someone else
  Repeat many times.

The material is well on its way to **long term memory**.
You will forget 40% if you do not do step 4. **GO TO STEP 4**

## Step 4: Track/Review

- Create a Revision Diary (one A4 page per day)
- Make a revision plan for the topic, e.g. 1 day later, 1 week later, 1 month later.
- Record your revision in your Revision Diary, e.g.
  Physics: Mechanics, 3rd March 25 minutes
  Physics: Mechanics, 5th March 15 minutes
  Physics: Mechanics, 3rd April 15 minutes
  ... and then at monthly intervals.

# Mechanics

## The following topics are covered in this chapter:

- Newton I and III
- Momentum and the second law
- Energy to work
- Motion in a circle

## 1.1 Newton I and III

**After studying this section you should be able to:**

- state Newton's first and third laws of motion
- apply the first law to situations where the forces acting on an object are balanced
- apply the third law to the forces acting on two objects that interact

**LEARNING SUMMARY**

### Newton's laws of motion

With his three laws of motion, Newton aimed to describe and predict the movement of every object in the Universe. Like many other 'laws' in physics, they give the right answers much of the time. The paths of planets, moons and satellites are all worked out using Newton's laws.

This section looks at formal statements of the first and third law and how they apply to everyday situations.

### The first law

| | | | |
|---|---|---|---|
| AQA A | AS | EDEXCEL B | AS |
| AQA B | AS | OCR A | M4 |
| EDEXCEL A | AS | OCR B | AS |
| NICCEA | AS | WJEC | M4 |

> **KEY POINT**
>
> Newton's first law states that:
> An object maintains its state of rest or uniform motion unless there is a resultant, or unbalanced, force acting on it.

The phrase 'uniform motion' means moving in a straight line at constant speed, i.e. moving at constant velocity.

friction and air resistance create a resultant force on this sliding object

At first, this seems contrary to everyday experience. Give an object a push and it slows down before coming to rest. This is not 'continuing in a state of uniform motion'. Newton realised that in this case the unseen resistive forces of friction and air resistance together act in opposition to the motion; as there is no longer a driving force after the object has been pushed, there is a resultant backwards-directed force acting on it.

Motion without resistive forces is difficult to achieve on Earth: there is always air resistance or friction from a surface that a moving object rests on. The nearest that we can get to modelling motion with no resistive forces is to study motion on ice or an air track.

The converse of Newton's first law is also true:

if an object is at rest or moving at constant velocity the resultant force on it is zero

### Key point from AS

- When the forces on an object are balanced, the vector diagram is a closed figure.
Revise AS section 1.1

This means that where there are two forces acting on an object that satisfies these conditions, they must be equal in size and opposite in direction. If there are three or more forces acting then the vector diagram is a closed figure.

One example of the first law is a vehicle moving at constant velocity; the driving force and resistive force are equal in size and act in opposite directions. The diagram shows an example of a situation where three forces sum to zero; it illustrates a vehicle parked on a hill at rest and a vector triangle that shows the forces.

> A normal reaction force acts on all four wheels; the arrow shown on the diagram represents the sum of these forces.

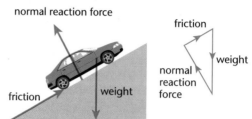

## The third law

| | | | |
|---|---|---|---|
| AQA A | AS | EDEXCEL B | AS |
| AQA B | AS | OCR A | M4 |
| EDEXCEL A | AS | OCR B | AS |
| NICCEA | AS | WJEC | AS |

> The laws were originally written in Latin. The third law is directly translated as 'to every action there is an equal and opposite reaction'. This was widely misinterpreted as meaning that two equal and opposite forces act on the same object, resulting in zero acceleration.

This is both the simplest in its statement and the most misunderstood of the three laws. The statement given here is not a direct translation of the original, but it helps to remove some of the misunderstanding.

> **KEY POINT**
>
> Newton's third law can be stated as:
>
> If object A exerts a force on object B, then B exerts a force equal in size and opposite in direction on A.

According to the third law, forces do not exist individually but in pairs. However, it is important to remember that:

- the forces are of the same type, i.e. both gravitational or electrical
- the forces act on different objects
- the third law applies to all situations.

## Some examples

| | | | |
|---|---|---|---|
| AQA A | AS | EDEXCEL B | AS |
| AQA B | AS | OCR A | M4 |
| EDEXCEL A | AS | OCR B | AS |
| NICCEA | AS | WJEC | AS |

Application of Newton's third law can lead to some surprising results. If you step off a wall the Earth pulls you down towards the ground. According to the third law, you also pull the Earth up with an equal-sized force. So why doesn't the Earth accelerate upwards to meet you instead of the other way round? The answer is it does, but if you apply $F = ma$ to work out the Earth's acceleration, it turns out to be minimal.

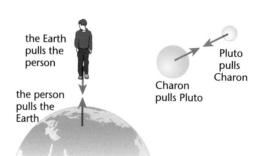

Newton's third law: pairs of forces that are equal in size and opposite in direction.

**Key point from AS**

• **force =
mass × acceleration**
*Revise AS section 1.6*

This type of situation is where the greatest misunderstanding of Newton's third law occurs. Remember, third law pairs of forces act on **different** objects.

Planets pull moons and, according to the third law, moons pull planets with an equal-sized force. Why does the moon go round the planet instead of the other way round?

Again, the answer is it does. In fact, they both rotate around a common centre of mass. In the case of the Earth and its Moon, the centre of mass is so close to the centre of the Earth that to all intents and purposes the Moon orbits the Earth.

The diagram shows another pair of forces that are equal in size and opposite in direction. This is an application of the **first** law – the vase is in a state of rest so there is no resultant force acting on it.

the Earth
pulls the
vase

the table
pushes
the vase

These forces are also equal in size and opposite in direction – an application of Newton's first law, not third.

## Progress check

1  Explain why a driving force is needed to maintain motion on the surface of the Earth.

2  Suggest why the driving force needed to maintain motion on the surface of the Moon is less than for the same vehicle on the surface of the Earth.

3  A foot kicks a ball. What is the force that makes up the pair in the sense of the third law?

3  The ball pushes the foot.
2  There is no air resistance on the Moon.
1  The driving force is needed to balance the resistive forces that oppose motion.

# 1.2 Momentum and the second law

**After studying this section you should be able to**

- apply the principle of conservation of momentum to collisions in one dimension
- state the relationship between the change of momentum of an object and the resultant force on it
- explain the difference between an elastic and an inelastic collision

**LEARNING SUMMARY**

## Momentum

| | | | |
|---|---|---|---|
| AQA A | AS | OCR A | M4 |
| AQA B | M4 | OCR B | M4 |
| EDEXCEL A | AS | NICCEA | M4 |
| EDEXCEL B | M4 | WJEC | M4 |

A collision between two objects need not involve physical contact. Imagine two positive charges approaching each other; the repulsive forces could cause them to reverse in direction without actually touching.

Which is more effective at demolishing a brick wall, a 1 kg iron ball moving at 50 m s$^{-1}$ or a 1000 kg iron ball moving at 1 m s$^{-1}$? Both the mass and the velocity need to be taken into account to answer this question.

Newton realised that what happens to a moving object involved in a collision depends on two factors:

- the mass of the object
- the velocity of the object.

He used the concept of **momentum** to explain the results of collisions between objects.

> **momentum = mass × velocity**
>
> $$p = m \times v$$
>
> Momentum is a vector quantity, the direction being the same as that of the velocity. It is measured in N s or kg m s$^{-1}$; these units are equivalent.
>
> **KEY POINT**

A ten-pin bowling ball has a mass of several kilogrammes, so even at low speeds it has much more momentum that a tennis ball (mass 0.06 kg) travelling as fast as you can throw it.

Everything that moves has momentum and exerts a force on anything that it interacts with. This applies as much to the light that is hitting you at the moment as it does to a collision between two vehicles.

No matter how hard you throw a tennis ball at a garden wall you will not knock the wall down; nor will you be very successful if you use a tennis ball to play ten-pin bowling!

## Conservation of momentum

| | | | |
|---|---|---|---|
| AQA A | AS | OCR A | M4 |
| AQA B | M4 | OCR B | M4 |
| EDEXCEL A | AS | NICCEA | M4 |
| EDEXCEL B | M4 | WJEC | M4 |

The phrase 'equal and opposite' is used as a shorthand way of writing 'equal in size and opposite in direction'.

When a force causes an object to change its velocity, there is also a change in momentum. When two objects interact or collide, they exert equal and opposite forces on each other and so the momentum of each one changes. This is illustrated in the diagram, where the blue ball loses momentum and the black ball gains momentum.

| before | | after | |
|---|---|---|---|
| 5.0 m s$^{-1}$ | 3.0 m s$^{-1}$ | 4.2 m s$^{-1}$ | 6.2 m s$^{-1}$ |
| 0.4 kg | 0.1 kg | | |

Like the forces they exert on each other during the collision, the changes in momentum of the balls are *equal in size and opposite in direction*.

It follows that:

- the combined momentum of the balls is the same before and after they collide.

To satisfy the conditions for conservation of momentum to apply, the only forces acting on the objects must be the ones between the objects themselves.

This is why conservation of momentum is usually demonstrated on an air track or other low friction surface.

This is an example of the principle of conservation of momentum.

> The principle of conservation of momentum states that:
>
> When two or more objects interact the total momentum remains constant provided that there is no external resultant force.

In the context of the colliding balls, an external resultant force could be due to friction or something hitting one of the balls. Either of these would result in a change in the total momentum.

## Different types of interaction

| AQA A | AS | OCR A | M4 |
|-------|----|-------|----|
| AQA B | M4 | OCR B | M4 |
| EDEXCEL A | AS | NICCEA | M4 |
| EDEXCEL B | M4 | WJEC | M4 |

### Objects moving in opposite directions

When two objects approach each other and 'collide' head-on, the result of the collision depends on whether they stick together or whether one or both rebound.

The diagram shows two 'vehicles' approaching each other on an air track. The vehicles join together and move as one.

$0.40\,\text{m s}^{-1}$     $0.30\,\text{m s}^{-1}$     velocity = ?

0.30 kg     0.20 kg     0.50 kg

before     after

To make two air track vehicles stick together, one has some Plasticine stuck to it and the other is fitted with a pin that sticks into the Plasticine.

The principle of conservation of momentum can be used to work out the combined velocity, $v$, after the collision.

Taking velocity from left to right as being positive:

total momentum before collision =
$0.3\,\text{kg} \times + 0.4\,\text{m s}^{-1} + 0.2\,\text{kg} \times - 0.3\,\text{m s}^{-1} = 0.06\,\text{N s}$.

This must equal the momentum after the collision, so $0.5\,\text{kg} \times v = 0.06\,\text{N s}$,

i.e. $v = 0.12\,\text{m s}^{-1}$. As this is positive, it follows that the 'vehicle' is moving from left to right.

### Rebound

When a light object collides with a heavier one, the light object may rebound, reversing the direction of its momentum. The diagram shows an example of such a collision.

before     after

A rebound collision can be modelled on an air track by fitting the vehicles with repelling magnets.

$10\,\text{m s}^{-1}$     $0\,\text{m s}^{-1}$     $v$     $3.33\,\text{m s}^{-1}$

2 kg     10 kg

The figures here are rounded. The 3.33 is a rounded value of 3⅓.

Application of the principle of conservation of momentum in this case gives:

$2\,\text{kg} \times 10\,\text{m s}^{-1} + 10\,\text{kg} \times 0\,\text{m s}^{-1} = 2\,\text{kg} \times v + 10\,\text{kg} \times 3.33\,\text{m s}^{-1}$

So $v = -6.7\,\text{m s}^{-1}$. The significance of the negative sign is that the green ball is now travelling from right to left.

When one object rebounds off another during a collision, both objects undergo a greater change in momentum than if they stick together. In the example above, the size of the change in momentum of the small ball is $2\,\text{kg} \times 10\,\text{m s}^{-1} - 2\,\text{kg} \times - 6.7\,\text{m s}^{-1} = 33.3\,\text{N s}$. Had it carried on moving from left to right, the change in momentum would have been much smaller than this.

## Recoil

In some situations two objects are stationary before they interact, having a combined momentum of zero. After they interact, both objects move off but the combined momentum is still zero. Examples of this include:

- a stationary nucleus undergoes radioactive decay, giving off an alpha particle
- a person steps off a boat onto the quayside
- two ice skaters stand facing each other; then one pushes the other
- two air track 'vehicles' fitted with repelling magnets are held together and then released.

In each case, both objects start moving from rest. For the combined momentum to remain zero, each object must gain the same amount of momentum but these momentum gains must be in *opposite directions*.

push        0.6 m s$^{-1}$        v = ?

The ice skater who does the pushing **recoils** as an equal and opposite force is exerted by the one who is pushed.

In the example shown in the diagram above, an 80 kg man pushes a 50 kg child. Their combined momentum is zero both before and after this interaction.

Taking velocity from left to right as positive, this means that:

$$80 \text{ kg} \times -0.6 \text{ m s}^{-1} + 50 \text{ kg} \times v = 0$$

So $v$, the velocity of the child after the interaction, is 0.96 m s$^{-1}$ to the right.

## Elastic or inelastic

In all three examples above, both momentum and energy are conserved. This is always the case. A collision where the total **kinetic** energy before and after the interaction is the same is called an **elastic** collision. Where there is a gain or loss of kinetic energy as a result of the interaction the collision is called **inelastic**.

In an elastic collision: momentum, kinetic energy and total energy are conserved.

In an inelastic collision: momentum and total energy are conserved, but the amount of kinetic energy changes.

Of the examples given, only the second is an elastic collision.

## Momentum and the second law

Newton's second law establishes the relationship between the resultant force acting on an object and the change of momentum that it causes. An important result that follows from the second law and the definition of the newton is $F = m \times a$.

> **Newton's second law states that:**
>
> The rate of change of momentum of an object is proportional to the resultant force acting on it and acts in the direction of the resultant force.
>
> $$\Delta p / \Delta t \propto F$$

KEY POINT

The definition of the size of the newton fixes the proportionality constant at one. This enables the second law to be written as:

> **Force = rate of change of momentum**
> $F = \Delta p / \Delta t$

**KEY POINT**

In this form, the second law is useful for working out the force in situations such as jet and rocket propulsion where the change in momentum each second can be calculated easily.

A rocket carries its own oxygen supply, so that it can fire the engines when it is travelling in a vacuum. A spacecraft travelling where it is not affected by any gravitational fields or resistive forces maintains a constant velocity, so it only needs to fire the rocket engines to change speed or direction.

Jet and rocket engines use the same principle; hot gases are ejected from the back of the engine to provide a force in the forwards direction. The difference is that jet engines are designed to operate within the Earth's atmosphere, so they can take oxygen from their surroundings.

The principle of rocket propulsion can be seen by blowing up a balloon and letting it go; the air squashed out of the neck of the balloon gains momentum as it leaves, causing the balloon and the air that remains inside to gain momentum at the same rate but in the opposite direction. The diagram shows the principle of rocket propulsion.

The unit kg m s⁻² is equivalent to the N, so a rate of change of momentum of 4 500 000 kg m s⁻² is another way of stating: a force of 4 500 000 N.

exhaust gases gain momentum in this direction

rocket gains momentum in this direction

conservation of momentum and rocket propulsion

In this example, the exhaust gases are emitted at a rate of 3000 kg s⁻¹. The speed of the exhaust gases relative to the rocket is 1500 m s⁻¹. The rate of change of momentum of these gases is therefore 4 500 000 kg m s⁻², i.e. 4 500 000 kg m s⁻¹ each second. This is the size of the force on both the gases and the rocket.

## Progress check

1 Two vehicles fitted with repelling magnets are held together on an air track. The masses of the vehicles are 0.10 kg and 0.15 kg.

   The vehicles are released. The 0.10 kg vehicle moves to the left with a speed of 0.24 m s⁻¹. Calculate the velocity of the heavier vehicle.

2 A ball of mass 0.020 kg hits a wall at a speed of 12.5 m s⁻¹ and rebounds at a speed of 10.0 m s⁻¹.
   a Calculate the change in momentum of the ball.
   b If the ball is in contact with the wall for 0.010 s, calculate the size of the force between the ball and the wall.
   c Explain whether this collision is elastic or inelastic.

   c Inelastic. The ball has less kinetic energy after rebound than before it hits the wall.
   b 45 N
   2 a 0.45 N s in the direction of the rebound
   1 0.16 m s⁻¹ to the right

# 1.3 Energy to work

*After studying this section you should be able to:*

- *explain the energy transfer that takes place when a force causes motion*
- *calculate kinetic energy and changes in gravitational potential energy*
- *apply the principle of conservation of energy*

**LEARNING SUMMARY**

## Working

| | | | |
|---|---|---|---|
| AQA A | AS | OCR A | M4 |
| AQA B | M4 | OCR B | M4 |
| EDEXCEL A | AS | NICCEA | M4 |
| EDEXCEL B | AS | WJEC | M4 |

Every event requires **work** to make it happen. Even if you relax and try to do nothing, your body is working just to keep you alive. Any force that causes movement is doing work. Pushing a supermarket trolley is working, as is throwing or kicking a ball. However, holding some weights above your head is not working; it may cause your arms to ache, but the force on the weights is not causing any movement!

How much work a force does depends on:

- the size of the force
- the direction of movement
- the distance that an object moves.

> The phrase 'distance moved in the direction of the force' is used here instead of 'displacement' to emphasise that the force and displacement it causes are measured in the same direction.

**KEY POINT**

The work done by a force that causes movement is defined as:

work = force × distance moved in the direction of the force

$$W = F \times s$$

Work is measured in joules (J) where 1 joule is the work done when a force of 1 N moves its point of application 1 m in the direction of the force.

Here are some examples of forces that are working and forces that are not working:

The shelf is not moving, so no work is being done.

The tension in the string is not working as there is no movement in the direction of the force.

> A pylon that supports electricity transmission cables is not working, but a wind that causes the cables to move is!

This force causes movement in the direction of the force.

The horizontal component of the force on the log is doing the work here.

> In circular motion, the direction of the velocity is along a tangent to the circle.

In the case of the stone being whirled in a horizontal circle, the stone's velocity is always at **right angles** to the force that maintains the motion, so the centripetal force is not working.

Although the force pulling the log and the movement it causes are not in the same direction, the force on the log has a component in the same direction as the log's movement. In calculating the work done the horizontal component of the force is multiplied by the horizontal distance moved by the log.

## Work and energy transfer

| | | | |
|---|---|---|---|
| AQA A | AS | OCR A | M4 |
| AQA B | M4 | OCR B | M4 |
| EDEXCEL A | AS | NICCEA | M4 |
| EDEXCEL B | AS | WJEC | M4 |

The physical processes of **working** and **energy transfer** are inseparable. Work requires an energy source, for example a fuel, an electric cell or a wound-up spring. As the work is done, energy leaves the source and ends up elsewhere, this is what is meant by **energy transfer**.

> An object has energy if it can exert a force that causes movement of the point of application of the force.

> **KEY POINT**
>
> Energy is the ability to do work.
> The energy transfer from a source is equal to the amount of work done.
> Like work, energy is measured in joules.

Here are some examples of everyday energy transfers.

### A bus accelerates away from a bus stop and then maintains a steady speed

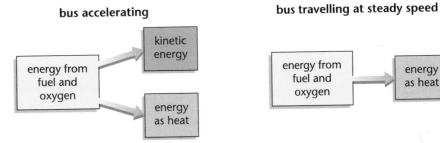

bus accelerating    bus travelling at steady speed

> A common error is to describe the energy transfer of a vehicle moving at a constant speed as 'energy from the fuel and oxygen being transferred to kinetic energy'.

As the bus speeds up, some of the energy from the fuel and oxygen is transferred to **kinetic energy** of the bus, the energy it has due to its movement. The rest of the energy is transferred to the surroundings. This takes place in a number of ways; the exhaust gases transfer energy as heat directly into the surrounding air and some energy is also transferred to heat wherever there are resistive forces.

Once the bus is maintaining a steady speed there is no increase in its kinetic energy, so all the energy transferred from the fuel and oxygen is going to the surroundings.

### An electric motor lifts a load at a steady speed

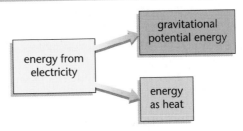

> The phrase 'gravitational potential energy' is often abbreviated to 'gpe' or just 'potential energy', although strictly speaking 'potential energy' can be any form of stored energy, for example in a spring or a fuel.

The load gains **gravitational potential energy** as it is lifted, the energy it has due to its position above the Earth's surface. All the movement is at a constant speed, so there is no change in the kinetic energy of the system. As with the bus travelling at a steady speed, work done against resistive forces causes heating, and this energy is transferred as heat to the surroundings.

## Energy and a filament lamp

| | | | |
|---|---|---|---|
| AQA A | AS | OCR A | M4 |
| AQA B | M4 | NICCEA | M4 |
| EDEXCEL A | AS | WJEC | M4 |

When a filament lamp is switched on, some energy is absorbed by the filament as it reaches its operating temperature. This takes a fraction of a second. Some energy is absorbed by the glass envelope and the gas inside. These take a little longer to reach a steady temperature, perhaps a few seconds.

Once the lamp has reached its steady state, of the 60 J of energy passing into the lamp each second from the electricity supply, typically 3 J passes out as light. This

In the steady state, the parts of the lamp are emitting energy and absorbing energy at the same rate, so there is no change in temperature.

Sankey diagrams are a useful way of showing the energy flow through a process where there are several stages, for example the energy flow through a power station.

is shown in the diagram. The Sankey diagram gives a visual indication of the relative proportions of energy transferred into a desirable output and wasted.

**Energy flow through filament lamp**

a Sankey diagram

The 'wasted' energy is transferred to the surroundings in two main ways:

- from the glass envelope to the surrounding air by conduction and convection
- from all parts of the lamp as non-visible electromagnetic radiation (mainly infra-red).

The light and the infra-red radiation all cause heating when they are absorbed, so all the energy input to the lamp ends up as heat!

These examples illustrate some important points about energy transfer:

- energy due to movement is called kinetic energy
- energy due to position above the Earth's surface is called gravitational potential energy, often abbreviated to potential energy
- work usually involves some transfer of energy to the surroundings.

## Conservation of energy

| AQA A | AS | OCR A | M4 |
| AQA B | M4 | OCR B | M4 |
| EDEXCEL A | AS | NICCEA | M4 |
| EDEXCEL B | AS | WJEC | M4 |

The examples of energy transfer identify some ways in which energy flows into and out of a process where work is done. Like mass and charge, energy is a conserved quantity.

Take care not to confuse 'conservation of energy' with 'conservation of energy resources'. Energy resources such as coal can be used up, and conservation in this context means preserving them as long as possible.

> **KEY POINT**
>
> The principle of **conservation of energy** states that:
>
> energy cannot be created or destroyed.

This simple statement means that energy is never used up, although often the energy output from a working process is in a form which is difficult to harness to do more work. The energy output from a filament lamp, for example, is effectively lost as heat in the surroundings. Like the energy from the bus, it raises the temperature of its surroundings by such a small amount that makes it difficult to recover.

In many energy transfer processes the amount of energy at each stage cannot easily be quantified; it is difficult to put a figure on the total kinetic and potential energy of the moving parts in an engine for example. There are simple formulas for calculating the kinetic energy and gravitational potential energy of individual objects.

The formula for change in gravitational potential energy is only valid for changes in height close to the Earth's surface, where g has a constant value.

> **KEY POINT**
>
> kinetic energy = ½ × mass × (speed)²
> $$E_k = \tfrac{1}{2} mv^2$$
> change in gravitational potential energy = weight × change in height
> $$\Delta E_p = mg\Delta h$$

The formula for calculating gravitational potential energy gives the change in energy because 'ground level' varies and so does not provide an accurate reference point for zero energy. This is not the case with kinetic energy as 'no movement relative to the Earth's surface' is the same the world over.

When a streamlined object is falling freely in the atmosphere at a low speed, the resistive forces are small and so little energy is transferred to the surroundings. In this case the principle of conservation of energy can be applied to the transfer of gravitational potential energy to kinetic energy. When a parachutist is falling at a constant speed conservation of energy still applies, but the energy transfer is from gravitational potential energy to heat in the atmosphere.

The principle of conservation of energy applies to all energy transfers.

**Energy transfer of falling objects**

A ball falling freely towards the Earth

gravitational potential energy → kinetic energy

A parachutist falling at constant speed

gravitational potential energy → heat in the surrounding air

## Progress check

A 2.0 kg mass is released from a height of 8.0 m above the Earth's surface and allowed to fall to the ground. Assuming $g = 10$ m s$^{-2}$, calculate:

a  the loss of gravitational potential energy
b  the gain in kinetic energy
c  the speed of the mass as it hits the ground.

a  160 J
b  160 J
c  12.6 m s$^{-1}$

# 1.4 Motion in a circle

*After studying this section you should be able to:*

- *use the radian to measure angular displacement*
- *describe the forces on an object moving in a circle*
- *recall and use the expressions for centripetal force and centripetal acceleration*

## Measuring angles

| | | | |
|---|---|---|---|
| AQA A | M4 | OCR A | M4 |
| AQA B | M4 | OCR B | M4 |
| EDEXCEL A | M4 | NICCEA | M4 |
| EDEXCEL B | M4 | WJEC | M4 |

When an object travels through a complete circle it moves through an angle of 360°, although the distance it travels depends on the radius of the circle. When a train or motor vehicle travels round a bend, the outer wheels have further to travel than the inner wheels. Similarly, when a compact disc is being played all points on the disc move through the same angle in any given time, but the linear speed at any point depends on its distance from the centre.

An alternative unit to the degree for measuring angular displacement is the radian, defined as the arc length divided by the radius of the circle. This is shown in the diagram.

**Measuring an angle in radians**

> To convert an angle in degrees to one in radians, multiply by $\pi$ and divide by 180.

It follows that movement through one complete circle, or 360°, is equal to a displacement of $\theta$ = circumference ÷ radius = $2\pi r \div r = 2\pi$ radians.

## Speed in a circle

| | | | |
|---|---|---|---|
| AQA A | M4 | OCR A | M4 |
| AQA B | M4 | OCR B | M4 |
| EDEXCEL A | M4 | NICCEA | M4 |
| EDEXCEL B | M4 | WJEC | M4 |

When an object moves in a circular path its velocity is always at right angles to a radial line, a line joining it to the centre of the circle.

> Angular velocity is a vector quantity that can have one of two directions: clockwise or anticlockwise.

Two objects, such as points on a CD, at different distances from the centre each complete the same number of revolutions in any specified time but they have different speeds. They have the same angular velocity but different linear velocities.

> **Angular velocity, $\omega$, is defined as:**
> *the rate of change of angular displacement with time.*
> It is represented by the gradient of an angular displacement–time graph.
> average angular velocity = angular displacement ÷ time
> $$\omega = \Delta\theta/\Delta t$$
> where $\Delta\theta$ is the angular displacement in time $\Delta t$.

The relationship between angular velocity and linear velocity for an object moving in a circle of radius $r$ is:

$$v = r\omega.$$

The time period for one revolution, $T$, is equal to the angular displacement ($2\pi$ radians) divided by the angular velocity:

$$T = 2\pi/\omega.$$

The frequency of a circular motion, $f$, is equal to $1/T$:

$$f = 1/T = \omega/2\pi.$$

## Centripetal acceleration

| | | | |
|---|---|---|---|
| AQA A | M4 | OCR A | M4 |
| AQA B | M4 | OCR B | M4 |
| EDEXCEL A | M4 | NICCEA | M4 |
| EDEXCEL B | M4 | WJEC | M4 |

Any object that changes its velocity is accelerating. Since velocity involves both speed and direction, a change in either of these quantities is an acceleration. An object moving at a constant speed in a circle is therefore accelerating as it is continually changing its direction. The vector diagram below shows the change in velocity, $\Delta v$, for an object following a circular path. This change in direction is always towards the centre of the circle.

> Note that $\Delta v$ is the change in velocity from $v_1$ to $v_2$, so
> $v_1 + \Delta v = v_2$

The change in direction of an object following a circular path

The acceleration of an object moving in a circle:

- is called **centripetal acceleration**
- is directed towards the centre of the circle.

> **KEY POINT**
>
> Centripetal acceleration, $a$, is the acceleration of an object moving in a circular path.
> $$a = v^2/r = r\omega^2$$
> where $v$ is the linear velocity, $\omega$ is the angular velocity and $r$ is the radius of the circle.

## The unbalanced force

| | | | |
|---|---|---|---|
| AQA A | M4 | OCR A | M4 |
| AQA B | M4 | OCR B | M4 |
| EDEXCEL A | M4 | NICCEA | M4 |
| EDEXCEL B | M4 | WJEC | M4 |

To make an object accelerate, there has to be an unbalanced force acting in the same direction as the acceleration. In the case of circular motion the unbalanced force is called the **centripetal force**.

> It is important to understand that the centripetal force is the resultant of all the forces acting on an object. A common misconception is that it is an extra force that appears when an object moves in a circular path.

When a train travels around a bend, the centripetal force comes from the push of the outer rail on the wheel. For a road vehicle such as a car, bus or bicycle it is the push of the road on the tyres; if there is insufficient friction due to ice or a slippery surface, the vehicle does not complete the turn and may leave the road.

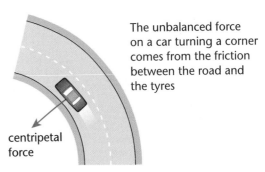

The unbalanced force on a car turning a corner comes from the friction between the road and the tyres

centripetal force

The size of the unbalanced force required to maintain circular motion can be calculated using $F = ma$:

> **KEY POINT**
>
> An object moving at constant speed in a circular path requires an unbalanced force towards the centre of the circle.
> centripetal force, $F = mv^2/r = mr\omega^2$

## The theme park ride

| AQA A | M4 | OCR A | M4 |
|-------|----|-------|----|
| AQA B | M4 | OCR B | M4 |
| EDEXCEL A | M4 | NICCEA | M4 |
| EDEXCEL B | M4 | WJEC | M4 |

Some theme park rides involve motion in a vertical circle. Part of the thrill of being on one of these rides is the apparent change in weight as you travel round the circle. This is because the force that you feel as you sit or stand is not your weight, but the normal contact force pushing up. If the forces on you are balanced, this force is equal in size to your weight, but it can become bigger or smaller if the forces are unbalanced, causing you to feel heavier or lighter.

The diagram below shows the forces acting on a person moving in a vertical circle when they are at the top and bottom of the circle.

> You feel the apparent change in weight when you travel in a lift. As the lift accelerates upwards you feel heavier because the normal contact force is greater than your weight.

### Key points from AS

- **Vector diagrams**
  *Revise AS section 1.1*

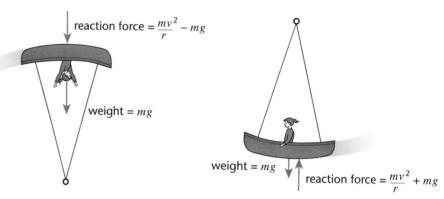

reaction force = $\frac{mv^2}{r} - mg$

weight = $mg$

weight = $mg$

reaction force = $\frac{mv^2}{r} + mg$

**The forces on a person moving in a vertical circle**

> Check that in each case the resultant of the forces is equal to $mv^2/r$ towards the centre of the circle.

**At the bottom of the circle:**

- The person feels heavier than usual as the size of the normal reaction force is equal to the person's weight plus the centripetal force required to maintain circular motion.

**At the top of the circle:**

- The person may feel lighter than usual as the person's weight is providing part of the centripetal force required to maintain circular motion. The rest comes from the normal reaction force, which is less than the person's weight if $mv^2/r < 2mg$.

> Another possibility is that the speed of the ride is such that the required centripetal force is less than the person's weight. In this case a safety harness is required to stop the person from falling out.

- If the speed of the ride is such that the required centripetal force is the same as the person's weight then he or she feels 'weightless' as the normal reaction force is zero.

## Progress check

1  **a**  Convert into radians:
   i   90°
   ii  720°

   **b**  An object is displaced through an angle of $8\pi$ radians.
        Explain why it has returned to its original position.

2  A stone is tied to a string and whirled round in a horizontal circle.
   **a**  What is providing the centripetal force?
   **b**  The stone moves at a horizontal speed of 3.5 m s⁻¹.
        The radius of the circle is 2.0 m.
        Calculate the centripetal acceleration.
   **c**  Calculate the size of the centripetal force if the mass of the stone is 0.25 kg.

1  **a**  i  $\pi/2$
        ii  $4\pi$
   **b**  It has travelled four complete revolutions.
2  **a**  The tension in the string.
   **b**  6.125 m s⁻²
   **c**  1.53 N

## Sample question and model answer

In this question, take the value of $g$, free fall acceleration, to be 10 m s$^{-2}$.

A toy car, mass 60 g, is at rest at the foot of a slope.

A second car, mass 90 g, which is moving at a speed of 1.20 m s$^{-1}$ hits the back of the stationary car and the cars stick together.

1.20 m s$^{-1}$

When objects stick together, there is always some energy transfer from kinetic energy to internal energy, as some deformation occurs.

(a) (i)  What type of collision occurs? [1]

Inelastic                                                                 1 mark

However, total energy is always conserved, no matter what type of collision.

(ii)  Which of the following quantities are conserved in this collision? [1]

                    kinetic energy    momentum    total energy

Momentum and total energy.                                                1 mark

Always write down the formula that you intend to use, and each step in the working. Otherwise, a simple error such as using the wrong mass results in no marks.

(b) (i)  Calculate the momentum of the larger car before the collision. [3]

momentum = mass × velocity                                               1 mark
          = 0.090 kg × 1.20 m s$^{-1}$                                    1 mark
          = 0.108 kg m s$^{-1}$                                           1 mark

(ii)  Calculate the combined speed of the cars after the collision. [3]

velocity = momentum ÷ mass                                               1 mark
         = 0.108 kg m s$^{-1}$ ÷ 0.150 kg                                 1 mark
         = 0.72 m s$^{-1}$                                                1 mark

The first mark here is for correct transposition of the equation.

(c)  Calculate the vertical height of the cars' centre of mass when they come to rest on the slope. [5]

As the cars rise up the slope, kinetic energy is transferred to
gravitational potential energy                                          1 mark
$\frac{1}{2}mv^2 = mg\Delta h$                                           1 mark
$\Delta h = v^2 \div 2g$                                                 1 mark
        = $(0.72 \text{ m s}^{-1})^2 \div (2 \times 10 \text{ m s}^{-2})$   1 mark
        = 0.026 m                                                        1 mark

It is not essential to include the units at each stage of the calculation, but it is good practice and you MUST include the correct unit with your answer to each part.

## *Practice examination questions*

Throughout this section, take the value of *g*, free fall acceleration, to be 10 m s$^{-2}$.

**1**

A child of mass 40 kg sits on a swing. An adult pulls the swing back, raising the child's centre of mass through a vertical height of 0.60 m. This is shown in the diagram.

(a) Calculate the increase in gravitational potential energy when the child moves from position B to position A. [3]

(b) The child is released from position A.
Calculate her speed as she passes through position B. [4]

(c) Explain why there must be a resultant (unbalanced) force on the child as she swings through position B and state the direction of this force. [4]

(d) Calculate the size of the resultant force that acts on the child as she swings through position B. [3]

(e) One force that acts on the child at position B is the upward push of the swing.
Write a description of the other force that acts on the child. [2]

(f) Write down the value of the upward push of the swing when the child is:
(i) stationary at B.
(ii) moving through B. [2]

**2**

The diagram shows two 'vehicles' on an air track approaching each other. After impact they stick together.

(a) Calculate the total momentum of the vehicles before the collision. [3]

(b) Calculate the velocity of the vehicles after the collision. State the direction of motion. [3]

(c) By performing appropriate calculations, explain whether the collision is elastic or inelastic. [3]

**3**

A communications satellite in a geostationary orbit (orbit time = 24 hours) goes round the Earth at a distance of $4.24 \times 10^7$ m from the centre of the Earth.

(a) Calculate the orbital speed of the communications satellite. [3]

(b) Calculate the centripetal acceleration of the satellite and state its direction. [4]

(c) What force causes this acceleration? [2]

**4**

A tennis ball has a mass of 60 g. A ball travelling at a speed of 15 m s⁻¹ hits a player's racket and rebounds with a speed of 24 m s⁻¹. The ball is in contact with the racket for a time of 12 ms.

(a) Calculate the change in momentum of the ball. [3]

(b) Calculate the average force exerted by the racket on the ball. [3]

(c) Calculate the average force exerted by the ball on the racket. [1]

(d) Describe the energy transfers that take place while the ball is in contact with the racket. [3]

**5**

The diagram shows a tractor being used to pull a log.

The tension in the steel rope is 600 N.

(a) Calculate the horizontal component of the forwards force on the log. [2]

(b) What other horizontal force acts on the log and what is the size of this force? [2]

(c) Calculate the work done on the log when it is pulled a distance of 250 m. [3]

(d) What happens to the energy used to pull the log? [2]

## *Practice examination questions* (continued)

**6**

A car of mass 800 kg travels around a circular corner of radius 100 m at a speed of 14 m s⁻¹.

(a) Calculate the centripetal acceleration of the car. [3]

(b) Calculate the size of the force needed to cause this acceleration. [3]

(c) (i)  Write a description of the force that causes this acceleration. [2]

    (ii) What is the other force that makes up the pair of forces in the sense of Newton's third law? [2]

(d) Suggest why, in wet or icy conditions, cars sometimes leave the road when travelling round a bend. [2]

**7**

A jet aircraft has a mass of 500 000 kg when fully laden.
It accelerates from rest to a take-off speed of 60 m s⁻¹ in 20 s.

(a) Calculate the momentum of the aircraft as it leaves the ground. [2]

(b) Calculate the mean force required to impart this momentum to the aircraft. [2]

(c) Suggest why the force produced by the engines during take-off is greater than the answer to (b). [2]

(d) How does the principle of conservation of momentum apply to this event? [2]

**8**

In a hydroelectric power station, water falls at the rate of $3.2 \times 10^5$ kg s⁻¹ through a vertical height of 180 m before entering a turbine.

(a) Calculate the loss in gravitational potential energy of the water each second. [3]

(b) Assuming that all this energy is transferred to kinetic energy, calculate the speed of the water as it enters the turbines. [3]

(c) The water leaves the turbines at a speed of 3 m s⁻¹. Calculate the maximum energy transfer to the turbine each second. [2]

# Waves

**The following topics are covered in this chapter:**

- Oscillations, resonance and damping
- Simple harmonic motion
- Wave properties
- Complex wave behaviour
- Quantum phenomena

## 2.1 Oscillations, resonance and damping

**After studying this section you should be able to:**

- explain the difference between a free and a forced vibration
- describe resonance and how it occurs
- distinguish between a damped oscillation and an undamped one and give examples of each

**LEARNING SUMMARY**

### Oscillations

| | | | |
|---|---|---|---|
| AQA A | M4 | OCR A | M4 |
| AQA B | M4 | OCR B | M4 |
| EDEXCEL A | M4 | NICCEA | M4 |
| EDEXCEL B | M5 | WJEC | M4 |

The terms 'oscillation' and 'vibration' have the same meaning. They are both used to describe a regular to-and-fro movement.

An **oscillation** is a repetitive to-and-fro movement. Loudspeaker cones oscillate, as do swings in a children's playground. However, there is an important difference between these oscillations.

The loudspeaker cone is an example of a **forced vibration**; the cone is forced to vibrate at the frequency of the current that passes in the coil. The amplitude of vibration depends on the size of the current. When reproducing speech or music, the frequency of vibration is constantly changing.

A child's swing, like a string on a piano or a guitar is an example of a **free vibration**. Once displaced from its normal position, it vibrates at its own **natural frequency**. The natural frequency of vibration of an object is the frequency it vibrates at when it is displaced from its normal, or rest, position and released. For a swing, this frequency depends only on the length of the chain or rope; in the case of a stringed instrument it is also affected by the tension and mass per unit length of the string.

The vibration of particles in a solid is a free vibration. Like a swing, the energy of the particles changes from kinetic to potential (stored) energy.

- In a free vibration there is a constant interchange between potential (stored) energy and kinetic energy.

- A swing has zero kinetic energy and maximum gravitational potential energy when at its greatest displacement; the kinetic energy is at a maximum and the gravitational potential energy at a minimum when the displacement is zero. This is shown in the diagram.

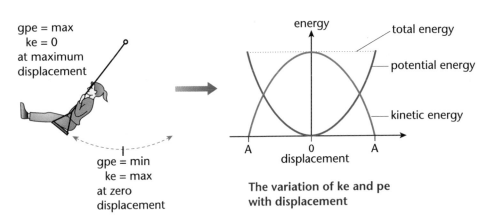

gpe = max
ke = 0
at maximum
displacement

gpe = min
ke = max
at zero
displacement

**The variation of ke and pe with displacement**

## Resonance

| AQA A | M4 | OCR A | M4 |
| AQA B | M4 | OCR B | M4 |
| EDEXCEL A | M4 | NICCEA | M4 |
| EDEXCEL B | M5 | WJEC | M4 |

There are many everyday examples of resonance. You may have noticed windows vibrating due to the sound from a bus waiting at a stop. Electrical resonance is used to tune in a radio or television to a particular station.

After a swing is given a push, it vibrates at its natural frequency. To make it swing higher, subsequent pushes need to coincide with the vibrations of the swing. This is an example of resonance, the large amplitude oscillation that occurs when the frequency of a forced vibration is equal to the natural frequency of the vibrating object.

The diagram shows how a vibration generator can be used to force a string to vibrate at any frequency, and the large amplitude vibration that occurs when the driving frequency coincides with the natural frequency of the string, also called its resonant frequency. At this frequency the string is able to absorb sufficient energy from the vibration generator to replace the energy lost due to resistive forces.

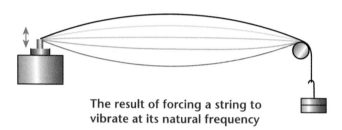

**The result of forcing a string to vibrate at its natural frequency**

If the frequency of the forced oscillation is increased slowly from zero, the amplitude of vibration of the string starts to increase at a frequency below the resonant frequency. It has its maximum value at this frequency and it then decreases again as the frequency continues to increase. The variation of amplitude around the resonant frequency is shown below.

The sharpness of the peak depends on the resistive forces acting. The greater the resistive force, the smaller the amplitude of vibration.

Vibration generator frequency

amplitude

natural frequency

These are some everyday examples of resonance.

* Microwave cookers heat food by causing water molecules in the food to resonate, absorbing the energy from the microwaves.

* Magnetic resonance imaging (MRI) is used in medicine to produce a detailed picture of the inside of the body by making the atomic nuclei resonate.

* Resonance in mechanical systems can be a nuisance or a danger. When the frequency of the engine vibrations in a vehicle matches the natural frequency of the exhaust the result is a rattle as the exhaust vibrates against the body of the vehicle.

* Resonance of parts of aircraft can cause failure due to excessive stress and several helicopter crashes have been attributed to resonance of the pilot's eyeballs resulting in the pilot being unable to see overhead power lines.

## Damping

| AQA A | M4 | OCR A | M4 |
| AQA B | M4 | OCR B | M4 |
| EDEXCEL A | M4 | NICCEA | M4 |
| EDEXCEL B | M5 | WJEC | M4 |

Just like linear motion, all mechanical oscillations are subject to resistive forces. The effect of resistive forces in removing energy from a vibrating object is known as damping.

A string or mass on a spring vibrating in air is lightly damped as the main resistive force is air resistance. The effect of light damping is to gradually reduce the energy, and therefore amplitude, of the vibrating object.

The body of a motor vehicle is connected to the wheels by springs. When the vehicle goes over a bump in the road the springs compress, giving the passengers a smoother ride. If motor vehicles were lightly damped they would continue to oscillate after going over a bump. Shock absorbers increase the resistive force, so that when the body of the vehicle has been displaced it returns to its original position without oscillating. A motor vehicle is critically damped so that it returns to the normal position in the minimum time to avoid vibrations that could cause 'car-sickness' and possibly loss of control by the driver.

> A car mechanic tests the shock absorbers by pushing down on the car bonnet. If the car oscillates, the shock absorbers need replacing.

Very large resistive forces result in heavy damping. Imagine a mass on a spring suspended so that the mass is in a viscous liquid such as syrup. If the mass is displaced, the force opposing its movement is very large and it takes a long time to return to the normal position.

> The effect of damping is to transfer the energy from the vibrating object to heat in the surroundings.

One effect of damping is to decrease the effects of resonance. The greater the damping, the smaller the amplitude of vibration at the resonant frequency. This is very important in mechanical structures such as bridges which can resonate due to the effects of wind. Resonance caused the suspension bridge built over the Tacoma Narrows, in America, to collapse in 1939. Since then, engineers have designed suspension bridges with shock absorbers built into the suspension to absorb the energy and prevent excessive vibration.

## Progress check

1 Describe the difference between a free vibration and a forced vibration.
2 Explain why a washing machine sometimes vibrates violently when the drum is spinning.
3 Explain how a free vibration is affected by the amount of damping.

3 Damping removes energy from a vibrating object. Increasing the damping reduces the amplitude of vibration and may stop any vibration from occurring.
2 The washing machine has a natural frequency of vibration. If the drum rotates at this frequency it causes the machine to resonate.
1 A free vibration occurs when an object is displaced and it vibrates at its natural frequency. A forced vibration can be at any frequency and is caused by an external oscillating force.

# 2.2 Simple harmonic motion

**After studying this section you should be able to:**

- *explain what is meant by simple harmonic motion*
- *apply the equations of simple harmonic motion*
- *sketch graphs showing the change in displacement, velocity and acceleration in simple harmonic motion*

**LEARNING SUMMARY**

## What is simple harmonic motion?

| | | | |
|---|---|---|---|
| AQA A | M4 | OCR A | M4 |
| AQA B | M4 | OCR B | M4 |
| EDEXCEL A | M4 | NICCEA | M4 |
| EDEXCEL B | M5 | WJEC | M4 |

> The resultant force on an object in simple harmonic motion is a restoring force since it is always directed towards the equilibrium position that the object has been displaced from.

As the name implies, it is the simplest kind of oscillatory motion. One set of equations can be used to describe and predict the movement of any object whose motion is simple harmonic.

The motion of a vibrating object is **simple harmonic** if:

- its acceleration is proportional to its displacement
- its acceleration and displacement are in opposite directions.

The second bullet point means that the acceleration, and therefore the resultant force, always acts towards the equilibrium position, where the displacement is zero.

Common examples of simple harmonic motion (often abbreviated to shm) include the oscillations of a simple pendulum (provided the amplitude is small, $\leq 10°$) and those of a mass suspended vertically on a spring.

The diagram shows the size of the acceleration of a simple pendulum and a mass on a spring when they are given a small displacement, $x$, from the equilibrium position.

$k$ = spring constant
$m$ = mass on spring

$g$ = gravitational field strength
$l$ = length of pendulum

> $k$, the spring constant, is defined as the force per unit extension
> $$k = F/x$$

In each case:

- the numerical value of the acceleration is equal to a constant multiplied by the displacement, showing that acceleration is proportional to displacement
- the negative value of the acceleration shows that it is in the opposite direction to the displacement, since acceleration and displacement are both vector quantities.

## Frequency of simple harmonic motion

| | | | |
|---|---|---|---|
| AQA A | M4 | OCR A | M4 |
| AQA B | M4 | OCR B | M4 |
| EDEXCEL A | M4 | NICCEA | M4 |
| EDEXCEL B | M5 | WJEC | M4 |

The constants in the equations for the acceleration of the mass on a spring (constant = $k/m$) and the pendulum bob (constant = $g/l$) give useful information about the frequency and time period of the oscillation.

> Since $(2\pi f)^2$ is always positive, $a$ and $x$ have the opposite signs so they are always in opposite directions.

**KEY POINT**

The acceleration of any object whose motion is simple harmonic is related to the frequency of vibration by the equation:

$$a = -(2\pi f)^2 x$$

where $a$ is the acceleration, $f$ is the frequency and $x$ is the displacement.

**Key points from AS**

- **The difference between distance and displacement**
  *Revise AS section 1.5*

A graph of acceleration against displacement is therefore a straight line through the origin, as shown in the diagram. The gradient of the graph has a negative value and is numerically equal to $(2\pi f)^2$.

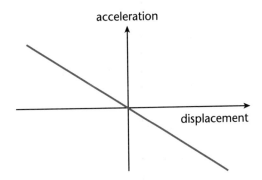

The relationship between the frequency, $f$, and the time period, $T$, is $f = 1/T$ so each of these quantities is the reciprocal of the other.

The frequency, $f$, and time period, $T$, are related to the constants in the equations. For a mass on a spring:

$$(2\pi f)^2 = k/m$$

So:

**KEY POINT**

$$f = \frac{1}{2\pi} \sqrt{\frac{k}{m}} \quad \text{and} \quad T = 2\pi \sqrt{\frac{m}{k}}$$

For a simple pendulum:

$$(2\pi f)^2 = g/l$$

So:

**KEY POINT**

$$f = \frac{1}{2\pi} \sqrt{\frac{g}{l}} \quad \text{and} \quad T = 2\pi \sqrt{\frac{l}{g}}$$

These equations for frequency and time period show that:

- increasing the mass on a spring causes the frequency to decrease and the time period to increase

The spring constant, $k$, is a measure of the stiffness of a spring.

- for a given mass, the stiffer the spring the higher the frequency of oscillation
- the frequency and time period of a pendulum do not depend on the mass of the pendulum bob
- increasing the length of a pendulum causes the frequency to decrease and the time period to increase.

## Displacement and time

| | | | |
|---|---|---|---|
| AQA A | M4 | OCR A | M4 |
| AQA B | M4 | OCR B | M4 |
| EDEXCEL A | M4 | NICCEA | M4 |
| EDEXCEL B | M5 | WJEC | M4 |

The displacement of an object in simple harmonic motion varies sinusoidally with time. This means that a displacement–time graph has the shape of a sine or cosine curve, and the displacement at any time, $t$, can be written in terms of a sine or cosine function.

The symbol $x_0$ is also used as an alternative to $A$ to represent amplitude.

**KEY POINT**

The displacement of an object in simple harmonic motion can be calculated using either of the equations:

$$x = A \sin 2\pi f t$$
$$x = A \cos 2\pi f t$$

where $A$ is the amplitude of the motion.

The graphs below show the variation of displacement with time for a simple harmonic motion.

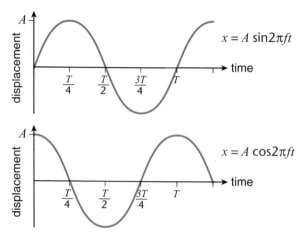

Which equation and graph apply to a simple harmonic motion depends on the displacement when the time is zero.

- If the displacement is zero at time $t = 0$, then the sine applies.
- If the displacement is at the maximum at time $t = 0$, then the cosine applies.

> Sine and cosine are identical functions with a 90°, or $\pi/2$ radians, phase difference.

## Displacement, velocity and acceleration

| | | | |
|---|---|---|---|
| AQA A | M4 | OCR A | M4 |
| AQA B | M4 | OCR B | M4 |
| EDEXCEL A | M4 | NICCEA | M4 |
| EDEXCEL B | M5 | WJEC | M4 |

Displacement, velocity and acceleration are linked graphically because:

- the gradient of a displacement–time graph represents velocity
- the gradient of a velocity–time graph represents acceleration.

> These bullet points apply to any motion.

The following diagram shows the variation of all three variables for a simple harmonic motion where the variation of the displacement with time is represented by a sine function.

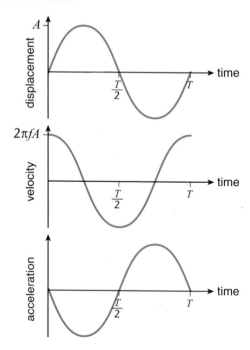

These graphs show that:

- when the displacement is zero the velocity is at a maximum and the acceleration is zero
- when the displacement has its maximum value the velocity is zero and the acceleration is a maximum in the opposite direction.

> These statements about the acceleration follow from the definition of simple harmonic motion.

Knowledge of the frequency and amplitude of a simple harmonic motion enables the maximum speed and the velocity at any displacement to be calculated.

> The ± symbol is used because at any value of the displacement the velocity could be in either direction.

**KEY POINT**

The maximum speed of an object in simple harmonic motion is:
$$v = 2\pi fA$$
The relationship between velocity and displacement is:
$$v = 2\pi \sqrt{(A^2 - x^2)}$$

## Progress check

When a 100 g mass is attached to the bottom of a vertical spring it causes it to extend by 5.0 cm.

**a** Calculate the value of the spring constant, $k$.
Assume $g = 10$ N kg$^{-1}$.
The mass is pulled down a further 3 cm and released.

**b** Calculate the frequency and time period of the resulting oscillation.

**c** **i** What is the value of the amplitude of the oscillation?
**ii** Calculate the maximum speed of the mass.

**c** **ii** 0.42 m s$^{-1}$
**i** 3 cm
**b** 2.25 Hz and 0.44 s
**a** 20 N m$^{-1}$

## 2.3 Wave properties

*After studying this section you should be able to:*

- *describe the difference between a longitudinal and a transverse wave*
- *recall and apply the wave equation to calculate speed, frequency and wavelength*
- *explain the meaning of polarisation and intensity*

## Wave motion

| AQA A | M4 | OCR A | AS |
| AQA B | AS | OCR B | AS |
| EDEXCEL A | M4 | NICCEA | AS |
| EDEXCEL B | AS | WJEC | AS |

*The waves on a water surface are almost transverse. The particles move in an elliptical path.*

Waves are used to transfer a signal or energy. They do this without an accompanying flow of material, although some waves, such as sound, can only be transmitted by the particles of a substance.

All waves consist of vibrations or oscillations. Sound and other compression waves are classified as longitudinal because the vibrations of the particles carrying the wave are along, or parallel to, the direction of wave travel. All electromagnetic waves are transverse; the vibrations of the electric and magnetic fields in these waves are at right angles to the direction of wave travel.

Waves transmitted through the body of water, or any other liquid, are longitudinal.

The diagram shows a longitudinal wave being transmitted by a spring and a transverse wave on a rope.

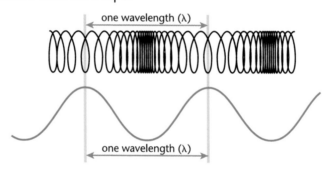

one wavelength (λ)

## Wave measurements

| AQA A | M4 | OCR A | AS |
| AQA B | AS | OCR B | AS |
| EDEXCEL A | M4 | NICCEA | AS |
| EDEXCEL B | AS | WJEC | AS |

*Most waves that you come across every day are progressive. Waves that are not progressive are discussed in the section 'complex wave behaviour'.*

These measurements apply to all progressive waves. A progressive wave is one that has a profile that moves through space.

- wavelength (symbol $\lambda$) is the length of one complete cycle; a compression (squash) and rarefaction (stretch) in the case of a longitudinal wave, a peak and a trough in the case of a transverse wave (see diagram above).
- amplitude (symbol $a$) is the maximum displacement from the mean position (see diagram page 51).
- frequency (symbol $f$) is the number of vibrations per second, measured in hertz (Hz)
- speed (symbol $v$) is the speed at which the profile moves through space
- period (symbol $T$) is the time taken for one vibration to occur. It is related to frequency by the equation $T = 1/f$

For all waves, the wavelength, speed and frequency are related by the equation:
$$speed = frequency \times wavelength$$
$$v = f \times \lambda$$

KEY POINT

## Wave graphs

| AQA A | M4 | OCR B | AS |
| EDEXCEL A | M4 | NICCEA | M5 |
| EDEXCEL B | AS | WJEC | M4 |
| OCR A | AS | | |

The definition of the size of an angle in radians is *arc length ÷ radius*, so it follows that an angle of $360° = 2\pi r/r = 2\pi$ radians.

Both transverse and longitudinal waves can be represented by graphs of displacement against position. The position along a wavefront can be measured in two ways:

- as a distance from a point in space
- as an angle.

In angular measure, one complete cycle of the wave is represented by an angle of 360°, or $2\pi$ radians. This is shown in the diagram.

A displacement–position or displacement–time graph does not distinguish between a transverse wave and a longitudinal one.

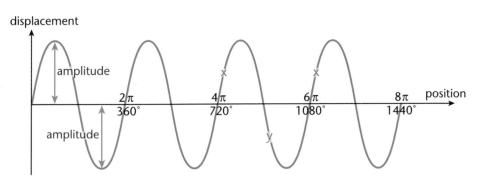

Using angular measure is a convenient way of comparing the **phase** of different parts of a wave. Two points on a wave are **in phase** if they have the same displacement and velocity. For progressive waves, this is only true for points exactly *n* wavelengths or $2\pi n$ radians or $360n°$ degrees apart, where *n* is an integer. Points marked x on the diagram above are in phase.

Try substituting values of *n*, for *n* = 0, 1, 2 etc. into the expression for 'an odd number of half wavelengths'. This will help you to understand the meaning of the phrase.

The point marked y is exactly **out of phase** with the points x as it has the opposite displacement and its velocity has the same value but in the opposite direction. This applies to any two points separated by an odd number of half wavelengths (i.e. 1/2, 3/2, 5/2 wavelengths, etc.). This can be expressed algebraically as $(2n + 1)\lambda/2$ or $(2n + 1)\pi$, where *n* is an integer.

The **phase difference** between two points on a wave describes their relative displacement and velocity and is normally expressed in degrees or wavelengths. A phase difference of $\lambda/2$ or 180° or $\pi$ radians describes two points separated by an odd number of half wavelengths (i.e. exactly out of phase), while a phase difference of $\lambda/4$ or 90° or $\pi/2$ radians compares two points that are separated by one-quarter, five-quarters, nine-quarters of a wavelength, etc.

A common error is to misinterpret a CRO display as a graph of displacement against distance.
This leads to a horizontal measurement, which represents time, being interpreted as a distance.

A **cathode ray oscilloscope** (CRO) can be used to display a different type of graph that represents wave motion. Rather than showing the displacement along a wave, the CRO plots the displacement of one point on the wave against time. The display on the CRO looks the same as that in the diagram, but the horizontal axis represents time rather than position on the wave.

To measure the **period** and **frequency** of a wave from a CRO display, the time-base control, which controls the rate at which the dot sweeps the screen horizontally, needs to be in the 'cal' (calibration) position. The diagram shows a CRO display of a sound wave, with the time–base set to 2 ms cm$^{-1}$.

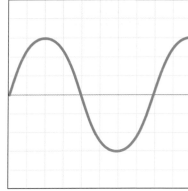

One cycle of the wave occupies a distance of 8 cm on the screen, so the period, $T = 8 \times 2$ ms = 16 ms = $1.6 \times 10^{-2}$ s.

The frequency of the wave, $f = 1/T = 1 \div 1.6 \times 10^{-2}$ s = 62.5 Hz.

## Polarisation

| | | | |
|---|---|---|---|
| AQA A | M4 | OCR A | AS |
| AQA B | AS | OCR B | AS |
| EDEXCEL A | M4 | NICCEA | AS |
| EDEXCEL B | AS | WJEC | AS |

Polarising material is used in some types of sunglasses and in filters for camera lenses. It cuts out reflected light from a flat surface such as an expanse of water.
Can you work out how it does this?

In order to receive a television or radio broadcast, the receiving aerial has to be lined up with that of the transmitter. This is because radio waves broadcast from aerials are polarised, the vibrations are only in one plane. For radio and television broadcasts the plane of polarisation is usually either vertical or horizontal.

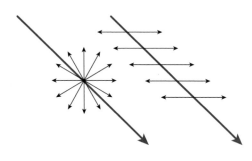

Light waves are not normally polarised, and nor are radio waves from stars. In an unpolarised wave the vibrations are in all planes at right angles to the direction of travel. Polarisation does not apply to longitudinal waves.

The diagram opposite shows the vibrations in polarised and unpolarised waves. Light waves can become polarised as they pass through some materials and they are partially polarised when reflected.

## Intensity

| | | | |
|---|---|---|---|
| AQA B | AS | OCR A | AS |
| EDEXCEL A | M4 | OCR B | AS |
| EDEXCEL B | AS | WJEC | AS |

A point source such as a star radiates energy in all directions. In this case the variation of intensity with distance follows an inverse square pattern, with:

$$I \propto 1/r^2$$

The signal strength of a radio broadcast decreases the further away the receiver is from the transmitting aerial. In a similar way, a sound appears fainter and a light seems dimmer further away from their origins. This is due to energy becoming spread over a wider area as the wave travels away from its source.

The intensity with which a wave is received depends on the power incident on the area of the detector.

> **KEY POINT**
>
> Intensity = power per unit area measured at right angles to the direction of travel.
> $$I = P/A$$
> Intensity is measured in $W\,m^{-2}$.

The diagram shows the sound spreading out from a loudspeaker. The ear that is further away detects a quieter sound because the power is spread over a wider area, so less enters the ear.

## Progress check

1. A television station broadcasts on a wavelength of 0.60 m. The speed of the waves is $3.0 \times 10^8$ m s$^{-1}$. Calculate the frequency of the waves.

2. Two waves have a phase difference of 60°. What is the phase difference in radians?

3. Explain why sound cannot be polarised.

3  Sound is a longitudinal wave, so there are no vibrations at right angles to the direction of travel.
2  $\pi/3$.
1  $5.0 \times 10^8$ Hz.

# 2.4 Complex wave behaviour

**After studying this section you should be able to:**

- describe how the diffraction of a wave at a gap depends on the wavelength and size of the gap
- explain how wave superposition causes interference patterns and describe the conditions for the patterns to be observable
- explain the formation of a stationary wave

**LEARNING SUMMARY**

## Diffraction

| | | | |
|---|---|---|---|
| AQA A | M4 | OCR A | AS |
| AQA B | AS | OCR B | AS |
| EDEXCEL A | M4 | NICCEA | AS |
| EDEXCEL B | AS | WJEC | AS |

Our everyday experience is that sound can travel round corners, but light travels in straight lines. The spreading out of waves as they pass through openings or the edges of obstacles is called diffraction. All waves can be diffracted, but the effect is more noticeable with long wavelength waves such as sound and radio waves than with short wavelength waves such as light and X-rays.

The diagrams show what happens when water waves pass through gaps of different sizes.

The amount of spreading when the wave has passed through the gap depends on the relative sizes of the gap and the wavelength.

> When answering questions about diffraction, always emphasise the size of the gap compared to the wavelength.

- For a gap that is many wavelengths wide, see diagram **a**, no detectable spreading takes place.

- Some spreading occurs when waves pass through a gap that is several wavelengths wide; diagram **b**.

- The maximum amount of spreading corresponds to a gap that is the same size as the wavelength as in diagram **c**.

Diagram **a** models what happens when light (wavelength approximately $5 \times 10^{-7}$ m) passes through a doorway, and diagram **c** models sound (wavelength approximately 1 m) passing through the same doorway.

> Diffraction explains why you can 'hear' round corners but you cannot 'see' round corners.

Diffraction also explains why long wavelength radio broadcasts can be detected in the shadows of hills and buildings, but short wavelength broadcasts cannot. Diffraction is also important in the design of loudspeakers to maximise the spreading for sounds of all wavelengths and in satellite transmissions to minimise the diffraction that occurs.

## Superposition and interference of sound

| | | | |
|---|---|---|---|
| AQA A | M4 | OCR A | AS |
| AQA B | AS | OCR B | AS |
| EDEXCEL A | M4 | NICCEA | AS |
| EDEXCEL B | AS | WJEC | AS |

The superposition of waves occurs when two or more waves cross. This is happening all the time, but the effects are so short-lived that we seldom notice them. One of the easiest effects of superposition to observe is two-source interference. This can be demonstrated with sound, light, surface water waves and 3 cm electromagnetic waves. The diagram shows the variation in the loudness of the sound in front of two loudspeakers vibrating in phase.

Superposition of sound waves

An observer walking along a line parallel to the loudspeakers notices that:

- there are regions where the sound is much louder than from one loudspeaker alone

- there are regions where the sound is barely audible

- in between these, the intensity of the sound varies from loud to quiet.

This is explained using the principle of superposition, which describes how two waves can reinforce each other or cancel each other out.

> **KEY POINT**
> When two or more waves cross, the resultant displacement is equal to the vector sum of the individual displacements.

This means that displacements in the same direction cause reinforcement, while displacements in opposite directions cause cancellation. The diagram below shows how a loud sound and no sound can be produced when two waves of equal amplitude meet at a point.

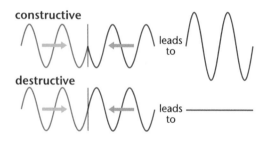

- If the waves are exactly in step (in phase) then constructive interference takes place, resulting in a loud sound

- If the waves are exactly out of step (out of phase) then destructive interference takes place, resulting in no sound.

## Conditions for interference to be observed

| AQA A | M4 | OCR B | AS |
|-------|-----|-------|-----|
| AQA B | AS | NICCEA | AS |
| EDEXCEL A | M4 | WJEC | AS |
| OCR A | AS | | |

The amplitudes do not need to be the same, but they should be comparable for the interference to be observed. If one wave has a much greater amplitude than the other then the effects of cancellation and reinforcement will not be noticeable.

The waves do not need to have the same amplitude for interference to take place. The effect of one wave having a greater amplitude than the other is that the destructive interference is not total. This happens when the observer is closer to one loudspeaker than to the other.

The diagram on page 55 represents the waves from two sources vibrating in phase. The solid and broken lines could represent peaks and troughs in the case of water waves or compressions and rarefactions in the case of sound waves. Some lines of constructive interference (c) and destructive interference (d) have been marked in.

If the sources in the diagram are in phase, then at any point equidistant from both sources the waves arrive in phase, giving constructive interference. This is the case for points along the central c line in this diagram. At all other points in the interference pattern, there is a path difference, i.e. the wave from one source has travelled further than that from the other source.

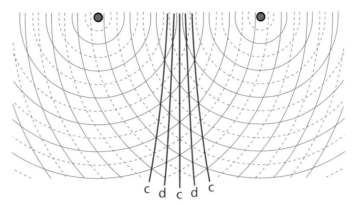

Try drawing similar patterns and investigating the effect of changing the wavelength and the separation of the sources.

c  d  c  d  c

**Constructive and destructive interference from two wave sources**

> **KEY POINT**
>
> A difference in the path length causes a difference in the phase of two waves arriving at a point. The relationship between phase difference and path difference is:
>
> $$\Delta(\text{phase}) = \frac{2\pi\Delta(\text{path})}{\lambda}$$

The conditions can be written as $n\lambda$ path difference for constructive interference and $(2n+1)\lambda/2$ path difference for destructive interference, where $n$ is an integer.

- **Constructive interference** takes place when the path difference is a whole number of wavelengths.

- **Destructive interference** takes place when the path difference is one and a half wavelengths, two and a half wavelengths, etc.

This is shown in the diagram.

the path difference here is half a wavelength, giving destructive interference

the path difference here is one wavelength, giving constructive interference

Clearly, for an interference pattern to be observed, the sources must not only be coherent but also of the same type.

These conditions for constructive and destructive interference only apply to two sources that are in phase. If the phase of one of the oscillators in the diagram is reversed so that the sources are in antiphase, then the conditions for constructive and destructive interference are also reversed. The effect is to shift the interference pattern so that lines of constructive interference (c) become destructive and *vice versa*. Any value of phase difference between the two sources gives an interference pattern that is stationary, provided that the phase difference is not changing. Two sources with a **fixed phase difference** are said to be **coherent**. Coherent sources must have the same wavelength and frequency.

## Interference of light

| | | | |
|---|---|---|---|
| AQA A | M4 | OCR A | AS |
| AQA B | AS | OCR B | AS |
| EDEXCEL A | M4 | NICCEA | AS |
| EDEXCEL B | AS | WJEC | AS |

Interference patterns are easy to set up with water waves and sound because two sources can be driven from the same oscillator, giving coherence. Because of the way in which light is emitted in random bursts of energy from a source, it is not possible to have two separate sources that are coherent. Instead, diffraction is used to obtain two identical copies of the light from a single source. This is done by illuminating two narrow slits with a lamp filament that is parallel to the slits so that the same wavefronts arrive at each slit. A suitable arrangement is shown in the diagram overleaf.

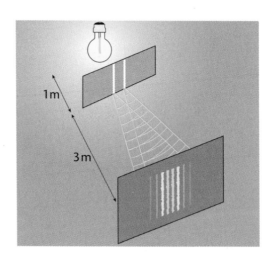

A colour filter can be used between the lamp and the slits to reduce the range of wavelengths interfering, but this also reduces the intensity of the interference pattern.

Bright and dark fringes on the screen are due to constructive and destructive interference between the two overlapping beams of light. The separation of the fringes depends on:

- the separation of the slits
- the wavelength of the light
- the distance between the slits and the screen.

This formula enables the wavelength of light to be measured from a simple experiment. The fringe spacing, $x$, should be obtained by measuring the separation of as many fringes as are visible and dividing by the number of fringes.

> **KEY POINT**
>
> The separation of the fringes is related to the other variables by the formula:
>
> $$\text{fringe spacing} = \frac{\text{wavelength} \times \text{distance from slits to screen}}{\text{slit separation}}$$
>
> $$x = \lambda D / a$$
>
> where $x$ is the distance between adjacent bright (or dark) fringes
> $\lambda$ is the wavelength of the light
> $D$ is the distance between the slits and the screen
> $a$ is the distance between the slits.

As the wavelength of light is very small, the slits need to be close together and separated from the screen by a large distance for the interference fringes to be seen.

## The diffraction grating

| | | | | | |
|---|---|---|---|---|---|
| AQA A | M4 | OCR B | AS | | |
| AQA B | AS | NICCEA | AS | | |
| EDEXCEL A | M4 | WJEC | M5 | | |

A diffraction grating consists of a large number of narrow slits that are ruled very close together on a glass slide. Diffraction takes place at each of the slits and results in an interference pattern that enables precise measurements of wavelength. The photograph shows the interference pattern obtained from a cadmium light source.

This photograph shows an emission spectrum, the wavelengths of light emitted by the source.

The bright lines in the photograph are due to constructive interference taking place between light diffracted at thousands of slits. The diagram on page 57 shows the path difference when the diffracted light is viewed at an angle $\theta$ to the normal.

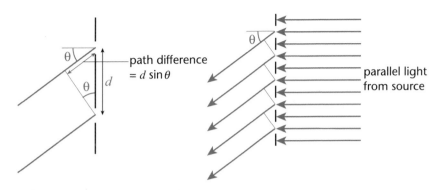

path difference
= $d \sin \theta$

parallel light from source

Bright lines are seen when waves from all the slits interfere constructively. For this to happen the path difference shown in the diagram has to be equal to a whole number of wavelengths.

> When $n = 0$, all wavelengths interfere constructively, so the bright central band consists of all wavelengths that are incident on the grating.

**KEY POINT**

The condition for constructive interference to occur when light is diffracted at a diffraction grating is:

$$d \sin \theta = n\lambda$$

where $d$ is the distance between adjacent slits
$\theta$ is the angle between the direction of the light and the normal
$n$ is an integer
$\lambda$ is the wavelength of the light.

When $n = 1$ the bright bands are known as the **first order spectrum**. The number of spectra that are visible depends on the separation of the slits.

Diffraction gratings are used to analyse the wavelengths of light emitted from gas discharge tubes and the light received from stars.

## Stationary (standing) waves

| | | | |
|---|---|---|---|
| AQA A | M4 | OCR A | AS |
| AQA B | AS | OCR B | AS |
| EDEXCEL A | M4 | NICCEA | AS |
| EDEXCEL B | AS | WJEC | AS |

A progressive wave is one whose profile moves through space, whereas a stationary wave has a static profile. The waves on the vibrating strings and in the vibrating air columns of musical instruments are stationary waves.

Stationary waves are caused by the superposition of two waves of the same wavelength travelling in opposite directions. They often arise when a wave is reflected at a boundary, but the waves can come from two separate coherent sources. The diagram shows the vibrations of a stationary wave on a string.

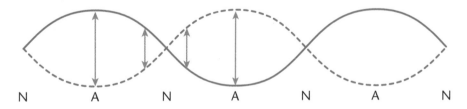

N    A    N    A    N    A    N

Stationary waves differ from progressive waves in a number of respects:

> All points on a progressive wave vibrate with the same amplitude, but each point is only in phase with points precisely $n$ wavelengths away.

- there is no flow of energy along a stationary wave, although stationary waves often radiate energy
- within each loop of a stationary wave, all particles vibrate in phase and exactly out of phase ($\pi$ radians or 180° phase difference) with the particles in adjacent loops
- the amplitude of vibration varies with position in the loop
- there are **nodes** (points marked N in the diagram), where the displacement is always zero and **antinodes** (marked A in the diagram) which vibrate with the maximum amplitude.

A convenient way of measuring the wavelength of a progressive wave is to use it to set up a stationary wave, for example by superposing the wave on its reflection from a barrier, and measuring the distance between a number of successive nodes or antinodes.

The wavelength of a stationary wave is twice the distance between two adjacent nodes or antinodes. The diagram below shows how a stationary wave (shown with a solid line) is formed from two progressive waves (shown with broken lines) travelling in opposite directions. In (a) the waves interfere constructively, so each point on the wave has its maximum displacement. In (b) each progressive wave has moved one quarter of a wavelength; the interference is now destructive, resulting in no displacement at all points on the wave.

formation of a stationary wave

## Progress check

1  Two waves arriving at a point interfere constructively. What are the possible values of the phase difference between the waves?

2  Two loudspeakers are placed 2.5 m apart and face in the same direction. They vibrate in phase, emitting sound with a wavelength of 0.3 m. An observer walks along a line parallel to the line joining the loudspeakers, 5.0 m in front of them.
   a  Calculate the distance between two adjacent loud sounds being heard.
   b  What difference would there be if the phase of one of the loudspeakers was reversed?

3  How is diffraction used to produce a two-source interference pattern using light?

# 2.5 Quantum phenomena

**After studying this section you should be able to:**

- describe the main features of the electromagnetic spectrum
- explain how the photoelectric effect gives evidence for the particulate nature of electromagnetic radiation
- explain line spectra and relate the spectral lines to energy level transitions

**LEARNING SUMMARY**

## The spectrum

| AQA B | AS | OCR B | AS |
|---|---|---|---|
| EDEXCEL A | M4 | NICCEA | AS |
| EDEXCEL B | AS | WJEC | AS |
| OCR A | AS | | |

The speed of all electromagnetic waves in a vacuum is $2.999 \times 10^8$ m s$^{-1}$. The value $3.00 \times 10^8$ m s$^{-1}$ is normally used for both a vacuum and air.

The different changes of speed when light consisting of a range of wavelengths is refracted is responsible for dispersion, the splitting of light into a spectrum.

Microwaves are given here as a separate part of the spectrum, although they can be considered as short-wavelength radio waves.

The **electromagnetic spectrum** consists of a whole family of waves, with some similarities and some differences in their behaviour.

Similar properties include:

- they consist of electric and magnetic fields oscillating at right angles to each other
- they travel at the same speed in a vacuum
- they are transverse waves, so they can be polarised
- they show the same pattern of behaviour in reflection, refraction, interference and diffraction.

The differences include:

- the shorter wavelength waves undergo a greater change of speed when being refracted
- wave-like properties such as diffraction and interference are more readily observable in the behaviour of the longer wavelength waves.

The diagram shows the range of wavelengths and frequencies of the waves that make up the spectrum.

| frequency / Hz | $10^{20}$ | | $10^{17}$ | | $10^{14}$ | | $10^{11}$ | | $10^8$ | | $10^5$ |
|---|---|---|---|---|---|---|---|---|---|---|---|

gamma rays     ultraviolet   infra-red      radio waves

X-rays      light     microwaves

| wavelength / m | $10^{-12}$ | | $10^{-9}$ | | $10^{-6}$ | | $10^{-3}$ | | 1 | | $10^3$ |
|---|---|---|---|---|---|---|---|---|---|---|---|

In general, the name given to an electromagnetic wave depends on its wavelength or frequency. Notable exceptions to this are X-rays and gamma rays, whose ranges of wavelength and frequency overlap. The difference here is in the origin of the waves, an X-ray is emitted when a high-speed electron is suddenly brought to rest and a gamma ray is emitted from an excited nucleus, usually following alpha or beta emission.

The table shows typical wavelengths, the origins of the waves and the uses of the main parts of the electromagnetic spectrum.

| name of radiation | typical wavelength/m | how produced | used for |
|---|---|---|---|
| gamma | $1 \times 10^{-12}$ | an excited nucleus releasing energy in radioactive decay | tracing the flow of fluids and treating cancer |
| X-rays | $1 \times 10^{-10}$ | high-speed electrons being stopped by a target | seeing inside the body |
| ultraviolet | $1 \times 10^{-8}$ | very hot objects and passing electricity through gases | suntan and security marking |

| | | | | |
|---|---|---|---|---|
| light | $5 \times 10^{-7}$ | hot objects and passing electricity through gases | vision and photography |
| infra-red | $1 \times 10^{-5}$ | warm and hot objects | heating and cooking |
| microwave | $1 \times 10^{-1}$ | microwave diode or oscillating electrons in an aerial | communications and cooking |
| radio | $1 \times 10^{2}$ | oscillating electrons in an aerial | communications |

All objects give out infra-red radiation, no matter what the temperature. The hotter the object, the more power is emitted and the greater the range of wavelengths.

## Waves or particles

| AQA A | AS | OCR A | AS |
| AQA B | M4 | OCR B | AS |
| EDEXCEL A | M4 | NICCEA A | AS |
| EDEXCEL B | AS | NICCEA B | AS |

The **photoelectric effect** provides evidence that electromagnetic waves have a particle-like behaviour which is more pronounced at the short-wavelength end of the spectrum. In the photoelectric effect electrons are emitted from a metal surface when it absorbs electromagnetic radiation.

The results of photoelectricity experiments show that:

The **threshold wavelength**, $\lambda_0$, is the wavelength of the waves that have the threshold frequency.
$$\lambda_0 = c \div f_0$$

- there is no emission of electrons below a certain frequency, called the **threshold frequency**, $f_0$, which is different for different metals

- above this frequency, electrons are emitted with a range of kinetic energies up to a maximum, $(\frac{1}{2}mv^2)_{max}$

The photoelectric effect can be demonstrated using a zinc plate connected to a gold leaf electroscope. An ultraviolet lamp discharges a negatively-charged plate but has no effect on a positively-charged plate.

- increasing the frequency of the radiation causes an increase in the maximum kinetic energy of the emitted electrons, but has no effect on the photoelectric current, i.e. the rate of emission of electrons

- increasing the intensity of the radiation has no effect if the frequency is below the threshold frequency; for frequencies above the threshold it causes an increase in the photoelectric current, so the electrons are emitted at a greater rate.

The wave model cannot explain this behaviour; if electromagnetic radiation is a continuous stream of energy then radiation of all frequencies should cause photoelectric emission. It should only be a matter of time for an electron to absorb enough energy to be able to escape from the attractive forces of the positive ions in the metal.

The word quantum refers to the smallest amount of a quantity that can exist. A quantum of electromagnetic radiation is the smallest amount of energy of that frequency.

The explanation for the photoelectric effect relies on the concept of a **photon**, a quantum or packet of energy. We picture electromagnetic radiation as short bursts of energy, the energy of a photon depending on its frequency.

A lamp emits random bursts of energy. Each burst is a photon, a quantum of radiation.

> **KEY POINT**
>
> The relationship between the energy, $E$ of a photon, or quantum of electromagnetic radiation, and its frequency, $f$, is:
> $$E = hf$$
> where $h$ is Planck's constant and has the value $6.63 \times 10^{-34}$ J s.

The energy of a photon can be measured in either joules or **electronvolts**. The electronvolt is a much smaller unit than the joule.

The conversion factor for changing energies in eV to energies in joules is $1.60 \times 10^{-19}$ J eV$^{-1}$

> **KEY POINT**
>
> One electron volt (1 eV) is the energy transfer when an electron moves through a potential difference of 1 volt.
> $$1 \text{ eV} = 1.60 \times 10^{-19} \text{ J}$$

Einstein's explanation of photoelectric emission is:

- An electron needs to absorb a minimum amount of energy to escape from a metal. This minimum amount of energy is a property of the metal and is called the work function, $\phi$.
- If the photons of the incident radiation have energy $hf$ less than $\phi$ then there is no emission of electrons.
- Emission becomes just possible when $hf = \phi$.
- For photons with energy greater than $\phi$, the electrons emitted have a range of kinetic energies, those with the maximum energy being the ones that needed the minimum energy to escape.
- Increasing the intensity of the radiation increases the number of photons incident each second. This causes a greater rate of emission of electrons, but does not affect their maximum kinetic energy.

The work function is the **minimum** energy needed to liberate an electron from a metal. Some electrons need more than this amount of energy.

Radiation below the threshold frequency, $f_0$, no matter how intense, does not cause any emission of electrons.

> **KEY POINT**
>
> Einstein's photoelectric equation relates the maximum kinetic energy of the emitted electrons to the work function and the energy of each photon:
>
> $$hf = \phi + (\tfrac{1}{2}mv^2)_{max}$$

At the threshold frequency, the minimum frequency that can cause emission from a given metal, $(\tfrac{1}{2}mv^2)_{max}$ is zero and so the equation becomes $hf_0 = \phi$.

## Stopping the emission

The work function of a metal is the minimum photon energy that causes photoelectric emission. This minimum energy can be increased by applying a positive potential, $V$, to the metal. An electron now needs an additional amount of energy $eV$ (in joules, or numerically equal to $V$ in eV) to escape.

A stopping potential is positive compared to earth.

> **KEY POINT**
>
> The **stopping potential** $V_s$, is the potential applied to a metal that just stops photoelectric emission. At the stopping potential:
>
> $$hf = \phi + eV_s$$

# *Particles or waves*

| AQA A | AS | OCR A | AS |
| AQA B | M4 | OCR B | AS |
| EDEXCEL A | M4 | NICCEA | M5 |
| EDEXCEL B | M4 | WJEC | M5 |

If waves can show particle-like behaviour in photoelectric emission, can particles also behave as waves? Snooker balls bounce off cushions in the same way that light bounces off a mirror, so reflection is not a test for wave-like or particle-like behaviour. Diffraction and interference are properties unique to waves, so particles can be said to have a wave-like behaviour if they show these properties. All particles have an associated wavelength called the de Broglie wavelength:

Try calculating the de Broglie (pronounced de Broy) wavelength of a moving snooker ball. Is it possible for the ball to show wave-like behaviour?

> **KEY POINT**
>
> The wavelength, $\lambda$, of a particle is related to its momentum, $p$, by the de Broglie equation:
>
> $$\lambda = h/p = h/mv$$
>
> where $h$ is the Planck constant.

The de Broglie wavelength of such an electron is of the order of $1 \times 10^{-10}$ m.

An electron that has been accelerated through a potential difference of a few hundred volts has a wavelength similar to that of X-rays and gamma rays.

This wavelength is also similar to the spacing of the atoms in crystalline materials, so these materials provide suitable sized 'gaps' to cause diffraction.

Diffraction patterns formed by a beam of electrons after passing through thin foil or graphite show a set of 'bright' and 'dark' rings on photographic film, similar to those formed by X-ray diffraction.

> You may have seen this pattern formed on a fluorescent screen in a vacuum tube.

Electrons can also be made to interfere when two coherent beams overlap. They produce an interference pattern similar to that of light, but on a much smaller scale.

Other 'particles' such as protons and neutrons also show wave-like behaviour.

*A diffraction pattern formed by passing a beam of electrons through graphite*

> Particles and waves are the models that we use to describe and explain physical phenomena. It is not surprising that the real world does not fit neatly into our models.

There are two separate models of how matter behaves. The particle model explains such phenomena as ionisation and photoelectricity, while the wave model explains interference and diffraction. It is not appropriate to classify matter as 'waves' or 'particles' as photons and electrons can fit either model, depending on the circumstances.

## Line spectra

| | | | |
|---|---|---|---|
| AQA A | M5 | OCR A | M5.1 |
| AQA B | M4 | OCR B | M5 |
| EDEXCEL A | M4 | NICCEA | AS |
| EDEXCEL B | AS | WJEC | AS |

> Gamma rays are the exception because they are a result of the nucleus losing energy.

With the exception of gamma rays, the emission of electromagnetic radiation is associated with electrons losing energy. A hot solid can radiate the whole range of wavelengths through the infra-red and part of the visible spectrum, the extent of the range depending on its temperature.

When an electric current is passed through an ionised gas, only a small number of wavelengths are emitted. The wavelengths are characteristic of the gas used and are called a line spectrum.

The diagram below shows some of the lines present in the hydrogen spectrum and their wavelengths.

> Not all of these lines lie in the visible spectrum, which extends from 0.4 μm to 0.65 μm.

wavelength/μm

The existence of line spectra provides evidence that the electrons in orbit around a nucleus can only have certain values of energy, the values being characteristic of an atom. Energy can only be emitted or absorbed in amounts that correspond to the differences between these allowed values.

An energy level diagram shows the amounts of energy that an electron can have. The diagram opposite is an energy level diagram for hydrogen.

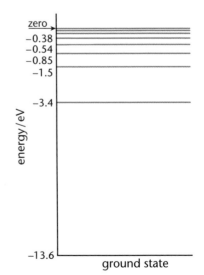

**Energy levels in a hydrogen atom**

Note that on an energy level diagram:

- the energies are measured relative to a zero that represents the energy of an electron at rest outside the atom, i.e. one that is just 'free'.
- an orbiting electron has less energy than a free electron, so it has negative energy relative to the zero
- an electron with the minimum possible energy is in the ground state; higher energy levels are called excited states.

Energy is emitted in the form of a photon when an electron moves from an excited state to a lower energy level. The energy of the photon is equal to the difference in the values of the energy levels.

> **KEY POINT**
>
> When an electron moves from an energy level $E_1$ to a lower energy level $E_2$, the energy of the photon emitted is given by
> $$hf = E_1 - E_2$$

For example, an electron moving from an energy level with –0.38 eV of energy to one with –0.85 eV loses energy equal to $(-0.38 - -0.85)$ eV = 0.47 eV. This corresponds to emitting a photon of frequency $1.13 \times 10^{14}$ Hz, which lies in the infra-red part of the electromagnetic spectrum.

Electrons can gain energy by absorbing photons. As with emission, the only photons that can be absorbed are those that correspond to allowed movements, or transitions, of electrons. An absorption spectrum is produced by shining white light through a sample of a gaseous element. The spectrum that emerges is the full spectrum with the element's emission spectrum 'missing' or of low intensity. This is due to the electrons absorbing photons of just the right energy to allow them to move to a more excited state.

> Movement of an electron to the ground state results in the emission of a photon with an energy in excess of 10 eV. Photons with this amount of energy give spectral lines in the ultraviolet region of the spectrum.

> The lines of the emission spectrum often appear black in the absorption spectrum. The photons of these energies are absorbed as the electrons move to more excited states, and then released when the electrons lose energy. The emitted photons are radiated in all directions, so very little energy is detected in any one direction.

## Electrons in atoms

The simple model of a hydrogen atom pictures a proton with a single electron in orbit. Since the electron is continually accelerating it ought to be radiating energy and spiralling in towards the nucleus. The fact that it does not do this is because of its wave-like behaviour. Each allowed orbit of the electron corresponds to a standing wave, with all the energy being contained within the wave and none being radiated.

## Progress check

1  The work function of potassium is 2.84 eV. Calculate:
   a  the minimum frequency of radiation that causes photoelectric emission.
   b  the maximum energy of the emitted electrons when ultraviolet radiation of frequency $8.40 \times 10^{14}$ Hz is absorbed by potassium.

2  Calculate the de Broglie wavelength of a neutron, mass = $1.67 \times 10^{-27}$ kg, travelling at a speed of 200 m s⁻¹.
   $h = 6.63 \times 10^{-34}$ J s

3  Calculate the energy of a photon that causes the excitation of an electron in a hydrogen atom from the ground state (–13.6 eV) to an excited state with an energy of –1.5 eV:
   a  in eV
   b  in J.

1  a  $6.85 \times 10^{14}$ Hz
   b  $1.03 \times 10^{-19}$ J
2  $1.99 \times 10^{-9}$ m
3  a  12.1 eV
   b  $1.94 \times 10^{-18}$ J

## Sample question and model answer

The diagram shows two similar loudspeakers that are driven from a common signal generator and vibrate in phase. The loudspeakers are 0.50 m apart and emit a note of frequency 1500 Hz.

An observer walks along a line 3.0 m from the loudspeakers, as shown in the diagram. He hears a loud sound at A, which decreases to a minimum at B and then increases in intensity as he walks towards C.

(a) Explain why the observer detects a loud sound at A. [2]

> A is equidistant from both loudspeakers, so as the waves are emitted in phase, they must arrive at A in phase.

The vibrations arrive at A in phase (1 mark) so they interfere constructively (1 mark).

(b) Explain why the loudness of the sound detected decreases as the observer walks from A to B and is a minimum at B. [3]

> The phase difference changes gradually from being totally in phase at A to being totally out of phase at B.

The phase difference increases from A to B (1 mark); the sound becomes quieter due to less constructive/more destructive interference (1 mark) and is a minimum at B where the vibrations are out of phase (1 mark).

(c) The speed of sound in air is 340 m s⁻¹.

Calculate the wavelength of the sound emitted by the loudspeakers. [3]

$\lambda = v/f$       1 mark
$= 340 \text{ m s}^{-1} \div 1500 \text{ Hz}$       1 mark
$= 0.227 \text{ m}$       1 mark

(d) Calculate the distance between A and C. [3]

> This relationship is valid for all examples of two–source interference, provided that the separation of the sources is small compared to the distance between the sources and the observer.

separation of maxima, $x = \lambda D/a$       1 mark
$= 0.227 \text{ m} \times 3.0 \text{ m} \div 0.50 \text{ m}$       1 mark
$= 1.36 \text{ m}$       1 mark

(e) The connections to one of the loudspeakers are reversed. What difference does the observer notice as he walks from A to C? Explain why he notices these differences. [2]

The points of maximum intensity become minima and vice versa (1 mark) as the loudspeakers are now vibrating out of phase (1 mark).

## Practice examination questions

Throughout this section use the following values of the electronic charge and Planck's constant:

$e = -1.60 \times 10^{-19}$ C

$h = 6.63 \times 10^{-34}$ J s

### 1

(a) Explain the meaning of the expression *natural frequency of vibration*. [2]

(b) In what circumstances does an object resonate? [2]

(c) The speed of sound in air is 340 m s$^{-1}$.

Sound waves with a wavelength of 0.15 m cause a glass tumbler to vibrate at its natural frequency.

Calculate the natural frequency of vibration of the tumbler. [3]

### 2

A mass is suspended on a spring. After it is given a small displacement its equation of motion is:

$$a = -kx/m$$

(a) State the meaning of the symbols used in this equation. [4]

(b) The diagram shows how the displacement of the mass changes during one cycle of oscillation.

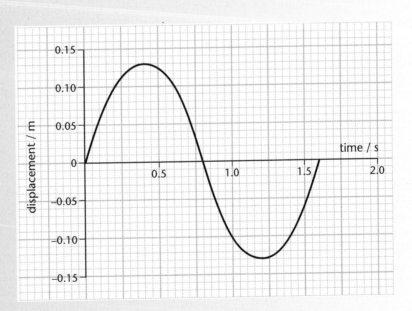

Calculate:

   (i) the amplitude of the oscillation [1]

   (ii) the frequency of the oscillation. [2]

(c) The mass on the spring is 0.50 kg. Calculate the spring constant. [3]

(d) (i) Calculate the maximum kinetic energy of the mass. [3]

   (ii) At what point on its motion does the mass have its maximum kinetic energy? [1]

**3**

The diagram represents some of the energy levels in a hydrogen atom.

(a) (i)  What is meant by the **ground state**? [1]

(ii) Explain why an electron in the ground state has a negative amount of energy. [2]

(b) An electron undergoes a transition from an energy level of –0.85 eV to the ground state.

(i)  Calculate the energy, in J, lost by the electron. [2]

(ii) Calculate the frequency of the photon that is emitted. [2]

(iii) In which part of the electromagnetic spectrum is the radiation emitted as a result of this transition? [1]

(iv) Give one danger and one use of radiation in this part of the spectrum. [2]

**4**

In a demonstration of two-source interference, a source of electromagnetic waves of wavelength 2.8 centimetres is directed at a metal barrier. There are two gaps in the barrier. The arrangement is shown in the diagram.

(a) (i)  Suggest why a single wave source is directed at two gaps in the barrier. [2]

(ii) Explain why 3.0 cm is a suitable size for the gaps. [2]

(b) Where is the strongest signal detected?
Explain why this is the strongest signal. [3]

(c) The detector is moved along the line AB.
What is the distance between the points where the strongest and weakest signals are detected? [4]

(d) The wave source is replaced by a lamp that emits light of one wavelength only.
The detector is replaced by a white screen placed along the line AB.
Explain why no interference pattern is observed. [2]

**5**

Einstein's equation for photoelectric emission is:

$$hf = \Phi + (\tfrac{1}{2}mv^2)_{max}$$

(a) (i) State the meaning of each term. [3]

(ii) What is the significance of the subscript 'max'? [1]

(b) Electromagnetic radiation of frequency $2.02 \times 10^{15}$ Hz is directed at a metal surface. The maximum kinetic energy of the electrons emitted is $4.05 \times 10^{-19}$ J. The intensity of the radiation is 4.5 $\mu$W m$^{-2}$.

Calculate:

(i) the energy of each photon [2]

(ii) the number of photons reaching the surface each second if the area is $2.5 \times 10^{-6}$ m$^2$ [2]

(iii) The work function of the metal. Give your answer in eV. [3]

**6**

The graph shows how the displacement of a pendulum bob changes over one cycle of its motion.

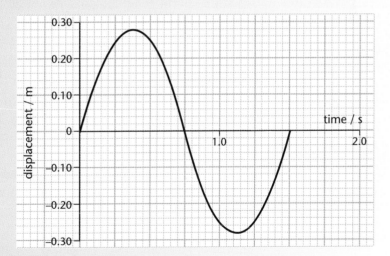

The equation that represents the variation of displacement with time is
$x = A \sin 2\pi ft$.

(a) Write down the values of $A$ and $f$. [2]

(b) Calculate the maximum velocity of the pendulum bob. [1]

(c) Calculate the velocity of the pendulum bob when its displacement is 0.20 m. [2]

## Chapter 3

# Fields

**The following topics are covered in this chapter:**

- Gravitational fields
- Electric fields
- Capacitors

- Magnetic fields
- Circular orbits
- Electromagnetic induction

## 3.1 Gravitational fields

**After studying this section you should be able to:**

- understand the concept of a field
- recall and use the relationship that describes the gravitational force between two masses
- describe the Earth's gravitational field and explain how the field strength varies with distance from the centre of the Earth

**LEARNING SUMMARY**

### Fields

| | | | |
|---|---|---|---|
| AQA A | M4 | OCR A | M4 |
| AQA B | M5 | OCR B | M4 |
| EDEXCEL A | M5 | NICCEA | M4 |
| EDEXCEL B | M5 | WJEC | M5 |

Radiant energy such as light has mass and so is affected by gravitational fields.

A field is a region of space where forces are exerted on objects with certain properties. In this and subsequent sections three types of field are considered:

- **gravitational fields** affect anything that has mass
- **electric fields** affect anything that has charge
- **magnetic fields** affect permanent magnets and electric currents.

These three types of field have many similar properties and some important differences. There are key definitions and concepts that are common to all three types of field.

### Gravitational fields

| | | | |
|---|---|---|---|
| AQA A | M4 | OCR A | M4 |
| AQA B | M5 | OCR B | M4 |
| EDEXCEL A | M5 | NICCEA | M4 |
| EDEXCEL B | M5 | WJEC | M5 |

The mass of the Earth is about $6 \times 10^{24}$ kg.

Newton realised that all objects with mass attract each other. This seems surprising, since any two objects placed close together on a desktop do not immediately move together. The attractive force between them is tiny, and very much smaller than the frictional forces that oppose their motion.

Gravitational attractive forces between two objects only affect their motion when at least one of the objects is very massive. This explains why we are aware of the force that attracts us and other objects towards the Earth – the Earth is very massive.

The diagram represents the Earth's gravitational field. The lines show the direction of the force that acts on a mass that is within the field.

This diagram shows that:

- gravitational forces are always attractive – the Earth cannot repel any objects
- the Earth's gravitational pull acts towards the centre of the Earth
- the Earth's gravitational field is radial; the field lines become less concentrated with increasing distance from the Earth.

The force exerted on an object in a gravitational field depends on its position. The less concentrated the field lines, the smaller the force. If the gravitational field strength at any point is known, then the size of the force can be calculated.

> **KEY POINT**
>
> The gravitational field strength (*g*) at any point in a gravitational field is *the force per unit mass* at that point:
> $$g = F/m$$
> Close to the Earth's surface, *g* has the value of $9.81\,\mathrm{N\,kg^{-1}}$, though the value of $10\,\mathrm{N\,kg^{-1}}$ is often used in calculations.

## Universal gravitation

| | | | |
|---|---|---|---|
| AQA A | M4 | OCR A | M4 |
| AQA B | M5 | OCR B | M4 |
| EDEXCEL A | M5 | NICCEA | M4 |
| EDEXCEL B | M5 | WJEC | M5 |

In studying gravitation, Newton concluded that the gravitational attractive force that exists between any two masses:

- is proportional to each of the masses
- is inversely proportional to the square of their distances apart.

> **KEY POINT**
>
> **Newton's law of gravitation** describes the gravitational force between two point masses. It can be written as:
> $$F = \frac{Gm_1 m_2}{r^2}$$
> where *G* is the universal gravitational constant and has the value $6.7 \times 10^{-11}\,\mathrm{N\,m^2\,kg^{-2}}$
> where $m_1$ and $m_2$ are the values of the masses and *r* is the separation of the centres of mass.

Although the Earth is a large object, on the scale of the Universe it can be considered to be a point mass. The gravitational field strength at its centre is zero, since attractive forces pull equally in all directions. Beyond the surface of the Earth, the gravitational force on an object decreases with increasing distance. When the distance is measured from the centre of the Earth, the size of the force follows an inverse square law; doubling the distance from the centre of the Earth decreases the force to one quarter of the original value. The variation of force with distance from the centre of the Earth is shown in the diagram.

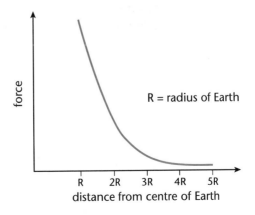

## g and G

| AQA A | M4 | OCR A | M4 |
| AQA B | M5 | OCR B | M4 |
| EDEXCEL A | M5 | NICCEA | M4 |
| EDEXCEL B | M5 | WJEC | M5 |

A negative sign is sometimes used on the right-hand side of this equation, following the convention that attractive forces are given negative values and repulsive forces positive values.

Newton's law of gravitation can be used to work out the value of the force between any two objects. It can also be used to calculate the strength of the gravitational field due to a spherical mass such as the Earth or the Sun.

> **KEY POINT**
>
> A small object, mass $m$, placed within the gravitational field of the Earth, mass $M$, experiences a force, $F$, given by
>
> $$F = \frac{GMm}{r^2}$$
>
> where $r$ is the separation of the centres of mass of the object and the Earth.

It follows from the definition of gravitational field strength as the *force per unit mass* that the field strength at that point, $g$, is related to the mass of the Earth by the expression:

> **KEY POINT**
>
> $$g = \frac{F}{m} = \frac{GM}{r^2}$$

Gravitational field strength is a property of any point in a field. It can be given a value whether or not a mass is placed at that point. Like gravitational force, beyond the surface of the Earth the value of $g$ follows an inverse square law. A graph of $g$ against distance from the centre of the Earth has the same shape as that shown in the diagram on page 69.

The radius of the Earth is about $6.4 \times 10^6$ m, so you would have to go much higher than aircraft-flying height for $g$ to change by 1%.

Because the inverse square law applies to values of $g$ when the distance is measured from the centre of the Earth, there is little change in its value close to the Earth's surface. Even when flying in an aircraft at a height of 10 000 m, the change in distance from the centre of the Earth is minimal, so there is no noticeable change in $g$.

The same symbol, $g$, is used to represent:

- gravitational field strength
- free-fall acceleration.

These are not two separate quantities, but two different names for the same quantity. Gravitational field strength, $g$, is defined as the force per unit mass, $g = F/m$. From Newton's second law and the definition of the newton, free-fall acceleration, $g$, is also equal to the gravitational force per unit mass. The units of gravitational field strength, N kg$^{-1}$, and free-fall acceleration, m s$^{-2}$, are also equivalent.

## Potential and potential energy

| AQA A | M4 | OCR A B | M4 |
| AQA B | M5 | WJEC | M5 |
| EDEXCEL A | M5 | | |

On an absolute scale of measurement, zero must be the smallest possible value and '20 units' must be twice as much as '10 units'.

When an object changes its position relative to the Earth, there is a change in potential energy given by $\Delta E_p = mg\Delta h$. It is not possible to place an absolute value on the potential energy of any object when $h$ is measured relative to the surface of the Earth. Two similar objects placed at the top and bottom of a hill have different values of potential energy, but relative to the ground the potential energy is zero for both objects, see diagram on the next page.

Absolute values of potential energy are measured relative to infinity. In this context, infinity means 'at a distance from the Earth where its gravitational field strength is so small as to be negligible'.

The car at the top of the hill has more potential energy than the one at the bottom, but relative to ground level they both have zero.

Work has to be done to move an object from within the Earth's gravitational field to infinity.

Using this reference point:

- all objects at infinity have the same amount of potential energy, zero
- any object closer than infinity has a negative amount of potential energy, since it would need to acquire energy in order to reach infinity and have zero energy.

Just as gravitational field strength is used to place a value on the gravitational force that would be experienced by a mass at any point in a gravitational field, the concept of gravitational potential is used to give a value for the potential energy.

> **KEY POINT**
>
> The gravitational potential at a point in a gravitational field is the potential energy per unit mass placed at that point, measured relative to infinity.

So if the potential at any point in a field is known, the potential energy of a mass placed at that point can be calculated by multiplying the potential by the mass.

## Calculating potential and potential energy

| | | | |
|---|---|---|---|
| AQA A | M4 | OCRA B | M4 |
| AQA B | M5 | WJEC | M5 |
| EDEXCEL A | M5 | | |

When an object is within the gravitational field of a planet, it has a negative amount of potential energy measured relative to infinity. The amount of potential energy depends on:

- the mass of the object
- the mass of the planet
- the distance between the centres of mass of the object and the planet.

The centre of mass of a planet is normally taken to be at its centre.

> **KEY POINT**
>
> The gravitational potential energy measured relative to infinity of a mass, $m$, placed within the gravitational field of a spherical mass $M$ can be calculated using:
>
> $$E_p = -\frac{GMm}{r}$$
>
> where $r$ is the distance between the centres of mass and $G$ is the universal gravitational constant.
>
> Gravitational potential energy is measured in joules (J).

Since gravitational potential is the gravitational potential energy per unit mass placed at a point in a field, it follows that:

> **KEY POINT**
>
> Gravitational potential, $V$, is given by the relationship:
>
> $$V = \frac{E_p}{m} = -\frac{GM}{r}$$
>
> Gravitational potential is measured in J kg$^{-1}$.

The relationship between potential and potential energy is similar to that between gravitational force and gravitational field strength:

- potential and field strength are properties of a point in a field
- potential energy and force are the corresponding properties of a mass placed within a field.

## Equipotential surfaces

| AQA B | M5 | OCR B | M4 |
|-------|-----|-------|-----|
| EDEXCEL A | M5 | | |

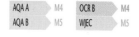 For a satellite in an elliptical orbit, there is an interchange between kinetic and potential energy as it travels around the Earth.

The potential energy of a satellite in a circular orbit around the Earth remains constant provided that its distance from the centre of the Earth does not change. To move to a higher or lower orbit the satellite must gain or lose energy. The satellite travels along an equipotential surface, the spherical shape consisting of all points at the same potential.

The diagram shows the spacing of equipotential surfaces around the Earth. The surfaces are drawn at equal differences of potential.

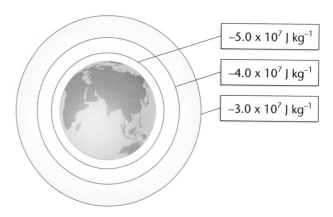

$-5.0 \times 10^7$ J kg$^{-1}$

$-4.0 \times 10^7$ J kg$^{-1}$

$-3.0 \times 10^7$ J kg$^{-1}$

The diagram shows that:

- equipotential surfaces around a spherical mass are also spherical
- the spacing between the equipotential surfaces increases with increasing distance from the centre of the Earth.

To move to a lower orbit, a rocket can lose energy by firing the rocket engines 'backwards' so that the exhaust gases are expelled in the direction of motion.

For a satellite to move from an orbit where the potential is $-4.0 \times 10^7$ J kg$^{-1}$ to one where the potential is $-3.0 \times 10^7$ J kg$^{-1}$, it needs to gain $1.0 \times 10^7$ J of gravitational potential energy for each kilogram of satellite. It does this by firing the rocket engines, transferring energy from its fuel supply.

## Calculating potential

| AQA A | M4 | OCR B | M4 |
|-------|-----|-------|-----|
| AQA B | M5 | WJEC | M5 |

The value of the gravitational potential at a point in a gravitational field depends on:

- the mass of the object causing the field
- the distance from the centre of mass of this object.

> The gravitational potential, $V$, due to a spherical mass, $M$, at a distance $r$ from its centre of mass is given by:
>
> $$V = -\frac{GM}{r}$$

**KEY POINT**

The diagram above shows that the rate of change of potential with distance, the potential gradient, decreases with increasing distance from the Earth.

As the potential gradient decreases, so does the gravitational field strength.

The concept of potential gradient is similar to that of the gradient of a hill or slope. The steeper the slope, the greater the acceleration of an object free to move down it.

> **KEY POINT**
>
> The relationship between gravitational field strength and potential gradient is:
>
> $$g = - \text{(potential gradient)} \quad \text{or} \quad g = -\frac{\Delta V}{\Delta x}$$
>
> where $\Delta V$ is the change in potential over a small distance $\Delta x$.

## Progress check

1  The radius of the Earth is $6.4 \times 10^6$ m and the gravitational field strength at its surface is 10 N kg⁻¹.
   At what height above the surface of the Earth is the gravitational field strength equal to 2.5 N kg⁻¹?

2  Two 2.5 kg masses are placed with their centres 10 cm apart.
   Calculate the size of the gravitational attractive force between them.
   $G = 6.7 \times 10^{-11}$ N m² kg⁻².

3  The mass of the Moon is $7.4 \times 10^{22}$ kg and its radius is $1.7 \times 10^6$ m.
   Using the value of $G$ from Q2, calculate the value of free-fall acceleration at the Moon's surface.

3  1.7 m s⁻²
2  $4.2 \times 10^{-8}$ N
1  $6.4 \times 10^6$ m

# 3.2 Electric fields

*After studying this section you should be able to:*

- *describe the electric field due to a point charge and between two charged parallel plates*
- *calculate the force on a charge in an electric field*
- *compare gravitational and electric fields*

LEARNING SUMMARY

## Charging up

EDEXCEL A | M5    OCR A | M4

Electrostatic phenomena such as attraction and repulsion are due to the forces between two charged objects. Similar charges repel and opposite charges attract.

Transfer of charge happens whenever two objects in physical contact move relative to each other. It is caused by electrons leaving one surface and joining another. This results in objects having one of two types of charge:

- an object that gains electrons has a negative charge
- an object that loses electrons has a positive charge.

In many cases any imbalance of charge on an object is removed by movement of electrons to and from the ground, but if at least one of the objects is a good insulator charge can build up.

A balloon is easily charged by rubbing but it is not possible to charge a hand-held metal rod since it is immediately discharged by electrons passing through the person holding it.

The quantum, or smallest unit, of charge is that carried by an electron. All quantities of charge must be a whole number multiple of $e = -1.6 \times 10^{-19}$ C.

## The electric field

AQA A | M4    OCR A | M4
AQA B | M5    OCR B | M5
EDEXCEL A | M5    NICCEA | M4
EDEXCEL B | M4    WJEC | M5

Unlike gravitational fields, which can only exert attractive forces, electric fields can attract or repel objects that are charged. When drawing field lines that represent the forces due to a charged object, the arrows show the direction of the force on a positive charge.

The diagram shows the electric fields due to a point charge and between a pair of oppositely charged parallel plates.

A pair of parallel plates become oppositely charged when connected to the positive and negative terminals of a d.c. supply.

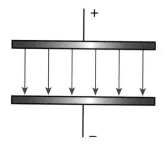

The field due to the point charge is radial, it decreases in strength with increasing distance from the charge. That between the parallel plates is uniform, it maintains a constant strength at all points between the plates.

Coulomb investigated the size of the force between two point charges and concluded that the force is:

- proportional to each of the charges
- inversely proportional to the square of their distances apart.

> **Permittivity** is a measure of the extent to which the medium reinforces the electric field. Water has a high permittivity due to its molecules being polarised.

**KEY POINT**

Coulomb's law states that the force between two point charges is given by

$$F = \frac{kQ_1 Q_2}{r^2}$$

where $Q_1$ and $Q_2$ represent the values of the charges, $r$ is their distance apart and $k$ has the value, in air or a vacuum, of $\frac{1}{4\pi\varepsilon_0} = 9.0 \times 10^9$ N m² C⁻².

The constant $\varepsilon_0$ is called the permittivity of free space.

## Field strength

| | | | |
|---|---|---|---|
| AQA A | M4 | OCR A | M4 |
| AQA B | M5 | OCR B | M5 |
| EDEXCEL A | M5 | NICCEA | M4 |
| EDEXCEL B | M4 | WJEC | M5 |

> The test charge has to be small enough to have no effect on the field.

Electric field strength:

- is defined *as the force per unit positive charge acting on a small charge placed within the field*
- is measured in N C⁻¹.

Coulomb's law can be used to express the field strength due to a point charge $Q$. Since the force between a charge $Q$ and a small charge $q$ placed with the field of $Q$ is given by $F = \frac{kQq}{r^2}$, it follows that:

**KEY POINT**

The electric field strength, $E$, due to a point charge $Q$ is given by the expression:

$$E = \frac{kQ}{r^2}$$

In a radial field, the field strength follows an inverse square law. This can be seen by the way in which the field lines spread out from a point charge. In a uniform field, like the one between two oppositely charged parallel plates, the field lines maintain a constant separation. The value of the electric field strength in a uniform field does not change.

> This gives an alternative unit for electric field strength, V m⁻¹, which is equivalent to the N C⁻¹.

**KEY POINT**

The electric field strength between two oppositely charged parallel plates is given by the expression:

$$E = \frac{V}{d}$$

where $V$ is the potential difference between the plates and $d$ is the separation of the plates.

## Potential

| | | | |
|---|---|---|---|
| AQA A | M4 | OCR B | M5 |
| AQA B | M5 | WJEC | M5 |
| EDEXCEL A | M5 | | |

The potential difference between the parallel plates shown in the diagram on page 74 represents the energy transfer per coulomb when charge moves between them. A charge $q$ moving between the plates would gain or lose energy $Vq$.

In a uniform field the potential changes by equal amounts for equal changes in distance.

The diagram here shows the variation in potential between two oppositely charged plates.

Potential could be measured relative to the upper plate, in which case the potentials would have negative values.

In this example:

- potential has been measured relative to the lower plate, which has been given the value 0
- the equipotential lines are parallel to the plates
- an equipotential surface, joining points all at the same potential, is simply a surface drawn parallel to the plates.

## Potential in a radial field

| AQA A | M4 | OCR B | M5 |
| AQA B | M5 | WJEC | M5 |

Unlike the potential in the gravitational field of a point mass, the electric potential in the field of a point charge is positive, since work has to be done to move a positive charge from infinity to any point in the field.

Like potential in a gravitational field, absolute potential in an electric field is measured relative to infinity. Infinity does not need to be a great distance, since it refers to a point where the field strength is negligibly small.

> **KEY POINT**
>
> The electric potential at a point, $V$, is the work done per unit positive charge in bringing a small charge from infinity to that point.
>
> In a radial field, $$V = \frac{1}{4\pi\varepsilon_0}\frac{Q}{r}$$

## Electron beams

| AQA A | M4 | EDEXCEL B | M4 |
| AQA B | M5 | NICCEA | M5 |
| EDEXCEL A | M5 | | |

Computer monitors, televisions and cathode ray oscilloscopes all use beams of electrons to produce a picture. The beam is produced by an electron gun, a device that accelerates and focuses the electrons given off by a hot wire. The diagram below shows an electron gun.

**An electron gun**

The filament is the cathode, or negative terminal. The anode is usually cylindrical and is connected to the positive terminal of the high voltage supply.

The low voltage supply heats the filament, causing it to emit electrons. The electrons are accelerated by the high voltage supply, gaining kinetic energy as they move towards the positively charged anode. Since the potential difference between the cathode and the anode represents the energy transfer per coulomb of charge, the kinetic energy of the accelerated electrons can be calculated.

> **KEY POINT**
>
> When an electron is accelerated through a potential difference $V$ it gains kinetic energy
> $$E_k = eV$$
> $$\tfrac{1}{2}m_e v^2 = eV$$
> where $m_e$ is the mass of an electron and $e$ is the electronic charge.

## Deflecting a beam of electrons

| AQA A | M4 | EDEXCEL B | M4 |
| AQA B | M5 | NICCEA | M5 |
| EDEXCEL A | M5 | | |

In a cathode ray oscilloscope, vertical deflection of the electron beam is achieved by passing the beam between a pair of oppositely charged parallel plates. The effect of this can be seen by studying the path of a beam of electrons in a deflection tube.

When passing between the plates, the electrons have a constant speed in the direction parallel to the plates. Perpendicular to the plates, the force on each electron is equal to *Ee*, so they accelerate in this direction. The consequent motion of the electrons is similar to that of a particle projected horizontally on the Earth; in each case the result is a parabolic path, shown in the diagram.

The path of the electrons before entering the field and after leaving the field is a straight line as there is no resultant force acting on them.

When moving between the plates:

- the electrons travel equal distances in successive equal time intervals in the direction parallel to the plates
- the electrons travel increasing distances in successive equal time intervals in the direction perpendicular to the plates.

## Comparing electric and gravitational fields

There are similarities and differences between electric and gravitational fields:

- electric field strength is defined as force per unit charge, gravitational field strength is defined as force per unit mass
- electric potential and gravitational potential are defined in similar ways
- the electric field due to a point charge is similar to the gravitational field of a point mass
- electric fields can attract or repel charged objects, gravitational fields can only attract masses.

## Progress check

1 Calculate the electric field strength due to a point charge of 3.0 μC at a distance of 0.10 m from the charge.
$k = 9.0 \times 10^9$ N m$^2$ C$^{-2}$

2 The potential difference between two parallel plates is 300 V. They are placed 0.15 m apart.
a Calculate the value of the electric field strength between the plates.
b Calculate the size of the force on an electron, charge $-1.6 \times 10^{-19}$ C, placed midway between the plates.
c Explain how the size of the force on the electron varies as it moves from the negative plate to the positive plate.

c It stays the same as the field strength does not vary.
b 3.2 × 10⁻¹⁶ N
2 a 2.0 × 10³ N C⁻¹
1 2.7 × 10⁶ N C⁻¹

# 3.3 Capacitors

**After studying this section you should be able to:**

- describe the action of a capacitor and calculate the charge stored
- relate the energy stored in a capacitor to a graph of charge against voltage
- explain the significance of the time constant of a circuit that contains a capacitor and a resistor

**LEARNING SUMMARY**

## The action of a capacitor

| | | | |
|---|---|---|---|
| AQA A | M4 | OCR A | M4 |
| AQA B | M4 | OCR B | M4 |
| EDEXCEL A | M5 | NICCEA | M5 |
| EDEXCEL B | M4 | WJEC | M4 |

Capacitors store charge and energy. They have many applications, including smoothing varying direct currents, electronic timing circuits and powering the memory to store information in calculators when they are switched off.

A capacitor consists of two parallel conducting plates separated by an insulator. When it is connected to a voltage supply charge flows onto the capacitor plates until the potential difference across them is the same as that of the supply. The charge flow and the final charge on each plate is shown in the diagram.

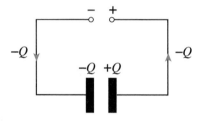

> When a capacitor is charging, charge flows in all parts of the circuit except between the plates.

As the capacitor charges:

- charge $-Q$ flows onto the plate connected to the negative terminal of the supply
- charge $-Q$ flows off the plate connected to the positive terminal of the supply, leaving it with charge $+Q$
- the capacitor plates always have the same quantity of charge, but of the opposite sign
- no charge flows between the plates of the capacitor.

## Capacitance

| | | | |
|---|---|---|---|
| AQA A | M4 | OCR A | M4 |
| AQA B | M4 | OCR B | M4 |
| EDEXCEL A | M5 | NICCEA | M5 |
| EDEXCEL B | M4 | WJEC | M4 |

The capacitor shown in the diagram above is said to store charge $Q$, meaning that this is the amount of charge on each plate.

When a capacitor is charged, the amount of charge stored depends on:

- the voltage across the capacitor
- its capacitance: i.e. the greater the capacitance, the more charge is stored at a given voltage.

> As the capacitor plates have equal amounts of charge of the opposite sign, the total charge is actually zero.
>
> However, because the charges are separated they have energy and can do work when they are brought together.

> The **capacitance** of a capacitor, $C$, is defined as:
> $$C = \frac{Q}{V}$$
> Where $Q$ is the charge stored when the voltage across the capacitor is $V$. Capacitance is measured in farads (F).
> 1 farad is the capacitance of a capacitor that stores 1 C of charge when the p.d. across it is 1 V.
>
> **KEY POINT**

One farad is a very large value of capacitance. Common values of capacitance are usually measured in picofarads (1 pF = $1.0 \times 10^{-12}$ F) and microfarads (1 µF = $1.0 \times 10^{-6}$ F).

## Combining capacitors

| AQA B | M4 | NICCEA | M5 |
|---|---|---|---|
| EDEXCEL A | M5 | WJEC | M4 |
| OCR A | M4 | | |

The effect of adding capacitors in series is to reduce the capacitance. When an additional capacitor is added, there is less p.d. across each one so less charge is stored.

Like resistors, capacitors can be connected in series or parallel to achieve different values of capacitance.

When capacitors in series are connected to a voltage supply:

- no matter what the value of its capacitance, each capacitor in the combination stores the same amount of charge, since any one plate can only lose or gain the charge gained or lost by the plate that it is connected to
- the total charge stored by a series combination is the charge on each of the two outer plates and is equal to the charge stored on each individual capacitor
- because the applied potential difference is shared by the capacitors, the total charge stored is less than the charge that would be stored by any one of the capacitors connected individually to the voltage supply.

The diagram shows the charge on the plates of three capacitors connected in series.

This results in the effective value of a series combination of capacitors being smaller than the lowest value capacitor in the combination.

A common error when using this relationship is to forget to carry out the final reciprocation, giving an answer which is equal to $\frac{1}{C}$ instead of C.

> **KEY POINT**
>
> The capacitance, C, of a number of capacitors connected in series is given by the expression:
> $$\frac{1}{C} = \frac{1}{C_1} + \frac{1}{C_2} + \frac{1}{C_3}$$

In contrast to this, the effect of connecting capacitors in parallel is to increase the capacitance so that the effective value of a number of capacitors in parallel is always greater than the largest value of the combination.

When capacitors are connected in parallel:

- all the capacitors are charged to the same potential difference
- each capacitor stores the same amount of charge as it would if connected on its own to the same voltage
- adding an additional capacitor increases the total charge stored.

The expressions for capacitors connected in series and parallel are similar to those for resistors, but the other way round.

> **KEY POINT**
>
> The capacitance, C, of a number of capacitors connected in parallel is given by the expression:
> $$C = C_1 + C_2 + C_3$$

## The energy stored in a capacitor

| AQA A | M4 | OCR A | M4 |
|---|---|---|---|
| AQA B | M4 | OCR B | M4 |
| EDEXCEL A | M5 | NICCEA | M5 |
| EDEXCEL B | M4 | WJEC | M4 |

Energy is needed from a power supply or other source to charge a capacitor. A charged capacitor can supply the energy needed to maintain the memory in a calculator or the current in a circuit when the supply voltage is too low.

The amount of energy stored in a capacitor depends on:

- the amount of charge on the capacitor plates
- the voltage required to place this charge on the capacitor plates, i.e. the capacitance of the capacitor.

The graph below shows how the voltage across the plates of a capacitor depends on the charge stored.

When a charge $\Delta Q$ is added to a capacitor at a potential difference $V$, the work done is $\Delta QV$. The total work done in charging a capacitor is $\Sigma\Delta QV$.

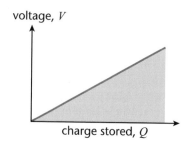

The shaded area between the graph line and the charge axis represents the energy stored in the capacitor.

> **KEY POINT**
>
> The energy, $E$, stored in a capacitor is given by the expression
> $$E = \tfrac{1}{2}\,QV = \tfrac{1}{2}\,CV^2$$
> where $Q$ is the charge stored on a capacitor of capacitance $C$ when the voltage across it is $V$.

## Charging and discharging a capacitor

| | | | |
|---|---|---|---|
| AQA A | M4 | OCR A | M4 |
| AQA B | M4 | OCR B | M4 |
| EDEXCEL A | M6 | NICCEA | M5 |
| EDEXCEL B | M4 | WJEC | M4 |

Having a resistor in the circuit means that extra work has to be done to charge the capacitor, as there is always an energy transfer to heat when charge flows through a resistor.

**Key points from AS**

- **The relationship between resistance, current and voltage.**
  *Revise AS section 2.1*

When a capacitor is charged by connecting it directly to a power supply, there is very little resistance in the circuit and the capacitor seems to charge instantaneously. This is because the process occurs over a very short time interval. Placing a resistor in the charging circuit slows the process down. The greater the values of resistance and capacitance, the longer it takes for the capacitor to charge.

The diagram below shows how the current changes with time when a capacitor is charging.

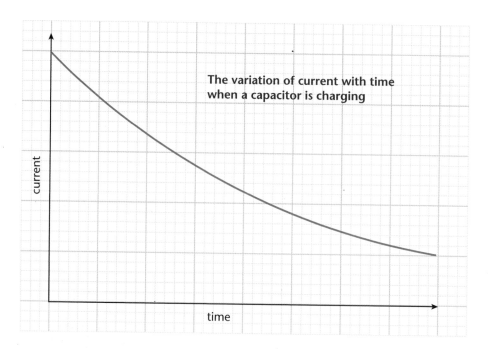

The variation of current with time when a capacitor is charging

This graph shows that:

- the charging current falls as the charge on the capacitor, and the voltage across the capacitor, rise
- the charging current decreases by the same proportion in equal time intervals.

The second bullet point shows that the change in the current follows the same pattern as the activity of a radioactive isotope. This is an example of an **exponential change**, the charging current decreases exponentially.

The graph shown above can be used to work out the amount of charge that flows onto the capacitor by estimating the area between the graph line and the time axis. Since *current = rate of flow of charge* it follows that:

> **KEY POINT**
>
> On a graph of current against time, the area between the graph line and the time axis represents the charge flow.

To calculate the charge flow:

- estimate the number of whole squares between the graph line and the time axis
- multiply this by the 'charge value' of each square, obtained by calculating $\Delta Q \times \Delta t$ for a single square.

> As the p.d. across the capacitor rises, that across the resistor falls, reducing the current.

> **Key points from AS**
>
> - **The relationship between current and charge flow.**
>   *Revise AS section 2.2*

## The time constant

| | | | |
|---|---|---|---|
| AQA A | M4 | OCR A | M4 |
| AQA B | M4 | OCR B | M4 |
| EDEXCEL A | M6 | NICCEA | M5 |
| EDEXCEL B | M4 | WJEC | M4 |

When a capacitor is charging or discharging, the amount of charge on the capacitor changes exponentially. The graphs in the diagram show how the charge on a capacitor changes with time when it is charging and discharging.

Graphs showing the change of voltage with time are the same shape.
Since $V = \dfrac{Q}{C}$, it follows that the only difference between a charge–time graph and a voltage–time graph is the label and scale on the *y*-axis.

> These graphs show the charge on the capacitor approaching a final value, zero in the case of the capacitor discharging, but never quite getting there.

**The charge on a capacitor during charging and discharging**

The rate at which the charge on a capacitor changes depends on the **time constant** of the charging or discharging circuit.

> **KEY POINT**
>
> The time constant, $\tau$, of a capacitor charge or discharge circuit is the product of the resistance and the capacitance:
>
> $$\tau = RC$$
>
> $\tau$ is measured in s.

The greater the values of *R* and *C* the longer the charge or discharge process takes. Knowledge of the values of *R* and *C* enables the amount of charge on a capacitor to be calculated at any time after the capacitor has started to charge or discharge. This is useful in timing circuits, where a switch is triggered once the charge, and therefore p.d., has reached a certain value.

e is a constant that has a value of 2.72 to 2 d.p. It is the base of the natural log and exponential functions.

The time constant $\tau$ represents:

- the time it takes for the charge on a capacitor to fall to $\frac{1}{e}$ of its initial value when a capacitor is discharging

- the time it takes for the charge on a capacitor to rise to $1-\frac{1}{e}$ of its final value when the capacitor is charging

The role of the time constant is similar to that of half-life in radioactive decay. When a capacitor is discharging, $\frac{1}{e^2}$ of the initial charge remains after time $2\tau$ and $\frac{1}{e^3}$ remains after $3\tau$.

The exponential function $e$ is used to calculate the charge remaining on a capacitor that is discharging.

Here $e$ is the exponential function, the inverse of natural log, ln. Take care not to confuse this with the EXP button on a calculator, which is used for entering powers of 10.

> **KEY POINT**
>
> The charge, $Q$, on a capacitor of capacitance $C$, remaining time $t$ after starting to discharge is given by the expression
> $$Q = Q_0 e^{-t/\tau}$$
> where $Q_0$ is the initial charge on the capacitor.

This expression shows that when $t$ is equal to $\tau$, i.e. after one time constant has elapsed, the charge remaining is equal to $Q_0 e^{-1}$, or $\frac{Q_0}{e}$

## Progress check

1 The charge on a capacitor is $3.06 \times 10^{-4}$ C when the p.d. across it is 6.5 V. Calculate the capacitance of the capacitor.

2 Three capacitors, of capacitance 2 μF, 3 μF and 6 μF, are connected together. Calculate the effective capacitance when they are connected:

a   in series

b   in parallel.

3 A 50 μF capacitor is charged to a p.d. of 360 V. Calculate the energy stored in the capacitor.

3   3.24 J
2   a   1 μF
    b   11 μF
1   $4.71 \times 10^{-5}$ F.

# 3.4 Magnetic fields

*After studying this section you should be able to:*

- *describe the magnetic fields inside a solenoid and around a wire when current passes in them*
- *understand the meaning of magnetic field strength*
- *calculate the size and direction of the force that acts on a current in a magnetic field*

## Magnetic fields

| EDEXCEL A | M5 | WJEC | M5 |

Magnetic fields:

- are due to permanent magnets and electric currents
- affect permanent magnets and electric currents.

> Although the field lines are often curves, the force at any point acts in a straight line.

Like other fields, magnetic fields are represented by lines with arrows. The arrows show the direction of the force at any point in the field. The convention when drawing magnetic fields is that the arrows show the direction of the force that would be exerted on the N-seeking pole of a permanent magnet placed at that point.

The diagrams show the field patterns around a bar magnet and between two different arrangements of pairs of magnets. Where equal-sized forces act in opposite directions, the result is a neutral point. At a neutral point, the resultant magnetic force is zero.

> How many forces act on a N-seeking pole placed at the neutral point shown in the middle diagram?

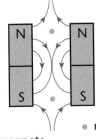

● neutral point

**Magnetic fields due to permanent magnets**

## Magnetic field strength

| AQA A | M4 | OCR A | M4 |
| AQA B | M5 | OCR B | M5 |
| EDEXCEL A | M5 | NICCEA | M5 |
| EDEXCEL B | M4 | WJEC | M5 |

> The force on any electric current that is parallel to a magnetic field is zero – there is no force.

A current-carrying wire placed in a magnetic field experiences a force provided that it is not parallel to the field. The force has its maximum value when the current is perpendicular to the field. Fleming's left hand rule shows the direction of the force on any current that has a component which is perpendicular to a magnetic field. The diagram illustrates Fleming's rule and its application to a simple motor.

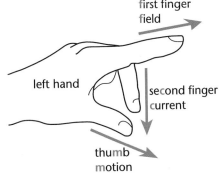

1 point your first finger in the direction of the magnetic field.

2 point your second finger in the direction of the current.

3 your thumb points in the direction of motion.

In electric and gravitational fields, the strength of the field is defined in terms of the force per unit mass or charge. There is no obvious equivalent 'unit of magnetism'.

The size of the force on a current-carrying conductor in a magnetic field depends on:

> These factors are used to define the 'unit' of magnetism as *current length*, the product of the current and the length of the conductor in the field.

- the size of the current
- the length of conductor in the field
- the orientation of the conductor relative to the field
- the strength of the magnetic field.

> **KEY POINT**
>
> These factors are used to define the strength of a magnetic field as:
> the force per unit (current × length) perpendicular to the field.
>
> Magnetic field strength is also known as *flux density*. Its symbol is $B$ and the unit is the tesla (T).

With the current and magnetic field directions shown in the diagram below the force is into the paper.

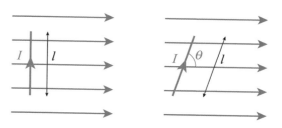

> You should always specify the directions of the quantities when using these relationships to define magnetic field strength.

> **KEY POINT**
>
> The force on a current-carrying conductor in a magnetic field is given by the expressions:
>
> when the conductor is perpendicular to the field lines:
> $$F = BIl$$
> when the angle between the conductor and the field lines is $\theta$,
> $$F = BIl\sin\theta$$

These relationships, together with the diagrams, can be used to define magnetic field strength.

## The force on a moving charge

| | | | |
|---|---|---|---|
| AQA A | M4 | OCR A | M4 |
| AQA B | M5 | OCR B | M5 |
| EDEXCEL A | M5 | NICCEA | M5 |
| EDEXCEL B | M4 | WJEC | M5 |

Any moving charge constitutes an electric current. In a vacuum tube, such as that in a television or a cathode ray oscilloscope, magnetic fields are used to exert forces on the electrons as they travel through the tube.

The size of the force on a charged particle moving in a magnetic field depends on:

- the charge on the particle
- the speed of the particle
- the strength of the magnetic field.

For a charge $Q$ moving with velocity $v$, the equivalent of (current × length) is (charge × velocity) or $Qv$.

> Remember when using Fleming's rule that the current direction is conventional current, taken to be from + to −. The current due to an electron beam is in the opposite direction to that of the electron movement.

> **KEY POINT**
>
> A charge $Q$ moving with velocity $v$ perpendicular to a magnetic field experiences a force:
> $$F = BQv$$

The direction of the force is given by Fleming's left hand rule.

## Fields due to currents

| AQA A | M4 | OCR A | AS |
| AQA B | M5 | OCR B | M5 |
| EDEXCEL A | M5 | NICCEA | M5 |
| EDEXCEL B | M4 | WJEC | M5 |

The diagram shows the magnetic field around a wire when a current passes in it.

The magnetic field pattern is a set of concentric circles around the wire. The change in the spacing of the circles shows that the magnetic field strength decreases with increasing distance from the wire. The magnetic field strength depends on:

- the size of the current
- the distance from the wire
- a constant called permeability, $\mu$, which is a measure of the extent to which the surrounding medium reinforces the magnetic field.

> Iron has a high permeability. This explains why a coil of wire wound on an iron core has a much greater magnetic field strength than a similar coil with an air core.

**KEY POINT**

The magnetic field strength, $B$, due to a current, $I$, passing in a wire is given by the expression:

$$B = \frac{\mu_0 I}{2\pi r}$$

where $\mu_0$ is the permeability of free space and has the value $4\pi \times 10^{-7}$ H m$^{-1}$ and $r$ is the distance from the wire.

> The symbol H represents the henry, a unit of electromagnetic induction.

The magnetic field pattern due to a current passing in a solenoid is shown in the diagram below.

On the outside of the solenoid, the field resembles that of a bar magnet. Inside, away from the ends of the solenoid, the parallel field lines show that the magnetic field strength is uniform.

The strength of this uniform field:

- is independent of the diameter of the solenoid
- depends on the number of turns of wire per metre of length, $n$.

**KEY POINT**

The magnetic field strength, $B$, inside a solenoid is given by the expression:

$$B = \mu_0 n I$$

where $n$ is the number of turns of wire per metre, $\mu_0$ is the permeability of free space and $I$ is the current in the solenoid.

## Progress check

1 Two parallel wires each carry a current in the same direction.
Use Fleming's rule to determine whether the force between them is attractive or repulsive.

2 The magnetic field strength between two permanent magnets is $5.6 \times 10^{-3}$ T.
A wire carrying a current of 3.5 A is placed at right angles to their magnetic field. The length of wire within the field is 0.1 m.
Calculate the size of the force on the wire.

3 An electron, charge $-1.6 \times 10^{-19}$ C, moves at a velocity of $2.4 \times 10^7$ m s$^{-1}$ perpendicular to a magnetic field of strength $4.5 \times 10^{-2}$ T.
Calculate the size of the force on the electron.

3 $1.73 \times 10^{-13}$ N
2 $1.96 \times 10^{-3}$ N
1 Attractive.

# 3.5 Circular orbits

**After studying this section you should be able to:**

- describe the forces acting on planets, moons and satellites
- explain how charged particles are accelerated in a cyclotron

## Movement in the Solar System

| | | | |
|---|---|---|---|
| AQA A | M4 | OCR A | MS.1 |
| AQA B | M4 | NICCEA | M4 |
| EDEXCEL A | M5 | | |

> Mercury and Pluto have highly elliptical orbits, the other planets follow paths that are very close to being circles.

All planetary movement in the Solar System is anticlockwise, when viewed from above the North Pole. The further a **planet** is from the Sun, the slower the speed in its orbit. Although the orbits of the planets are ellipses, for most planets they are so close to circles that our understanding of circular motion can be applied.

Planets can be considered to be:

- moving at constant speed in a circle around the Sun
- accelerating towards the Sun with centripetal acceleration $v^2/r$.

In space there are no resistive forces since the planets move through a vacuum. The only forces acting on them are gravitational. Gravitational attraction between a planet and the Sun provides the unbalanced force required to cause the centripetal acceleration.

The diagram shows the attractive forces between the Sun and a planet.

> This diagram shows that the gravitational force on a planet acts at its centre of mass and is directed towards the Sun's centre of mass.

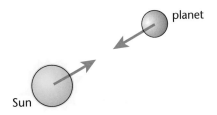

planet

Sun

The force on the planet is:

- equal in size and opposite in direction to that on the Sun
- at right angles to its direction of motion
- the unbalanced, centripetal force required to maintain circular motion.

> Asteroids in the asteroid belt, between Mars and Jupiter, have a wide range of masses but similar orbit times.

By equating the gravitational force to $mv^2/r$, it emerges that the orbital speed depends only on the orbital radius and not on the mass of the planet.

> The centripetal force required to keep a planet in a circular orbit is the gravitational force between the planet and the Sun:
>
> $$\frac{M_p v^2}{r} = \frac{GM_s M_p}{r^2}$$
>
> so
>
> $$v^2 r = GM_s$$
>
> where $M_s$ is the mass of the Sun and $M_p$ is the mass of the planet.

KEY POINT

## Satellites

| | | | |
|---|---|---|---|
| AQA A | M4 | OCR A | MS.1 |
| AQA B | M4 | NICCEA | M4 |
| EDEXCEL A | M5 | | |

> The relationship also applies to the Earth's natural satellite, the Moon.

The relationship between the orbital speed and radius of a planet can be applied to the orbit of a **satellite** around the Earth by replacing the mass of the Sun, $M_s$, with that of the Earth, $M_E$. This enables the speed of a satellite to be calculated at any orbital radius.

Some communications satellites occupy **geo-synchronous** orbits. A satellite in a geo-synchronous orbit:

- orbits above the equator
- remains in the same position relative to the Earth's surface
- has an orbit time of 24 hours.

The radius of a geo-synchronous orbit can be calculated from $v^2r = GM_E$. As there are two unknowns in this equation, $v$ can be written as $2\pi r/t$ to work out the value of $r$.

## Circular orbits in magnetic fields

| AQA A | M4 | OCR A | M4 |
| AQA B | M5 | OCR B | M5 |
| EDEXCEL B | M4 | NICCEA | M5 |

When applying Fleming's rule to electrons, remember that the direction of the current is opposite to that of the electrons' motion.

When a charged particle moves at right angles to a magnetic field, the magnetic force on the particle is perpendicular to both its direction of motion and the magnetic field. This can result in circular motion.

The diagram shows the path and the force on an electron moving in a magnetic field directed into the paper.

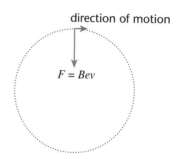

direction of motion

$F = Bev$

The electron follows a circular path, the magnetic force being the unbalanced force required to cause acceleration towards the centre of the circle. The radius of the circular path is proportional to the speed of the electron.

For an electron, $Q = e$, so the relationship is

$$Be = \frac{mv}{r}.$$

When a charge $Q$ moves in a circular path in a magnetic field of strength $B$:

$$BQv = \frac{mv^2}{r}$$

so

$$BQ = \frac{mv}{r}$$

## The cyclotron

| AQA A | M4 | EDEXCEL B | M4 |
| AQA B | M5 | OCR B | M5 |
| EDEXCEL A | M6 | NICCEA | M6 |

Particles accelerated in a cyclotron are used to probe atomic nuclei and for treating some cancers.

A **cyclotron** uses a magnetic field to force charged particles to move in a circular path, and an electric field to accelerate them as they travel around the circle. As the charged particles accelerate, the increase in speed results in an increase in the radius of the circle, so they spiral outwards.

A cyclotron consists of two D-shaped halves called dees. A magnetic field acting at right angles to the plane of the dees causes a beam of charged particles to follow a circular path. Particles such as protons and alpha particles are both suitable for use in cyclotrons.

The diagram shows the path of protons produced at the centre of the cyclotron.

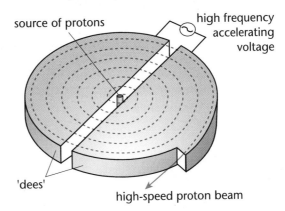

source of protons

high frequency
accelerating
voltage

'dees'

high-speed proton beam

In a cyclotron:

- the beam of charged particles is accelerated as it passes from one dee into the other
- this occurs because of the alternating electric field which changes polarity so that it attracts the particles as they enter a dee
- the frequency of the alternating voltage must be equal to the frequency of rotation of the particles
- the radius of orbit increases as the particles accelerate.

The frequency of rotation of the charged particles in a cyclotron matches that of the accelerating voltage and does not depend on the speed of the particles. The value of the magnetic field strength can be adjusted to achieve the desired frequency.

> If the frequency of the accelerating voltage is fixed, each orbit takes the same time. With an increase in the radius of successive orbits, the particles travel increasing distances in a given time period.

---

**KEY POINT**

The frequency of rotation of a charged particle in a cyclotron, $f$, is related to the magnetic field strength, $B$, by the expression:

$$f = \frac{BQ}{2\pi m}$$

where $Q$ is the charge on a particle of mass $m$.

---

## Progress check

1 The Moon orbits the Earth, mass $6.0 \times 10^{24}$ kg, at a radius of $3.84 \times 10^8$ m.
$G = 6.7 \times 10^{-11}$ N m² kg⁻².
Calculate:
a the speed of the Moon in its orbit.
b the time it takes for the Moon to complete one orbit.

2 An electron of mass $m_e = 9.10 \times 10^{-31}$ kg and charge, $e = 1.60 \times 10^{-19}$ C travels at a speed of $2.10 \times 10^7$ m s⁻¹ in a circular orbit at right angles to a magnetic field.
The magnetic field strength, $B = 6.5 \times 10^{-6}$ T.
Calculate the radius of the electron orbit.

3 In the diagram of the cyclotron what is the direction of the magnetic field?

3 Upwards.

2 18.4 m

b 2.36 × 10⁶ s
1 a 1.02 × 10³ m s⁻¹

# 3.6 Electromagnetic induction

LEARNING SUMMARY

**After studying this section you should be able to:**

- calculate the flux linkage through a coil of wire in a magnetic field
- explain how electromagnetic induction occurs due to changes in flux linkage
- apply Faraday's law and Lenz's law

## Flux and flux linkage

| | | | |
|---|---|---|---|
| AQA A | M4 | OCR A | M4 |
| AQA B | M5 | OCR B | M5 |
| EDEXCEL A | M5 | NICCEA | M5 |
| EDEXCEL B | M4 | WJEC | M5 |

Flux provides a useful model for explaining the effects of magnetic fields.

The current view is that these forces can be attributed to 'exchange particles'.

Almost everything we do, apart from sleeping in the dark, relies on **electromagnetic induction**. Induction is used to generate electricity in power stations and to transform its voltage as it passes through the distribution system.

The effects of induction are explained by using the concept of **flux**. Although the existence of flux has long been discredited, an awareness of its meaning is useful to understand the laws of induction as set out by Faraday and Lenz.

Like gravitational and electric fields, magnetic fields act at a distance. Magnetic field patterns are used to show the forces that are exerted around a magnet or electric current. These forces are exerted without any physical link between the magnet or current that causes the field and a magnetic material or current placed within the field. In the days of Faraday and Lenz, they were attributed to the effects of flux.

When drawing magnetic field patterns:

- the relative strength at different points in the field is shown by the separation of the field lines
- the closer the lines are together, the stronger the field
- these field lines represent **magnetic flux**, which is imagined as occupying the space around a magnet and being responsible for the effect of a magnet field.

To integrate the flux model with today's explanation of magnetic effects in terms of magnetic field strength, this can be thought of in terms of a flux density, being represented by the concentration of magnetic field lines. Flux density is the flux per unit area so flux is now defined in terms of the magnetic field strength and the area that the field permeates.

This definition relates the equivalence of the modern concept of magnetic field strength to that of the older 'flux density' concept.

> **KEY POINT**
>
> The magnetic flux, $\Phi$, through an area, $A$, is defined as the product of the magnetic field strength and the area normal to the field.
>
> $$\Phi = B \times A$$
>
> Magnetic flux is measured in webers (Wb) where 1 Wb is the flux through an area of 1 m² normal to a uniform field of strength 1 T.

The diagram shows the flux through a rectangular coil in a uniform magnetic field.

coil of $N$ turns and area $A$

magnetic field of strength $B$, represented by lines of magnetic flux

**The flux through a rectangular coil**

When the coil is rotated, it 'cuts' through the flux, or field lines and an e.m.f. is induced.

The size, or magnitude, of the induced e.m.f. depends on:

- the amount of flux through the coil
- the speed of rotation
- the number of turns on the coil.

Each turn on the coil has a flux linkage which changes as the coil rotates.

The flux linkage of a coil of $N$ turns is $N\Phi$, where $\Phi$ is the flux through the coil.

> Movement of the coil parallel to the field does not induce an e.m.f., since no field lines are being 'cut'.
>
> The induced e.m.f. has its greatest value when the movement of the coil is perpendicular to the field.

## Faraday's law

| | | | |
|---|---|---|---|
| AQA A | M4 | OCR A | M4 |
| AQA B | M5 | OCR B | M5 |
| EDEXCEL A | M5 | NICCEA | M5 |
| EDEXCEL B | M4 | WJEC | M5 |

Electromagnetic induction occurs whenever the magnetic field through a conductor changes. This can be due to a conductor moving through a magnetic field or a conductor being in a fixed position within a changing magnetic field, such as that due to an alternating current. Both of these result in an e.m.f. being induced in the conductor.

Examples of electromagnetic induction include:

- moving a magnet inside a wire coil
- generating the high voltage necessary to ionise the vapour in a fluorescent tube and cause the spark needed to ignite the explosive mixture in a petrol engine
- changing the voltage of an alternating current, using a transformer.

The diagram below shows the difference in the size of the e.m.f. when a magnet is moved at different speeds in a coil.

> In a power station, electricity is generated by an electromagnet spinning inside copper coils.

slow movement
produces a small e.m.f.

faster movement
produces a bigger e.m.f.

Faraday's law relates the size of the induced e.m.f. to the change in flux linkage.

> **KEY POINT**
>
> Faraday's law states that:
>
> *the size of the induced e.m.f. is proportional to the rate of change of flux linkage.*
>
> As the proportionality constant is equal to 1, for a uniform rate of change of flux linkage this can be written as:
>
> $$\text{magnitude of induced e.m.f.} = N\frac{\Delta\Phi}{\Delta t}$$
>
> where $\Delta\Phi$ is the change of flux in time $\Delta t$.

> To generate the high voltage needed to cause a spark, the flux has to change rapidly. This happens when the current in an electromagnet is switched off.

## What direction?

| | | | |
|---|---|---|---|
| AQA A | M4 | OCR A | M4 |
| AQA B | M5 | OCR B | M5 |
| EDEXCEL A | M5 | NICCEA | M5 |
| EDEXCEL B | M4 | WJEC | M5 |

Faraday's law can be used to work out the size of an induced e.m.f. such as that across the wingtips of an aircraft flying in the Earth's magnetic field. In Britain the Earth's field makes an angle of 20° with the vertical, see the following diagram.

20°

An aircraft flying in a North–South direction is cutting across the vertical component only, while flying East–West involves cutting across the horizontal component in addition.

The induced e.m.f. arises as a consequence of the force on the free electrons in the metal of the aircraft frame. As the aircraft travels through the air, the movement of these electrons forms a current in the opposite direction to that of flight. Fleming's left hand rule can be used to work out the direction of the force on the electrons and hence the direction of the induced e.m.f.

In the case of an aircraft flying from North to South:

- the current is South–North
- the magnetic field being 'cut' is vertically downwards
- the force on the free electrons is towards the East.

This results in a charge imbalance and a voltage across the wingtips.

The direction of the e.m.f. induced in the aircraft and when a magnet moves into a coil of wire can be worked out using **Lenz's law**.

> **KEY POINT**
>
> Lenz's law states that:
>
> *the direction of an induced e.m.f. is always in opposition to the change that causes it.*

The diagram below shows that when the North pole of a magnet is moved into one end of a coil, the induced e.m.f. causes an induced current in an anticlockwise direction. When current passes in a coil, the magnetic field is similar to that of a bar magnet, the North pole being the end where the current passes anticlockwise.

The induced current when a magnet enters a coil of wire

How to identify the poles of an electromagnet

Lenz's law is a re-statement of the principle of conservation of energy; the induced current opposes the motion of the magnet so work has to be done to move the magnet against the induced magnetic field. This work is the energy transfer to the circuit needed to cause a current.

Combining Faraday's and Lenz's laws gives the equation for induced e.m.f.:

> **KEY POINT**
>
> $$\varepsilon = -N\frac{\Delta\Phi}{\Delta t}$$
>
> Where $\varepsilon$ is the induced e.m.f. The negative sign shows that the induced e.m.f. is in opposition to the change of flux causing it.

## The transformer

| | | | | |
|---|---|---|---|---|
| AQA B | M5 | OCR B | M5 | |
| EDEXCEL A | M5 | NICCEA | M5 | |
| EDEXCEL B | M4 | | | |

Transformers use changing magnetic fields to change the size of an alternating voltage. An alternating current passing in one coil (the primary) induces an e.m.f. in an adjacent coil (the secondary).

The diagram below shows the flux when the two coils are wound on an iron core.

> The e.m.f. is induced whether or not there is a secondary circuit. If there is a complete circuit, there is also an induced current.

**Flux linking two transformer coils**

In a transformer:

> Iron is easily magnetised; its magnetic domains contribute to the strength of the magnetic field.

- alternating current in the primary produces an alternating magnetic field
- this is reinforced by the high-permeability iron core
- the flux concentrates in the iron
- an e.m.f. is induced in the secondary because of the changing flux linkage.

It follows from the last bullet point that the induced e.m.f. is proportional to the number of turns on the secondary coil.

> A transformer constructed from low-resistance coils on a laminated iron core is close to ideal.

**KEY POINT**

The relationship between the voltages and numbers of turns for an ideal transformer is:

$$\frac{V_p}{V_s} = \frac{N_p}{N_s}$$

This states that the voltages are in the same ratio as the numbers of turns. In an ideal transformer there is no energy loss in the wires or the core so the power output from the secondary is equal to the power input to the primary and the currents are in the inverse ratio to the voltages.

## Progress check

1 A rectangular coil has 25 turns and an area of $2.5 \times 10^{-4}$ m$^2$.
   It is placed in a magnetic field of strength $6.8 \times 10^{-6}$ T.
   Calculate the flux linkage when the plane of the coil is
   a parallel to the magnetic field
   b perpendicular to the magnetic field.

2 An aircraft flies from East to West.
   In what direction is the induced e.m.f. due to the 'cutting' of the Earth's horizontal magnetic field?

3 The flux linking a coil of 60 turns changes at the rate of $4.0 \times 10^{-3}$ Wb s$^{-1}$.
   Calculate the size of the induced e.m.f.

3 0.24 V
2 The top of the aircraft is positive and the bottom negative
   b 4.25 × 10⁻⁸ Wb
1 a 0

## Sample question and model answer

The diagram shows a capacitor connected in series with a 12.0 V power supply and a 500 Ω resistor.

The switch is closed.

(a) What is the initial voltage across:

At the instant that the switch is closed there is no charge on the capacitor, so there is no voltage across it. All the voltage is therefore across the resistor.

    (i) the capacitor?                                      [1]

        0 V                                            1 mark

    (ii) the resistor?                                        [1]

        12 V                                         1 mark

(b) Calculate the value of the current in the circuit immediately after the switch is closed.     [3]

The value of the resistor determines the initial current in the circuit.

    $I = V \div R$                           1 mark

        $= 12.0 \text{ V} \div 500 \text{ } \Omega$       1 mark

        $= 0.024 \text{ A}$             1 mark

(c) When the capacitor is fully charged, calculate:

    (i) the charge stored on the capacitor     [3]

        $Q = C \times V$         1 mark

            $= 200 \text{ } \mu\text{F} \times 12.0 \text{ V}$   1 mark

            $= 2.4 \times 10^{-3} \text{ C}$     1 mark

    (ii) the energy stored by the capacitor.     [3]

Alternatively, $\frac{1}{2}QV$ should give the same answer, provided that the charge, Q, has been calculated correctly.

        $E = \frac{1}{2} C \times V^2$       1 mark

           $= \frac{1}{2} \times 200 \text{ } \mu\text{F} \times 12.0^2$   1 mark

           $= 1.44 \times 10^{-2} \text{ J}$     1 mark

(d) The capacitor is discharged and a second 200 μF capacitor is connected in parallel with the capacitor shown in the diagram. The switch is then closed. Without doing any further calculations, explain how this affects:

    (i) the final charge stored     [2]

        This is doubled (1 mark) since the potential difference is unchanged and the capacitance is doubled (1 mark).

As the question states that further calculations are not required, give an answer based on the relationships used to calculate these quantities, with the reasons.

    (ii) the final energy stored.     [2]

        This is also doubled (1 mark) as there is double the capacitance (and double the amount of charge stored) for the same potential difference (1 mark).

## Practice examination questions

Throughout this section, use the following values of constants:
electronic charge, $e = -1.6 \times 10^{-19}$ C
universal gravitational constant, $G = 6.7 \times 10^{-11}$ N m² kg⁻²
$k = 1/4\pi\varepsilon_0 = 9.0 \times 10^9$ N m² C⁻²
gravitational field strength at the surface of the Earth, $g = 10.0$ N kg⁻¹

### 1

A hydrogen atom consists of a proton and an electron at an average separation of $5.2 \times 10^{-11}$ m.

(a) Calculate the size of the force between them. [3]

(b) (i) Calculate the electric field strength that each experiences due to the other. [3]

    (ii) What is the direction of the electric field between the proton and the electron? [1]

(c) Assuming that the electron orbits the proton, calculate the speed of its orbit. The mass of an electron, $m_e = 9.1 \times 10^{-31}$ kg. [3]

### 2

The diagram shows a pair of parallel plates connected to 2500 V supply.

An oil drop of mass 0.050 g between the plates carries a charge of $10e$.

(a) Calculate:

    (i) the voltage between C and D [1]

    (ii) the energy transfer when the drop moves from C to D. [3]

(b) Calculate the energy transfer when the drop moves from A to B. [1]

(c) The voltage is changed so that the drop is stationary between the plates.

    (i) What must the polarity of the plates be to achieve this? [1]

    (ii) Calculate the voltage required. [3]

### 3

Use this data to answer the questions:
The mass of the Earth = $6.0 \times 10^{24}$ kg.
The mass of the Moon = $7.4 \times 10^{21}$ kg.
The distance between the Earth and the Moon = $3.8 \times 10^8$ m.

(a) (i) Calculate the strength of the Earth's gravitational field at the Moon's orbit. [3]

    (ii) Calculate the size of the Earth's pull on the Moon. [2]

    (iii) Use the answer to (ii) to calculate the Moon's period of rotation around the Earth.

(b) Tides are due to the combined effect of the Moon and the Sun.

(i) Calculate the size of the Moon's pull on 1.0 kg of water. [3]

(ii) Suggest why the pull of the Sun has much less effect than the pull of the Moon on tides. [3]

(iii) The diagram shows two positions of the Moon relative to the Earth and the Sun.

Suggest why tides are higher when the Moon is in position A than when it is in position B. [2]

**4**

A moving coil loudspeaker consists of a cylindrical permanent magnet and an electromagnet. The electromagnet is positioned between the poles of the fixed magnet. The diagram shows the arrangement.

(a) What is the direction of the magnetic field between the poles of the fixed magnet? [1]

(b) The current in the electromagnet passes in a clockwise direction.
What is the direction of the force on the electromagnet? [1]

(c) Explain why the electromagnet vibrates when an alternating current passes in it. [2]

(d) The value of the magnetic field strength due to the fixed magnet is 0.85 T at the position of the electromagnet.
The electromagnet consists of a coil of wire of 150 turns and diameter 5.0 cm.
Calculate the size of the force on the coil when a current of 0.055 A passes in it. [3]

## Practice examination questions (continued)

5 Electrons can be made to move in a circular path by firing them into a region where there is a magnetic field acting at right angles to their velocity.

The diagram shows a device for producing high-speed electrons.

(a) Calculate:

    (i) the kinetic energy of an electron in the accelerated beam     [3]

    (ii) the speed of an electron in the accelerated beam.
       The mass of an electron, $m_e = 9.1 \times 10^{-31}$ kg.     [3]

(b) The electron beam passes into a magnetic field directed into the paper.

    (i) What is the direction of the force on the electron beam as it enters the magnetic field?     [1]

    (ii) Explain why the electrons follow a circular path.     [2]

    (iii) The strength of the magnetic field is 2.0 mT.
       Calculate the radius of the circular path of the electrons.     [3]

6

A rectangular coil of wire is placed so that its plane is perpendicular to a magnetic field of strength 0.15 T. This is shown in the diagram.

(a) Calculate the magnetic flux through the coil:

    (i) when the coil is in the position shown in the diagram     [3]

    (ii) when the coil is turned through 90°.     [1]

(b) The coil is connected to a high-resistance voltmeter.

    (i) Describe and explain how the reading on the voltmeter changes as the coil is rotated through 360°.     [4]

    (ii) What device makes use of this effect?     [1]

(c) State three ways of increasing the size of the induced voltage.     [3]

# Particle physics

**The following topics are covered in this chapter:**

- Thermal physics
- Radioactive decay
- Energy from the nucleus
- Probing matter

## 4.1 Thermal physics

### Energy of individual particles

| | | | |
|---|---|---|---|
| AQA A | AS | OCR A | M4 |
| AQA B | M4 | OCR B | M4 |
| EDEXCEL A | AS | NICCEA | M4 |
| EDEXCEL B | M5 | WJEC | M4 |

All materials are made up of particles, and particles have energy. The particles of an ideal gas have only kinetic energy since forces between the particles are negligible. However, this cannot be true of a solid or a liquid or the particles would not stay in that phase, and the particles in real gases exert both attractive and repulsive forces on each other during collisions.

Imagine a collision between two particles in a real gas.

- As the particles approach, the attractive forces cause an increase in the speed and kinetic energy.
- The particles then get closer and the speed decreases, being momentarily zero before they start to separate.
- The process is then reversed.

> The speed of colliding particles is momentarily zero as their direction of travel is reversed.

There is an interchange between potential (stored) energy and kinetic energy during the collision, as shown in the diagram.

> As the separation between two particles decreases, the net force changes from attractive to repulsive. The equilibrium position is where these forces are equal in size, so the resultant force is zero.

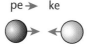

particles speed up due to
short-range attractive forces

particles slow down due to
close-range repulsive forces

The particles of real gases, like those of solids and liquids, have both potential and kinetic energy. The distribution of energy between kinetic and potential is constantly changing and so is said to be random.

The total amount of energy of the particles in an object is known as its **internal energy**. At a constant temperature the internal energy of an object remains unchanged, but the contributions of individual particles to that energy change due to the transfer of energy during interactions between particles.

## Measuring temperature

| AQA B | M4 | OCR B | M4 |
| EDEXCEL B | M5 | NICCEA | M4 |
| OCR A | M4 | WJEC | M4 |

Suitable thermometric properties include volume, resistance and length.

The Celsius scale, based on the ice-point and the steam-point, is a useful scale for everyday measurements of temperature. However, there are some drawbacks:

- thermometers which use different thermometric properties only agree at the points of calibration
- the Celsius scale is not an absolute scale, so 20°C is not twice as hot as 10°C.

Since temperature is related to the internal energy of a material, on an absolute scale of temperature:

- the zero should correspond to the minimum internal energy of the particles
- there should be a unique value for any temperature, which does not depend on the instrument being used to measure it
- 20 units should be twice as hot as 10 units.

The thermodynamic scale of temperature, based on the behaviour of an ideal gas, satisfies these conditions.

Note that the kelvin is not called a degree and does not have a ° in the unit.

> **KEY POINT**
>
> The absolute scale of temperature is called the **kelvin** scale; its unit is the kelvin (K) and its relationship to the Celsius scale is:
>
> $$T/K = \theta/°C + 273$$

The diagram shows the relationship between Celsius and kelvin temperatures.

## Being specific

| AQA A | AS | OCR A | M4 |
| AQA B | M4 | OCR B | M4 |
| EDEXCEL A | AS | NICCEA | M4 |
| EDEXCEL B | AS | WJEC | M4 |

Changing the temperature of an object involves a change in its internal energy, the total potential and kinetic energy of the particles.

The energy transfer required to change the temperature of an object depends on:

- the temperature change
- the mass of the object
- the material the object is made from.

These are all taken into consideration in the concept of specific heat capacity. The term 'specific' means 'for each kilogram' so any physical measurement that is described as 'specific' refers to 'per kilogram of material'.

> **KEY POINT**
>
> The specific heat capacity of a material, $c$, is defined as:
>
> the energy transfer required to change the temperature of 1 kg of the material by 1°C.
>
> $$\Delta E = mc\,\Delta\theta$$
>
> where $c$, specific heat capacity, is measured in $J\ kg^{-1}\ °C^{-1}$ or $J\ kg^{-1}\ K^{-1}$

When using this relationship the temperatures can be in either Celsius or kelvin since the temperature change is the same in each case.

## Changing phase

| AQA A | AS | OCR A | M4 |
| AQA B | M4 | OCR B | M4 |
| EDEXCEL A | AS | | |

If you try to squash a solid or a liquid, the particles are pushed closer together and they repel each other. Stretching has the opposite effect; when the separation of the particles is increased the forces are attractive. At increased separations the particles have increased potential energy and this energy has to be supplied for any process that involves expansion to take place.

The term phase means whether the substance is a solid, liquid or gas.

When a substance changes phase there is a change in the potential energy of the particles.

For most substances there is a small increase in potential energy when changing from solid to liquid and a much bigger change from liquid to gas.

The increase in particle separation during a change of phase from solid to liquid is small, but that during a change of phase from liquid to gas is large.

solid      liquid      gas

On each horizontal part of the curve the substance exists in two phases. Can you identify them?

The diagram below shows a typical temperature–time graph as a solid is heated at a constant rate and passes through the three phases. Where the curve is horizontal it shows that energy is being absorbed with no change in temperature. This corresponds to a change of phase. Note that much more energy is absorbed during the change from liquid to gas than during the change from solid to liquid.

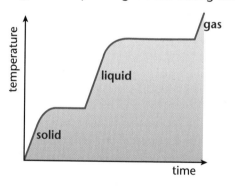

## An ideal gas

| AQA A | AS | OCR A | M4 |
| AQA B | M4 | OCR B | M4 |
| EDEXCEL A | AS | NICCEA | M4 |
| EDEXCEL B | M5 | WJEC | M4 |

The concept of an ideal gas is useful because the behaviour of all gases is close to ideal provided that the gas is neither at a high pressure nor close to its boiling point. The ideal gas equation gives the relationship between the three variables pressure, volume and temperature.

For an ideal gas these variables are related by the single equation:

$$pV/T = \text{constant}$$

1 mole of gas consists of the Avogadro constant, $N_A$, of particles. The value of the Avogadro constant is $6.02 \times 10^{23}$ mol$^{-1}$.

The value of the constant depends only on the amount of gas. For 1 mole of any gas under ideal conditions it has the value of 8.3 J mol$^{-1}$ K$^{-1}$. This constant is called the molar gas constant and has the symbol $R$. Doubling the number of particles by considering two moles of gas has the effect of doubling the value of the constant, so:

The equation of state for an ideal gas is:
$$pV = nRT$$
where $n$ is the number of moles of gas and $R$ is the molar gas constant.

In an ideal gas, all the internal energy exists as kinetic energy of the particles. The internal energy is proportional to thermodynamic temperature.

The mean kinetic energy of the particles of an ideal gas is proportional to the kelvin temperature.

This means that doubling the thermodynamic temperature of a gas results in the mean (speed)$^2$ being doubled.

## Progress check

1 1.5 kg water at 10°C is placed in a kettle. Calculate the energy transferred to the water in bringing it to the boil. The specific heat capacity of water, $c = 4.2 \times 10^3$ J kg$^{-1}$ °C$^{-1}$.

2 Calculate the molar volume of an ideal gas at atmospheric pressure ($1.01 \times 10^5$ Pa) and a temperature of 0°C.

3 The mean kinetic energy of the particles of an ideal gas is $6.07 \times 10^{-21}$ J at a temperature of 20°C. At what temperature is their mean kinetic energy equal to $12.14 \times 10^{-21}$ J?

3  586 K or 313°C
2  $2.24 \times 10^{-2}$ m$^3$
1  $5.67 \times 10^5$ J

# 4.2 Radioactive decay

*After studying this section you should be able to:*

- describe the main types of radioactive emission
- explain the effects of radioactive decay on the nucleus
- calculate the half-life of a radioactive isotope from an activity–time graph

## Radiation all around us

| | | | |
|---|---|---|---|
| AQA A | M5-9 | EDEXCEL B | AS |
| AQA B | AS | OCR A | M4 |
| EDEXCEL A | AS | WJEC | M5 |

The term 'background radiation' is also used to describe the microwave radiation left over from 'Big Bang'. This is a different type of radiation to nuclear radiation.

Radioactive decay occurs when an atomic nucleus changes to a more stable form. This is a random event that cannot be predicted. The emissions from these nuclei are collectively called radioactivity or radiation.

We are subjected to a constant stream of radiation called background radiation. Most of this is 'natural' in the sense that it is not caused by the activities of people living on Earth. Sources of background radiation include:

- the air that we breathe; radioactive radon gas from rocks can concentrate in buildings
- the ground and buildings; all rocks contain radioactive isotopes
- the food that we eat; the food chain starts with photosynthesis. Radioactivity enters the food chain in the form of carbon-14, an unstable form of carbon that is continually being formed in the atmosphere
- radiation from space, called cosmic radiation
- medical and industrial uses of radioactive materials.

## The emissions

| | | | |
|---|---|---|---|
| AQA A | M5-9 | OCR A | M4 |
| AQA B | AS | OCR B | M5 |
| EDEXCEL A | AS | NICCEA | M5 |
| EDEXCEL B | AS | WJEC | M5 |

The range of a beta-plus particle is effectively zero, since it is annihilated when it collides with an electron.

Alpha radiation is the most intensely ionising and can cause a lot of damage to body tissue.

Although it cannot penetrate the skin, alpha emitters can enter the lungs during breathing.

Radioactive emissions are detected by their ability to cause ionisation, creating charged particles from neutral atoms and molecules by removing outer electrons. This results in a transfer of energy from the emitted particle, which is effectively absorbed when all its energy has been lost in this way. The four main emissions are alpha ($\alpha$), beta-plus ($\beta^+$), beta-minus ($\beta^-$) and gamma ($\gamma$). Of these, alpha radiation is the most intensely ionising and has the shortest range, with the exception of beta-plus, while gamma radiation is the least intensely ionising and has the longest range.

- In alpha emission the nucleus emits a particle consisting of two protons and two neutrons. This has the same make-up as a helium nucleus.
- Beta-minus emission occurs when a neutron decays into a proton, emitting an electron in the process.
- In beta-plus emission a proton changes to a neutron by emitting a positron, an anti-electron.
- Gamma emission is short wavelength electromagnetic radiation.

| radioactive emission | nature | charge/e | symbol | penetration | causes ionisation | affected by electric and magnetic fields |
|---|---|---|---|---|---|---|
| **alpha** | two neutrons and two protons | +2 | $^4_2\text{He}$ or $^4_2\alpha$ | absorbed by paper or a few cm of air | intensely | yes |
| **beta-minus** | high-energy electron | –1 | $^0_{-1}\text{e}$ or $^0_{-1}\beta$ | absorbed by 3 mm of aluminium | weakly | yes |
| **beta-plus** | positron (antielectron) | +1 | $^0_{+1}\text{e}$ or $^0_{+1}\beta$ | annihilated by an electron | | yes |
| **gamma** | short-wavelength electromagnetic radiation | none | $^0_0\gamma$ | reduced by several cm of lead | very weakly | no |

Notice that the electron emitted in beta-minus decay and the positron emitted in beta-plus decay have been allocated the atomic numbers –1 and +1. This is because of the effect on the nucleus when these particles are emitted.

Other types of nuclear decay include:

- **electron capture**, where a proton in the nucleus captures an orbiting electron and becomes a neutron. The energy lost by the electron is emitted as an X-ray
- **nucleon emission**; some artificially-produced isotopes decay by emitting a proton or a neutron.

Alpha and beta radiations are the most intensely ionising but they are readily absorbed and so easy to shield. Gamma radiation is very penetrating. Although it only reacts weakly with matter, its effects can be devastating. One of the safest ways of protecting against danger from gamma radiation is to maximise the distance from the source to any people. The radiation from a gamma source is emitted equally in all directions, so the intensity decreases as an inverse square law.

> In this context, intensity means the number of photons detected per square metre.

> **KEY POINT**
>
> The intensity, $I$, of gamma radiation detected from a source is related to the intensity of the source, $I_0$, by the equation:
>
> $$I = \frac{kI_0}{x^2}$$
>
> where $k = 1/4\pi$ and $x$ is the distance from the source.

## Balanced equations

| | | | |
|---|---|---|---|
| AQA A | M5–9 | OCR A | M4 |
| AQA B | AS | OCR B | M5 |
| EDEXCEL A | M4 | NICCEA | M5 |
| EDEXCEL B | M4 | WJEC | M5 |

When a nucleus decays by alpha or beta emission, the numbers of protons and neutrons are changed. Gamma emission does not change the make-up of the nucleus, but corresponds to the nucleus losing excess energy. Gamma emission often occurs alongside alpha and beta emissions, though some artificial radioactive isotopes emit gamma radiation only.

The changes that take place due to alpha and beta emissions are:

> Technetium-99 is an artificial isotope that emits gamma radiation only. It is used as a tracer in medicine. The gamma radiation can be detected outside the body and there is less risk of cell damage than with an isotope that also emits alpha or beta radiation.

- **alpha**; the number of protons decreases by 2 and the number of neutrons also decreases by two
- **beta-minus**; the number of neutrons decreases by one and the number of protons increases by one
- **beta-plus**; the number of neutrons increases by one and the number of protons decreases by one.

In writing equations that describe nuclear decay, both charge (represented by the atomic number, $Z$) and the number of nucleons (represented by the mass number, $A$) are conserved. The table summarises these changes and gives examples of each type of decay.

> Check that the equations given as examples are balanced in terms of charge and number of nucleons.

| particle emitted | effect on A | effect on Z | example |
|---|---|---|---|
| alpha | –4 | –2 | $^{226}_{88}Ra \rightarrow ^{222}_{86}Rn + ^{4}_{2}He$ |
| beta-minus | unchanged | +1 | $^{14}_{6}C \rightarrow ^{14}_{7}N + ^{0}_{-1}e$ |
| beta-plus | unchanged | –1 | $^{11}_{6}C \rightarrow ^{11}_{5}B + ^{0}_{+1}e$ |

## Rate of decay and half-life

| | | | |
|---|---|---|---|
| AQA A | M5–9 | OCR A | M4 |
| AQA B | AS | OCR B | M4, M5 |
| EDEXCEL A | M6 | NICCEA | M5 |
| EDEXCEL B | M5 | WJEC | M5 |

> **KEY POINT**
>
> The activity, or rate of decay, of a sample of radioactive material is measured in **becquerel** (Bq). An activity of 1 Bq represents a rate of decay of 1 $s^{-1}$.

Radioactive decay is a random process and the decay of an individual nucleus cannot be predicted. However, given a sample containing large numbers of undecayed nuclei, then statistically the rate of decay should be proportional to the number of undecayed nuclei present. Double the size of the sample and, on the average, the rate of decay should also double.

There are only two factors that determine the rate of decay of a sample of radioactive material. They are:

- the radioactive isotope involved
- the number of undecayed nuclei.

> Unlike chemical reactions, radioactive decay is not affected by changes in temperature.

**KEY POINT**

The relationship between the rate of decay, or activity, of a radioactive isotope and the number of undisclosed nuclei is:

$$\text{activity, } A = \lambda N$$

Where the activity is measured in becquerel, $N$ is the number of undecayed nuclei present and $\lambda$ is the **decay constant** of the substance. $\lambda$ has units of $s^{-1}$.

When the activity of a radioactive isotope (after deducting the average background count) is plotted against time, the result is a curve that shows the activity decreasing as the number of undecayed nuclei decreases. A decay curve is shown in the diagram.

> The graph shown is a plot of activity against time. A plot of number of undecayed nuclei against time would be identical but with a different scale on the y-axis.

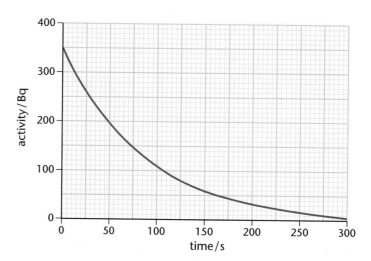

The shape of the curve is the same for all radioactive substances, but the activities and time scales depend on the size of the sample and its decay constant.

> The negative sign is needed here because the *rate of change of undecayed nuclei* is always negative due to the decrease in the numbers of undecayed nuclei with increasing time.

**KEY POINT**

An alternative way of expressing activity is *the rate of change of undecayed nuclei with time*:

$$\frac{\Delta N}{\Delta t} = -\lambda N$$

this leads to the relationship between the number of undecayed nuclei, $N$, and the number at time $t = 0$, $N_0$:

$$N = N_0 e^{-\lambda t}$$

The curve on the graph above shows exponential decay. In an exponential decay curve:

- the rate of change of $y$, represented by the gradient of the curve, is proportional to $y$
- in equal intervals of $x$ the value of $y$ always changes in the same ratio.

This second point means that it always takes the same time interval for the activity to decrease to a given fraction of any particular value. The time for the activity to halve is a measure of the **half-life** of the substance.

Note that although half-life is defined in terms of the number of undecayed nuclei, it is usually measured as the average time for the activity of a sample to halve.

> **KEY POINT**
>
> The half-life of a radioactive isotope, $t_{1/2}$ is the average time taken for the number of undecayed nuclei of the isotope to halve.

Because radioactive decay is a random process, the results of any experiment to measure activity do not fit the curve exactly and there is some variation in the change of activity when identical samples of the same material are compared. This is why the term *average* is used in the definition of half-life.

The values of half-lives range from tiny fractions of a second to many millions of years. Starting with $N$ nuclei of a particular isotope, the number remaining after one half-life has elapsed is $N/2$, after two half-lives $N/4$ and after $n$ half-lives it is $N/2^n$.

There is a relationship between the half-life and the decay constant of any particular radioactive isotope. The shorter the half-life, the greater the rate of decay and the decay constant.

When using this relationship, $\lambda$ and $t_{1/2}$ must be measured in consistent units, e.g. $\lambda$ in $s^{-1}$ and $t_{1/2}$ in s.

> **KEY POINT**
>
> The half-life of a radioactive isotope and its decay constant are related by the equation:
> $$\lambda t_{1/2} = \ln 2 = 0.69$$

## Stable and unstable nuclei

AQA A    M5    OCR B    M5
EDEXCEL A    AS

Carbon-11, carbon-12 and carbon-14 are three isotopes of carbon. Of these, only carbon-12 is stable. It has equal numbers of protons and neutrons. The graph shows the relationship between the number of neutrons ($N$) and the number of protons ($Z$) for stable nuclei.

Remember, isotopes of an element all have the same number of protons but different numbers of neutrons in the nucleus.

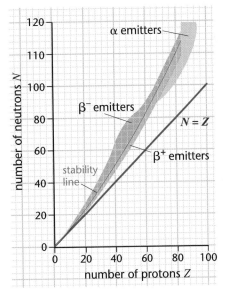

It can be seen from this graph that the condition for a nucleus to be stable depends on the number of protons:

- for values of $Z$ up to 20, a stable nucleus has equal numbers of protons and neutrons
- for values of $Z$ greater than 20, a stable nucleus has more neutrons than protons.

For stable nuclei with more than 20 protons, the neutron:proton ratio increases steadily to a value of around 1.5 for the most massive nuclei.

Unstable nuclei above the stability line in the diagram are **neutron-rich**; they can become more stable by decreasing the number of neutrons. They decay by $\beta^-$

emission; this leads to one less neutron and one extra proton and brings the neutron–proton ratio closer to, or equal to, one. An example is:

$$^{24}_{11}Na \rightarrow \, ^{24}_{12}Mg + \, ^{0}_{-1}e$$

Unstable nuclei below the stability line decay by $\beta^+$ emission; this increases the neutron number by one at the expense of the proton number. An example is:

$$^{11}_{6}C \rightarrow \, ^{11}_{5}B + \, ^{0}_{+1}e$$

The emission of an alpha particle has little effect on the neutron–proton ratio for isotopes that are close to the $N=Z$ line and is confined to the more massive nuclei. For these nuclei, emission of an alpha particle changes the balance of the proton–neutron ratio in the favour of the neutrons. In the decay of thorium-228 shown below, the neutron–proton ratio increases from 1.53 to 1.55.

$$^{228}_{90}Th \rightarrow \, ^{224}_{88}Ra + \, ^{4}_{2}He$$

The $N=Z$ line corresponds to a neutron:proton ratio of 1. As an alpha particle consists of two neutrons and two protons, its emission would hardly affect a neutron:proton ratio that is nearly 1.

## Progress check

1 What is the effect on atomic and mass number of decay by:
   a an alpha particle only
   b a beta-minus particle only
   c a gamma photon only?

2 Technetium-99 is a gamma emitter with a half-life of 6 hours.
   A fresh sample is prepared with an activity $A$.
   a Calculate the activity of the sample when 24 hours have elapsed since its preparation.
   b Calculate the decay constant of technetium-99.

3 Complete the equation for the decay of phosphorus–32 by beta-minus emission.
   $$^{32}_{15}P \rightarrow S+$$

3 $^{32}_{15}P \rightarrow \, ^{32}_{16}S + \, ^{0}_{-1}e$

b $3.2 \times 10^{-5} \, s^{-1}$

2 a $A/16$

c Both the atomic number and the mass number are unchanged.

b The atomic number increases by 1 and the mass number is unchanged.

1 a The atomic number decreases by 2 and the mass number decreases by 4.

# 4.3 Energy from the nucleus

**After studying this section you should be able to:**

- *represent nuclear reactions by nuclear equations*
- *explain the equivalence of mass and energy*
- *describe how energy from the nucleus is released in a fission reactor*

**LEARNING SUMMARY**

## The nucleus

| AQA A | AS | OCR A | M4 |
|---|---|---|---|
| AQA B | AS | OCR B | M5 |
| EDEXCEL A | AS | NICCEA | M6 |
| EDEXCEL B | M5 | WJEC | AS |

> You will not find El in the periodic table – it is fictitious.

If the size of an atom is compared to that of a cathedral or a football stadium, the **nucleus** is about the size of a tennis ball. Evidence of the size of the nucleus comes from alpha particle scattering experiments (see section 4.4).

There are two types of particle in the nucleus, **protons** and **neutrons**.

> **KEY POINT**
>
> The nucleus of an element is represented as $^A_Z$El.
>
> $Z$ is the atomic number, the number of protons.
>
> $A$ is the mass number, the number of nucleons (protons and neutrons).

The element is fixed by $Z$, the number of protons. A neutral atom has equal numbers of protons in the nucleus and electrons in orbit. Different atoms of the same element can have different values of $A$ due to having more or fewer neutrons. As this does not affect the number of electrons in a neutral atom, the chemical properties of these atoms are the same. They are called **isotopes** of the element.

The most common form of carbon, for example, is carbon-12, $^{12}_6$C, which has six protons and six neutrons in the nucleus. Carbon-14, $^{14}_6$C, has the same number of protons but two extra neutrons. These two forms of carbon are isotopes of the same element.

## Atomic mass and energy conservation

| AQA A | M4 | OCR A | M4 |
|---|---|---|---|
| AQA B | M5 | OCR B | M5 |
| EDEXCEL A | AS | NICCEA | M5 |
| EDEXCEL B | M5 | WJEC | M5 |

> The phrase 'relative to the value' means compared to the actual amount of charge, ignoring the sign.

The charges and relative masses of atomic particles are shown in the table. The masses are in atomic mass units (u), where $1u = 1/12$ the mass of a carbon-12 atom $= 1.661 \times 10^{-27}$ kg. The charges are relative to the value of the electronic charge, $e = 1.602 \times 10^{-19}$ C.

| atomic particle | mass | charge |
|---|---|---|
| proton | 1.0073 | +1 |
| neutron | 1.0087 | 0 |
| electron | $5.49 \times 10^{-4}$ | –1 |

A carbon-12 nucleus consists of six protons and six neutrons. The mass of the atom is precisely 12u, by definition, so after taking into account the mass of the electrons, that of the nucleus is 11.9967u. The mass of the constituent neutrons and protons is:

$$6\,m_p + 6\,m_n = 6(1.0073\,u + 1.0087\,u) = 12.0960\,u$$

The nucleus has less mass than the particles that make it up. This appears to contravene the principle of conservation of mass. Einstein established that *energy has mass*. The mass that you gain due to your increased energy when you walk upstairs is infinitesimally small, but at a nuclear level this mass cannot be ignored. Any change in energy is accompanied by a change in mass, and *vice versa*.

> Try using Einstein's equation to calculate the increase in your mass when you climb upstairs to see how small it is.

> **KEY POINT**
>
> Einstein's equation relates the change in energy, $\Delta E$, to the change in mass, $\Delta m$
>
> $$\Delta E = \Delta mc^2$$
>
> where $c$ is the speed of light.
>
> It applies to **all** energy changes.

When dealing with the nucleus and nuclear particles, energy and mass are so closely linked that their equivalence, and that of their units, has been established.

> **KEY POINT**
>
> $$1\,u = 931.5\ \text{MeV}$$
>
> where 1 eV (one electron volt) is the energy transfer when an electron moves through a potential difference of 1 volt.

Using this relationship, the separate conservation rules regarding mass and energy can be combined into one so that (mass + energy) is always conserved in nuclear interactions.

To split a nucleus up into its constituent nucleons requires energy. It follows that a nucleus has less energy than the sum of the energies of the corresponding number of free neutrons and protons. So the fact that a nucleus has less energy than its nucleons would have in isolation means that it also has less mass.

> A common misconception is that the binding energy is the energy that holds a nucleus together. It is the energy needed to split it up.

> **KEY POINT**
>
> The difference between the sum of the masses of the individual nucleons and the mass of the nucleus is called the **mass defect** or **nuclear binding energy**. It represents the energy required to separate a nucleus into its individual nucleons.

In the case of carbon-12 the mass defect, or nuclear binding energy, is equal to $0.093\,u = 89.3$ MeV.

As would be expected, the greater the number of nucleons, the greater the binding energy. The diagram shows how the binding energy per nucleon varies with nucleon number. The most stable nuclei have the greatest binding energy per nucleon.

# Joining and splitting

| AQA A | M4 | OCR B | M5 |
| AQA B | M5 | NICCEA | M5 |
| EDEXCEL B | M5 | WJEC | M5 |
| OCR A | M4 | | |

The nuclear reaction that releases energy in stars like our Sun is fusion.

In a fusion reaction:

- small nuclei join together to form larger, more massive ones
- the nucleus formed in fusion has less mass than those that fuse together, due to its greater binding energy per nucleon
- the mass difference is released as energy.

Despite some scientists' claims to have carried out 'cold fusion' reactions, these have not been verified.

Fusion reactions can only take place at high temperatures where the nuclei have enough thermal energy to come together despite the repulsive forces that act as they approach each other.

Large nuclei release energy and become more stable by fission.

In fission:

- a large nucleus splits into two smaller ones and two or three neutrons
- the particles formed from fission have less mass than the original nucleus, due to the greater binding energy per nucleon
- the mass difference is released as kinetic energy of the fission fragments.

## Induced fission

| | | | |
|---|---|---|---|
| AQA A | M4 | OCR B | M5 |
| AQA B | M5 | NICCEA | M5 |
| EDEXCEL B | M5 | WJEC | M5 |

In a neutron-rich nucleus the neutron–proton ratio is greater than that required for stability.

A thermal neutron has energy similar to the mean kinetic energy of neutrons at room temperature.

Fission can occur spontaneously in a neutron-rich nucleus or it can be caused to take place by changing the make-up of a nucleus so that it is neutron-rich. This is known as induced fission.

Induced fission is triggered when a large nucleus absorbs a neutron. The nucleus formed has a greater mass than the original nucleus and neutron put together, so the neutron must supply the difference in the form of kinetic energy. A slow-moving, or thermal, neutron has the right amount of energy to cause the fission of uranium-235.

When uranium-235 undergoes fission:

- a nucleus of $^{235}_{92}U$ absorbs a neutron to become $^{236}_{92}U$
- the resulting nucleus is unstable and splits into two approximately equal-sized smaller nuclei, together with two or three neutrons
- a large amount of energy, approximately 200 MeV, is released as kinetic energy.

The fission of uranium-235 is illustrated in the diagram.

The fission of uranium-235

Check that this equation is balanced in terms of both mass and charge.

When uranium-235 fissions, the nuclei formed are not always the same. The equation represents one possible reaction, the products being barium and krypton.

$$^{235}_{92}U + ^{1}_{0}n \rightarrow ^{144}_{56}Ba + ^{90}_{36}Kr + 2^{1}_{0}n$$

## Releasing the energy

| | | | |
|---|---|---|---|
| AQA A | M4 | NICCEA | M5 |
| AQA B | M5 | WJEC | M5 |

If the neutrons released from fission go on to cause further fissions, then a chain reaction can build up. This is illustrated in the diagram overleaf.

If every neutron produced were to cause the fission of another uranium-235 nucleus, the reaction would quickly go out of control. This does not happen with a small quantity of uranium, since the neutrons produced in the fission process are moving at high speeds. These high-speed neutrons:

- are not as likely to be absorbed by uranium-235 nuclei as thermal neutrons are

- have a good chance of leaving the material altogether due to the relatively large surface of a small mass of material.

**A chain reaction**

The atomic bomb dropped on Hiroshima was detonated by combining three small masses of uranium into one that was above the critical mass.

For a chain reaction to be sustained then, on average, the neutrons released from each fission must go on to cause one further fission. The minimum mass of material needed for this to happen is called the critical mass. For uranium-235, the critical mass is 15 kg. This amount of uranium-235 in a spherical shape (so that it has the minimum surface area) can just sustain a reaction without it dying out.

## Harnessing the energy

| AQA A | M4 | NICCEA | M5 |
| AQA B | M5 | WJEC | M5 |

In a nuclear reactor, energy released from fission is removed and used to generate electricity. Three important features of a reactor are the moderator, control rods, and coolant. These are shown in the diagram.

This type of reactor is called a thermal reactor as it uses thermal neutrons to cause fission.

**A nuclear reactor**

The moderator:

- slows down the high-speed neutrons released by fission so that they are more likely to cause further fissions
- should not absorb neutrons.

Deuterium, $_1^2H$, is an isotope of hydrogen that contains an extra neutron.

Graphite and heavy water, similar to normal water but using deuterium in place of hydrogen, are commonly used as moderators. Normal water is unsuitable as the hydrogen nuclei absorb neutrons.

The control rods:

- are used to control the neutron concentration by absorbing neutrons
- can be raised or lowered within the moderator material to increase or decrease the rate at which fissions take place
- must be able to absorb neutrons without becoming unstable.

The material used for control rods needs to have nuclei that are successful at capturing neutrons. Cadmium and boron are both used as they can capture neutrons over a large cross-section of their nuclei.

The coolant:

- should flow easily
- should not corrode the metal casings of the moderator and fuel rods
- removes energy to a heat exchanger, where steam is generated to turn turbines as in a coal-fired power station.

The coolant used depends on the temperature of the reactor core. Coolants in use include water, carbon dioxide and liquid sodium.

## Nuclear fuels

*Fast-breeder reactors use **fast** neutrons and **breed** their own fuel.*

The first nuclear power stations to be built used natural uranium as a fuel. Natural uranium contains less than 1% uranium-235, the remainder being uranium-238 which does not fission readily. Enriched uranium fuel contains a higher proportion of uranium-235, up to 3%, so it has a longer usable lifetime than natural uranium. Fast-breeder reactors use the plutonium, $^{239}_{94}Pu$, formed when uranium-238, $^{238}_{92}U$, captures a neutron and then decays by $\beta^-$ emission (twice). Plutonium needs fast neutrons to fission, but the spare neutrons left over from the fission process are used to create more plutonium fuel from uranium-238, so they make their own fuel from the relatively abundant isotope of uranium, uranium-238.

## Safety

No nuclear reactor is totally safe. The coolant and other materials used in building the reactor become radioactive. The by-products of fission are also radioactive. Removal of the control rods results in a melt-down, as happened at Chernobyl. This released vast quantities of radioactive material into the atmosphere, having disastrous effects in the immediate vicinity and affecting farmland thousands of miles away.

In normal use a thick concrete wall built around the reactor shields the outside environment by absorbing much of the radiation emitted within. In an emergency, a reactor can be shut down by lowering the control rods fully so that they absorb more neutrons, leaving insufficient neutrons to maintain the chain reaction.

The way in which active waste materials are disposed of depends on the level of radioactivity.

- Low-level waste, such as laboratory clothing and packaging, is buried underground or at sea.
- Medium-level waste, such as empty fuel casing and parts of the reactor fabrication, is kept in a concrete-lined store or underground cavern.
- High-level waste, such as the radioactive fission by-products and spent fuel rods, is kept in steel tanks of water to keep it cool.

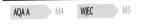

*Spent fuel rods contain highly radioactive materials with long half-lives. There is no safe way of disposing of these.*

## Artificial nuclides

Many naturally-occurring radioactive isotopes have long half-lives and so are not suitable for use as tracers in the environment or in the body. Man-made radioactive isotopes can be produced in two ways:

- in a nuclear reactor, by irradiating a stable isotope with neutrons
- by firing charged particles at a stable isotope using a particle accelerator.

The resulting change in the nucleus is known as artificial transmutation. An isotope used extensively in medicine is a form of technetium-99, $^{99}Tc^m$, which emits gamma radiation only and has a half-life of 6 hours. It is formed when molybdenum-99, artificially created in a reactor, decays by beta-emission. This form of technetium-99 is particularly useful in medical diagnosis because:

- the gamma radiation can be detected outside the body using a gamma camera
- it does not emit the more intensely-ionising alpha and beta radiations which could cause cell damage
- the half-life is long enough for it to be detected for several hours after being injected and allowed to circulate the body
- the half-life is short enough for a minimum dose to be used
- by attaching the technetium-99 to different substances, it can be targeted towards specific body organs.

> The shorter the half-life, the greater the rate of decay so the smaller the dose needed to be detectable.

## Progress check

1 $4.0 \times 10^5$ J of energy is transferred to some water in a kettle to bring it to the boil.
Calculate the increase in mass of the water due to this energy transfer.

2 Describe the difference between nuclear *fusion* and nuclear *fission*.

3 Explain why a fast-breeder reactor does not need a moderator.

3 Fission is caused by fast neutrons, so they do not need to be slowed down.
Fission is the splitting up of a nucleus.
2 Fusion is the joining together of two nuclei.
1 $4.4 \times 10^{-12}$ kg

# 4.4 Probing matter

**After studying this section you should be able to:**

- describe how the scattering of alpha particles and high-energy electrons gives evidence for the atomic model
- understand how electron diffraction enables measurements of atomic spacing to be made
- understand how the diffraction of X-rays and neutrons gives evidence for crystal structures

*LEARNING SUMMARY*

## Evidence for the atomic model

| | | | |
|---|---|---|---|
| AQA A | M5–9 | OCR A | M4 |
| AQA B | AS | OCR B | M5 |
| EDEXCEL A | AS | NICCEA | M5 |
| EDEXCEL B | M4 | WJEC | M5 |

*An alpha particle is a positively-charged particle consisting of two protons and two neutrons.*

*The experiments were carried out in a vacuum so that the alpha particles were not scattered by air particles.*

The simple atomic model pictures the atom as a tiny, positively-charged nucleus surrounded by negatively-charged particles (electrons) in orbit. Evidence for this model comes from the scattering of alpha particles as they pass through a thin material such as gold foil. The results of these experiments, first carried out under the guidance of Rutherford in 1911, can be summarised as:

- most of the alpha particles travel straight through the foil with little or no deflection
- a small number are deflected by a large amount
- a tiny number are scattered backwards.

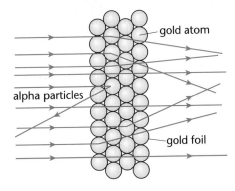

alpha particles / gold atom / gold foil

Rutherford concluded that the atoms of gold are mainly empty space, with tiny regions of concentrated charge. This charge must be the same sign as that of alpha particles (positive) to explain the back-scattering as being due to the repulsion between similar-charged objects.

## The size of the nucleus

| | | | |
|---|---|---|---|
| AQA A | M5–9 | OCR B | M5 |
| EDEXCEL A | AS | NICCEA | M5 |
| OCR A | M4 | WJEC | M5 |

*It is a maximum value as the nucleus cannot be larger than this but is probably smaller, as the repulsive force reverses the direction of motion of the alpha particles before they reach the nucleus.*

Rutherford's experiments established the existence of a nucleus that is very small compared to an atom. They also enabled the first estimates of the size of a nucleus to be made. When an alpha particle is scattered back along its original path:

- as it approaches the nucleus its kinetic energy is transferred to potential energy due to its position in the electric field
- at the closest distance of approach all its energy is potential
- the energy transfer is reversed as it moves away from the nucleus.

The closest distance of approach to the nucleus is therefore calculated from the kinetic energy of the alpha particle, assuming that Coulomb's law (see section 3.2) applies to the nucleus. This gives a maximum value for the size of the gold nucleus of about $10^{-14}$ m.

## Electron diffraction

| | | | |
|---|---|---|---|
| AQA A | M5-9 | OCR B | AS |
| EDEXCEL A | AS | NICCEA | MS, M6 |
| OCR A | M4 | WJEC | MS |

More precise evidence comes from the scattering of electrons, known as electron diffraction. Moving electrons show wave-like behaviour (see section 2.5) with a wavelength that depends on their speed.

Electrons accelerated through a potential difference of a few hundred volts have a wavelength similar to that of X-rays and gamma rays.

This wavelength is also similar to the spacing of the atoms in crystalline materials, so these materials provide suitable sized 'gaps' to cause diffraction.

Diffraction patterns formed by a beam of electrons after passing through thin foil or graphite show a set of 'bright' and 'dark' rings on photographic film, similar to those formed by X-ray diffraction (see below).

Analysis of these patterns is used to calculate the spacing between rows of atoms in the sample. This is typically around $10^{-10}$ m.

Higher energy electrons are used to estimate the size of a nucleus. Electrons accelerated through a voltage of a few hundred million volts have a wavelength comparable to that of a nuclear diameter.

When probing matter, the choice of particle used depends on its interaction with the matter being probed.

Unlike alpha particles, electrons are not repelled as they approach a nucleus. Instead, they are diffracted in the same way that light is diffracted around a circular or spherical obstacle and a diffraction pattern similar to that shown in the diagram above emerges, with the rings having fuzzy edges.

Analysis of the diffraction pattern gives a nuclear size of about $10^{-15}$ m and the fuzziness of the rings shows that the nucleus does not have a perfectly circular cross-section.

Unlike their nuclei, the atoms of different elements can have very different densities.

The results of electron diffraction experiments show that:

- all nuclei have approximately the same density, about $1 \times 10^{17}$ kg m$^{-3}$
- the greater the number of nucleons, the larger the radius of the nucleus
- the nucleon number, $A$, is proportional to the cube of the nuclear radius, $R$.

An alternative way of writing this last point is:

the nuclear radius, $R$, is proportional to the cube root of the nucleon number, $A$.

> **KEY POINT**
>
> The relationship between nucleon number, $A$, and nuclear radius, $R$, is:
> $$R = r_0 A^{1/3}$$
> where $r_0$ has the value $1.2 \times 10^{-15}$ m.

This relationship means that if the nuclei of ⁶⁴Cu and ³²S are compared, the radius of the copper nucleus is $2^{1/3} = 1.26$ times that of the sulphur, as the nucleon number of copper is twice that of sulphur. Similarly, the copper nucleus has a radius which is $4^{1/3} = 1.59$ times that of ¹⁶O.

## X-ray diffraction

OCR A · M4 · WJEC · M5
NICCEA · M5

Metals are made up of small crystals. You can see these clearly on a clean metal surface such as the zinc plating on the inside of a can used for pineapple.

The diffraction of X-rays by crystalline materials such as metals shows the arrangement of atoms and allows measurements of their size. The regular arrangement of atoms in parallel planes causes a diffraction pattern when the waves used are comparable to the spacing. X-rays with a wavelength of the order of $1 \times 10^{-10}$ m (0.1 nm) are suitable for this use. The diagram shows X-rays reflected at three consecutive planes within a crystal.

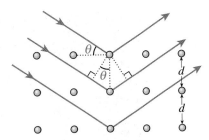

Constructive interference between the reflected X-rays is detected on photographic film. The angles, $\theta$, at which this occurs are related to the wavelength of the X-rays and the spacing of the atomic planes, $d$, allowing the spacing to be calculated.

X-rays are absorbed by some metals due to their interaction with electrons.

X-rays do not interact with the nuclei of atoms in a material as they are uncharged and are not affected by electric or nuclear forces.

## Neutron diffraction

OCR A · M4 · WJEC · M5

Neutrons moving at a speed of around $4 \times 10^4$ m s⁻¹ have a similar wavelength to that of X-rays and therefore should show a similar diffraction pattern of a crystal. This turns out to be the case, but neutron diffraction can give different information about a crystal structure to that obtained from X-ray diffraction. This is because:

The magnetic field is due to the spin of the neutrons and the charges contained within them.

- moving neutrons have a magnetic field
- this magnetic field interacts with the magnetic fields of orbiting electrons, revealing information about the magnetic structure of a material
- some metals absorb X-rays but transmit neutrons.

The diagrams on the next page compare X-ray and neutron diffraction patterns for the same metal alloy. Note that each one shows the molecules arranged in a regular pattern and reveals the structure of the crystal.

These diffraction patterns are due to a metal oxide, $ZrTiO_4$.

The contrasts in the diffraction patterns are:

- both show the arrangement of four oxygen atoms around the metal atoms
- the X-ray diffraction pattern shows the metals zirconium (Zr) and titanium (Ti) very clearly
- the neutron diffraction pattern shows the oxygen atoms more clearly than the metal atoms.

Neutron diffraction is used in preference to X-ray diffraction when examining the structure of nuclear fuel rods and some archaeological artefacts.

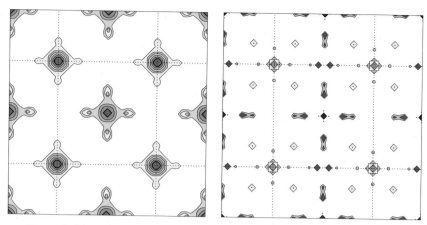

**X-ray (left) and neutron (right) diffraction patterns for a metal alloy**

## Progress check

1 In alpha particle scattering by a gold nucleus, what feature of the results of the experiment shows that:
   **a** most of the atom is empty space
   **b** the nucleus has the same charge as the alpha particles?

2 What is the speed of an alpha particle at its closest distance of approach to a nucleus?

3 Explain why it is more difficult to accelerate neutrons than it is to accelerate electrons.

3 Electrons are charged so they can be accelerated by an electric field. Neutrons are uncharged.

2 Zero.

1 **a** Most of the particles pass through undeviated.
  **b** Some alpha particles are back-scattered.

# Sample question and model answer

Radon-220 decays by *alpha-emission* to polonium-216.

(a) Explain what is meant by *alpha-emission*.                    [2]

Note that the two marks awarded here are for describing the structure of an alpha particle and describing the emission.

The nucleus emits a particle (1 mark) which consists of two protons and two neutrons (1 mark).

(b) Complete the decay equation.                    [2]

$$^{220}_{86}Rn \rightarrow Po + He$$

The nuclear equation should be balanced in terms of both mass (top number) and charge (lower number).

$^{220}_{86}Rn \rightarrow ^{216}_{84}Po + ^{4}_{2}He$ (1 mark for each correct symbol)

(c) The graph shows how the activity of a sample of radon-220 changes with time.

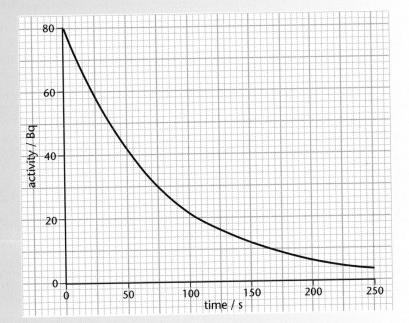

(i) Use the graph to determine the half-life of radon-220.                    [2]

It is important to take at least two readings here to determine the *average* time for the rate of decay to halve. Had the two readings been different, a third one would be necessary.

time taken for activity to halve from 80 Bq to 40 Bq = 52.5 s   1 mark
time taken for activity to halve from 40 Bq to 20 Bq = 52.5 s
half-life = average of these readings = 52.5 s                  1 mark

(ii) The relationship between half-life, $t_{1/2}$ and decay constant, $\lambda$, is:

$$\lambda \, t_{1/2} = \ln 2$$

If these relationships are not given in the question, they will be provided on a data sheet. Always check a data sheet, it may contain the formula that you cannot remember!

Calculate the decay constant of radon-220.                    [2]

$\lambda = \ln 2 \div t_{1/2}$ (1 mark) $= \ln 2 \div 52.5$ s $= 1.32 \times 10^{-2}$ s$^{-1}$   1 mark

(iii) The activity, *A*, of a radioactive sample is related to the number of undecayed nuclei, *N*, by the formula:

$$A = \lambda N$$

Calculate the number of undecayed nuclei at time = 0 on the graph.   [2]

activity at time $t = 0$ is 80 Bq                                       1 mark
$N = A \div \lambda = 80$ Bq $\div 1.32 \times 10^{-2}$ s$^{-1}$ = 6060   1 mark

## Practice examination questions

**1**

A sample of an ideal gas has a volume of $2.20 \times 10^{-6}$ m³ at a temperature of 25°C and a pressure of $2.02 \times 10^5$ Pa.

(a) Calculate the temperature of the gas in kelvin. [1]

(b) Calculate the number of moles of gas in the sample. [3]

$R = 8.3$ J mol⁻¹ K⁻¹

(c) The absolute temperature of the gas is doubled, while the pressure remains constant.

(i) What is meant by the *internal energy* of the gas? [1]

(ii) In what form is this energy in an ideal gas? [1]

(iii) What happens to the internal energy of the gas when its absolute temperature is doubled? Explain your answer. [2]

(iv) Calculate the new volume occupied by the gas after its absolute temperature has doubled. [2]

**2**

In a nuclear reactor uranium-238, $^{238}_{92}$U, undergoes fission when it captures a neutron.

(a) (i) Write down the symbol for the isotope formed when uranium-238 captures a neutron. [1]

(ii) Explain what happens when a nucleus of this isotope undergoes fission. [3]

(iii) Explain how fission can lead to a chain reaction. [2]

(b) (i) Explain why nuclear fission results in the release of energy. [2]

(ii) In what form is the energy released? [2]

(iii) Outline how the energy released in fission is used to generate electricity. [3]

(c) State the purpose of each of the following parts of a nuclear reactor:

(i) control rods [1]

(ii) moderator. [1]

**3**

A kettle is filled with 1.50 kg of water at 10°C.
The kettle is fitted with a 2.40 kW heater.
The specific heat capacity of water is $4.2 \times 10^3$ J kg⁻¹ K⁻¹.

(a) Calculate the time it takes for the water to boil after the kettle is switched on. [3]

(b) Suggest three reasons why the time taken is longer than the answer to (a). [3]

**4**

Iodine-123 and iodine-131 are two radioactive isotopes of iodine.
Iodine-123 emits gamma radiation only with a half-life of 13 hours.
Iodine-131 emits beta and gamma radiation and has a half-life of 8 days.

Both can be used to monitor the activity of the thyroid gland by measuring the count rate using a detector outside the body.

(a) Which of the radioactive emissions from iodine-131 can be detected outside the body? Give the reason for your answer. [2]

(b) In terms of the radiation emitted, explain why iodine-123 is often preferred to iodine-131. [2]

## Practice examination questions *(continued)*

(c) Explain why a smaller dose is needed when iodine-123 is used than for iodine-131. [2]

(d) A patient is given a dose of iodine-123. After what time interval can medical staff be sure that the activity of the patient's thyroid gland is less than 2% of its greatest value? [2]

**5**

Information about the size of a nucleus can be obtained by firing high-energy electrons at the nucleus and examining the subsequent diffraction pattern.

(a) (i) Why are electrons suitable for this purpose? [2]
    (ii) Explain why the electrons need to be high-energy. [3]

(b) The relationship between the radius, $R$, of a nucleus and its atomic number, $A$ is:
$$R = r_0 A^{1/3}$$
where $r_0$ has the value $1.20 \times 10^{-15}$ m.

Calculate the radius of the nuclei of:

    (i) $^4_2\text{He}$ [2]

    (ii) $^{32}_{16}\text{S}$ [1]

(c) Explain what the answers to (b) show about the densities of the helium and sulphur nuclei. [3]

**6**

The diagram shows some of the results of firing alpha particles at gold foil.

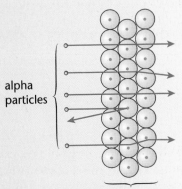

alpha particles

atoms in gold foil

Explain how each of the following provides evidence for the atomic model.

(a) Most alpha particles pass through the foil undeviated. [2]

(b) A very small number of alpha particles are scattered backwards. [3]

# Medical physics

*The following topics are covered in this chapter:*

- *Body mechanics*
- *Using lenses*

- *Hearing*
- *Medical imaging and treatment*

## 5.1 Body mechanics

**After studying this section you should be able to:**

- *understand how bones and muscles enable body movement*
- *apply the principle of moments to bones and muscles*
- *explain how the force between the body and the ground changes when the body moves*

**LEARNING SUMMARY**

## Body structure

OCR A ▶ M5.2

Animal cells do not have cell walls to give them rigidity, instead they rely on a system of **bones** which are connected together to make a **skeleton**. In the skeleton:

- bones meet at different types of **joint** which allow restricted movement
- at a joint, **ligaments** connect the bones together
- **muscles** are responsible for moving bones, they are attached to the bones by **tendons**.

The diagram below shows the ligaments, muscles and tendons at a hip joint.

Muscles can only pull. They exert a force by contracting and release it when they relax. The two sets of muscle tissue shown in the diagram allow movement both ways. They are known as antagonistic muscles.

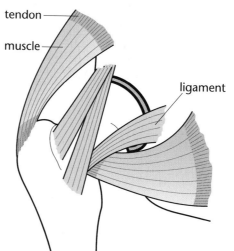

Bones are moved when muscles contract. The movement of a bone by a muscle is an example of a **lever** being used to magnify distance. A small contraction of the muscle causes a much larger movement at the far end of the bone that it connects to. The diagram below shows the turning forces involved when the biceps contract to raise the forearm from a horizontal position in order to raise a hand-held load.

Use your own arm to estimate the distances between the forces and the pivot.

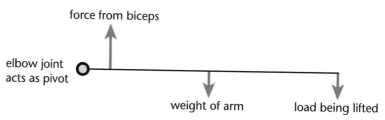

For the system to be balanced, the moment of the force from the biceps must be equal to the sum of the moments of the weight of the arm and the load held in the hand – this is an application of the **principle of moments**. Because of the relative distances, this means that the force from the biceps is much greater than the sum of the other two forces.

The **mechanical advantage** of a lever is a measure of how much easier it enables a job to be done.

> Most pulley systems have mechanical advantages greater than 1, they enable a small input force to move a large load.

**KEY POINT**

mechanical advantage = load/effort
where the effort is the input force and the load is the output force.
Mechanical advantage is a ratio so it has no units.

Joints in the body have mechanical advantages less than one; their effect is to enable a small movement to cause a much larger movement.

## Forces and the body

OCR A ▷ MS.2

When standing still, the upward force from the ground that acts on a body is equal to the weight of the body. Walking and running both require additional forces to accelerate and decelerate the body both vertically and horizontally.

When a foot leaves the ground:

- an additional upward force acts to accelerate the body upwards in order to raise the centre of mass
- a forwards-directed force due to friction acts to accelerate the body forwards.

> When you walk or run, you push down on the ground so that it pushes back up on you.

When a foot hits the ground:

- an additional upward force acts to decelerate the body vertically
- a backwards-directed force due to friction acts to decelerate the body in the forwards direction.

The resultant of the forces acting on a foot is called the **ground force**; it acts parallel to the leg. The diagram shows the ground force, $G$, when a foot hits and leaves the ground during walking.

> The reaction force, $R$, is equal to the body weight when standing still and greater than the body weight when walking and running.

## Bending and lifting

OCR A    M5.2

Back injuries are a major cause of time lost at work due to illness.

When a person bends, the upper body turns around the base of the spine. The muscles attached to the spine have to produce a moment to counteract that of the weight of the upper body (about 60% of the total body weight). This is shown in the diagram below.

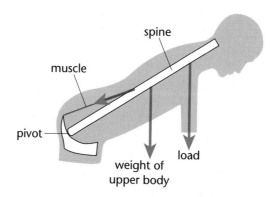

Because of the angles involved, the force from the muscle must be very much greater than the weight of the upper body. The diagram also shows what happens when a load is lifted. This puts a large stress on the muscle and the disc at the base of the spine. To avoid damage to these, heavy loads should be lifted by crouching, bending the knees and keeping the spine vertical.

## Progress check

1 At a joint, what:
   a connects the bones together
   b attaches the muscles to the bones
   c contracts to cause movement?

2 The data in the table refers to the diagram of a forearm on page 120.

| force | value of force | distance between force and pivot |
|---|---|---|
| biceps | | 0.04 m |
| weight of arm | 20 N | 0.18 m |
| load | 50 N | 0.36 m |

Calculate the force in the biceps.

2 540 N
   c muscle
   b tendon
1 a ligament

# 5.2 Using lenses

*After studying this section you should be able to:*

- *use ray diagrams and the lens formula to describe image formation by lenses*
- *identify, and state the function of, the main parts of the eye*
- *explain how the quality of vision depends on the intensity of the ambient light*

**LEARNING SUMMARY**

## The action of a lens

| AQA A | M5, M6 | OCR B | AS |
|-------|--------|-------|-----|
| EDEXCEL B | AS | NICCEA | AS |
| OCR A | M5.2 | | |

Light from a source converges after passing through a concave lens if the source is at a greater distance from the lens than the focal length.

The dotted lines in the diagram of a concave lens show where light **appears** to have come from.

The image is virtual in the case of a concave lens, and real for a convex lens.

When light leaves a lamp or other source of light it spreads out or diverges, reducing in intensity with increasing distance from the source. A concave or diverging lens exaggerates this effect, making the light more divergent. A convex or converging lens acts against this effect, so that the light either converges or becomes less divergent.

The diagrams show the effect of these lenses on a parallel beam of light.

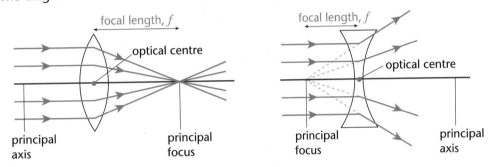

The diagrams also show some important features of the lenses:

- the **optical centre** is at the centre of the lens
- the **principal axis** is a line drawn through the optical centre, perpendicular to the plane of the lens
- the **principal focus** is the image position for light that is parallel to the principal axis
- the **focal length** is the distance between the optical centre and the principal focus.

The shorter the focal length of a lens, the greater the effect on converging or diverging light that passes through it. This effect is measured by the power of the lens.

> **Lens power** is calculated using the relationship:
>
> power = 1/focal length
>
> The unit of power is the dioptre (D) when the focal length is measured in m.

**KEY POINT**

## The lens formula

| AQA A | M5, M6 | OCR B | AS |
|-------|--------|-------|-----|
| EDEXCEL B | AS | NICCEA | AS |
| OCR A | M5.2 | | |

Concave lenses form **virtual images**. Convex lenses can form either **real** or virtual images, depending on the distance of the object from the optical centre of the lens compared to its focal length. In each case, the position, nature and size of the image can be determined by a ray diagram.

The diagram below shows ray diagrams for an object placed at a distance of 8 cm from converging and diverging lenses each of focal length, $f = 5$ cm.

A real image is formed if the light converges after passing through the lens. If the light is still diverging, then the image is virtual.

The position, nature and size of the images can be determined from the diagrams. They can also be calculated using the **lens formula**:

The magnification, i.e. the size of the image compared to the object, is equal to $v/u$.

> **K E Y   P O I N T**
>
> The lens formula applies to both convex and concave lenses. It relates the positions of the object and image to the focal length of the lens.
>
> $$\frac{1}{u} + \frac{1}{v} = \frac{1}{f}$$
>
> Where $u$ is the distance between the object and the lens, $v$ is the distance between the image and the lens, and $f$ is the focal length of the lens.

This is known as the 'real is positive' sign convention.

When using the lens formula:

- real objects, images and the focal length of a convex lens have positive values
- virtual images, and the focal length of a concave lens, have negative values.

## Structure of the eye

| AQA A | M6 | NICCEA | AS |
| OCR A | M5.2 | | |

The eye uses two converging lenses, the cornea and the eye lens, to produce a real image on the retina. The diagram shows the main parts of the eye and the focusing action of these lenses.

The cornea also acts as a protective covering, so that no dust or other objects can enter the eye.

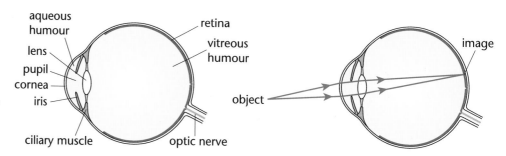

- The **cornea** is a fixed lens that carries out most of the convergence of light required for image formation.
- The **aqueous humour** and **vitreous humour** supply the eye with essential nutrients and exert pressure to maintain the shape of the eyeball.
- The **iris** adjusts the size of the pupil to control the amount of light entering the eye.
- The **lens** is of variable focal length and acts as a fine-focus control.
- The **ciliary muscles** are connected to the lens by ligaments: they can relax and contract to change the shape of the lens.
- The **retina** consists of nerve-endings which are sensitive to the colour and brightness of light and produce electrical signals.
- The **optic nerve** transfers these signals to the brain.

## Focusing the eye

AQA A · M6 · NICCEA · AS
OCR A · M5.2

For a normal eye which does not require spectacles, the near point is at a distance of about 25 cm and the far point is at infinity.

The **near point** is the closest distance that an eye can focus on. The eye lens is at its most powerful when focusing on the near point and at its least powerful when focused on the **far point**, the greatest distance from the eye at which objects can be seen clearly. Whatever the distance the object is that the eye is focused on, objects nearer and further than this can also be seen clearly. The range of distances over which objects can be seen in focus is called the **depth of field**.

Adjustment of the eye to focus on objects at different distances is called **accommodation**; this is achieved by changing the shape of the lens.

## Defective vision

AQA A · M6 · NICCEA · AS
OCR A · M5.2

An eye lens is elastic tissue which becomes more difficult to stretch with increased use.

The common defects in vision are:

- **short sight** or **myopia** when an eye cannot focus on distant objects but can focus on near objects. It is caused by the eyeball being too long or the lens, at its weakest, being too strong.
- **long sight** or **hypermetropia** is the opposite of short sight; distant objects can be seen clearly but near ones cannot. It is caused by the eyeball being too short or the lens, at its strongest, being too weak.
- **presbyopia** is the condition of being both long-sighted and short-sighted. It often occurs in older people due to the lens becoming stiffer with age.
- **astigmatism** is caused by the surface of the cornea not being spherical. This results in objects in one direction being sharp and in focus while those in a different direction are blurred.

The diagrams show the conditions of short sight and long sight and their correction with converging and diverging lenses respectively.

These diagrams shows light from a distant object arriving at a short-sighted eye.

short sight

correcting short sight with a diverging lens

These diagrams show light from a near object arriving at a long-sighted eye.

long sight

correcting long sight with a converging lens

The lens formula can be used to calculate the focal length, and hence power, of the correcting lens needed. The object distance, *u*, is the distance of the actual object and the image distance, *v*, is the desired distance for the eye to focus on it.

People who suffer from presbyopia often wear bifocal spectacles. The lower half of each lens is convex for reading and the upper half is concave for distant viewing.

To correct astigmatism cylindrical lenses are used. These are only curved in one direction to compensate for the unsymmetrical curvature of the cornea.

## Colour vision

AQA A · M6 · NICCEA · AS
OCR A · M5.2

The light-sensitive cells that make up the retina have different sensitivities:

- **rods** are sensitive to light intensity only
- there are three types of **cones**, called red, green and blue. Each of these responds to a different range of wavelengths and frequencies.

This explains why, in low-intensity light such as moonlight, we see in monochrome.

Cones enable colour vision but they do not respond to low-intensity light. The graph shows the relative sensitivities of the three types of cone.

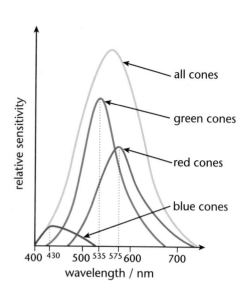

The diagram shows why blue-coloured lamps always appear dim and yellow-coloured lamps appear bright.

The low sensitivity of the blue cones explains why we never perceive this colour as bright as red or yellow. An advertising hoarding designed to catch the eye would not use blue to do so. However, it is a more acceptable colour for 'low-key' signs and buildings.

Vision in low-intensity light is scotopic – we cannot detect colour but only differences in brightness. In scotopic vision the cones are not responsive so the rods are the only detectors operating.

Vision is most acute at the fovea, or yellow spot. This is the point on the retina where the light-sensitive cells are most closely-packed.

In more intense light the cones take over and vision is photopic. Photopic vision allows differentiation of colour in addition to intensity. The ability of the eye to distinguish between two objects that are close together, called its spatial resolution, is also greater in photopic vision. Spatial resolution requires at least one unstimulated cell between those that are stimulated. Resolution is greatest at the centre of the retina where the cells are most concentrated. It is worst at the edge of the retina where the cells are more widespread and several rods share a single nerve fibre, so the brain cannot distinguish between individual rods.

## Making still pictures move

AQA A    M6

In film animation, a series of photographs is taken, with each one showing a small change in position from the previous photograph.

We 'see' moving pictures when watching a film at a cinema or watching television. In reality there are no moving pictures.

- At the cinema, 24 still pictures appear on the screen each second.
- On the television, there is only a 'splodge' of light, consisting of three coloured patches, at any instant. The splodge sweeps the screen 25 times each second, with each sweep consisting of 625 lines.

A succession of still pictures merge into each other because the light-sensitive cells on the retina remain in an excited state for a short time after the image has been removed. This is called persistence of vision and it also explains why we see a moving dot on a CRO screen as a line.

## Progress check

1 An object of height 2.0 cm is placed on the principal axis at a distance of 12.0 cm from a convex lens of focal length 5.0 cm.
By drawing or calculation, work out the position, size and nature of the image.

2 Which TWO parts of the eye converge light as it passes through them?

3 Explain why the brain does not perceive colour in low-intensity ambient light.

3 The cones do not function in low-intensity light, and the rods only detect brightness.

2 The cornea and the eye lens.

1 The image is real, 8.6 cm on the other side of the lens and 1.4 cm high.

# 5.3 Hearing

**After studying this section you should be able to:**

- *describe how sound travels through the ear and is detected by the brain*
- *explain that the ear is sensitive to a range of frequencies and intensities of sound*
- *compare sound intensities in W m⁻² and in decibels*

## Ear structure

AQA A  M6    OCR A  M5.2

The ear receives sound in the form of compression waves and produces an electrical signal that represents these waves. This is the action of a transducer, a device that transfers a signal from one form to another.

The diagram shows the main parts of the ear.

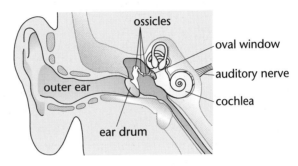

> The three ossicles are the malleus (hammer), incus (anvil) and stapes (stirrup).

In the ear:

- sound waves in the outer ear cause the ear drum to vibrate

> The ear drum is also known as the tympanic membrane. It is very thin and vibrates with a typical amplitude of $10^{-9}$ m.

- the ossicles (ear bones) transmit the vibrations to the oval window of the cochlea. In doing so they reduce the amplitude of the vibrations and maximise the energy transfer

- vibrations of the fluid in the cochlea are detected by tiny hairs which produce electrical impulses

- these impulses are transmitted along the auditory nerve to the brain.

## Sensitivity of the ear

AQA A  M6    NICCEA  AS
OCR A  M5.2

The ear responds to a range of frequencies and a range of intensities.

> **KEY POINT**
>
> The intensity of a wave is defined as the power incident per unit area perpendicular to the direction of wave travel.
>
> $$I = P/A$$
>
> where intensity, $I$ is measured in W m⁻². Intensity can also be measured in **decibels** (dB) – see overleaf.

Although increasing the intensity of a sound at a particular frequency makes it sound louder, we perceive sounds of different frequencies and the same intensity as varying in loudness. This is because of the different sensitivity of the ear at different frequencies, shown in the graph overleaf.

127

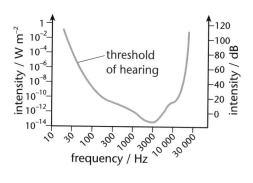

This graph shows that:

- the frequency range of a human ear is approximately 20 Hz to 20 000 Hz, but the upper limit falls with increasing age and exposure to noise due to wear of the ossicles

- the minimum intensity required to detect a sound, $I_0$, depends on frequency

- the ear is at its most sensitive at frequencies around 3000 Hz.

Logarithmic scales are used on the graph because the ear responds to relative changes in intensity, it has a logarithmic response. This means that doubling the intensity doubles the perceived loudness, no matter what the initial intensity of the sound was. This leads to the decibel scale for measuring intensity.

$$\text{intensity in decibels} = 10 \log (I/I_0)$$
where $I_0$, the threshold of hearing, is taken to be $1.0 \times 10^{-12}$ W m$^{-2}$.

On the decibel scale, an increase of 10 dB corresponds to the intensity becoming ten times as great.

A sound of intensity 120 dB or 1 W m$^{-2}$ causes discomfort as it causes feeling in the ear; one of 140 dB or 100 W m$^{-2}$ causes extreme pain in the ear.

## Loudness

AQA A ▸ M6    OCR A ▸ M5.2

Because our ears are most sensitive to frequencies around 3000 Hz, higher and lower frequencies need a greater intensity to sound equally loud. Loudness is measured in phons, where 1 phon is equivalent to an intensity of 1dB at a frequency of 1kHz. The graph shows the variation in intensity required to cause equal loudness at four loudness levels.

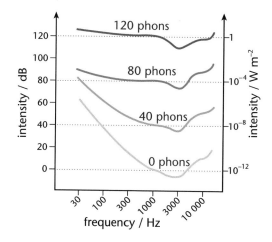

The dBA scale of intensity takes into account the fact that mid-range frequency sounds seem louder than those at the extreme frequencies.

## Progress check

1 When light passes through a slit which is 2 mm wide, it produces a narrow beam. Explain why there is no observable diffraction.

2 For an observable interference pattern between two sources, the sources must be **coherent**.
   a What is meant by two coherent sources?
   b What TWO other conditions should be met for an observable interference pattern?

3 In a two-slit interference experiment using light, the slits are separated by a distance of 1.2 mm. The distance from the slits to the screen is 3.0 m and the separation of the bright fringes is 1.4 mm.
   Calculate the wavelength of the light.

3 $5.6 \times 10^{-7}$ m
   b They should be the same type of wave and similar in amplitude.
2 a Coherent means they have a fixed phase relationship.
1 The width of the slit is very large compared to the wavelength of light.

# 4.4 Wave properties across the spectrum

*After studying this section you should be able to:*

- *describe the main features of the electromagnetic spectrum*
- *explain how the photoelectric effect gives evidence for the particulate nature of electromagnetic radiation*
- *explain line spectra and relate the spectral lines to energy level transitions*

**LEARNING SUMMARY**

## The spectrum

| | | | |
|---|---|---|---|
| AQA B | M2 | OCR A | M2 |
| EDEXCEL A | A2 | OCR B | M1, M2 |
| EDEXCEL B | M1 | WJEC | M2 |
| NICCEA | M2 | | |

The speed of all electromagnetic waves in a vacuum is $2.999 \times 10^8$ m s$^{-1}$. The value $3.00 \times 10^8$ m s$^{-1}$ is normally used for both a vacuum and air.

The different changes of speed when light consisting of a range of wavelengths is refracted is responsible for dispersion, the splitting of light into a spectrum.

Microwaves are given here as a separate part of the spectrum, although they can be considered as short-wavelength radio waves.

The electromagnetic spectrum consists of a whole family of waves, with some similarities and some differences in their behaviour.

Similar properties include:

- they consist of electric and magnetic fields oscillating at right angles to each other
- they travel at the same speed in a vacuum
- they are transverse waves, so they can be polarised
- they show the same pattern of behaviour in reflection, refraction, interference and diffraction.

The differences include:

- the shorter wavelength waves undergo a greater change of speed when being refracted
- wave-like properties such as diffraction and interference are more readily observable in the behaviour of the longer wavelength waves.

The diagram shows the range of wavelengths and frequencies of the waves that make up the spectrum.

| frequency / Hz | $10^{20}$ | | $10^{17}$ | | $10^{14}$ | | $10^{11}$ | | $10^{8}$ | | $10^{5}$ |
|---|---|---|---|---|---|---|---|---|---|---|---|

**gamma rays**　**ultraviolet**　**infra-red**　**radio waves**

**X-rays**　**light**　**microwaves**

| wavelength / m | $10^{-12}$ | | $10^{-9}$ | | $10^{-6}$ | | $10^{-3}$ | | 1 | | $10^{3}$ |
|---|---|---|---|---|---|---|---|---|---|---|---|

In general, the name given to an electromagnetic wave depends on its wavelength or frequency. Notable exceptions to this are X-rays and gamma rays, whose ranges of wavelength and frequency overlap. The difference here is in the origin of the waves: an X-ray is emitted when a high-speed electron is suddenly brought to rest and a gamma ray is emitted from an excited nucleus, usually along with alpha or beta emission.

The table shows typical wavelengths, the origins of the waves and the uses of the main parts of the electromagnetic spectrum.

| name of radiation | typical wavelength/m | how produced | used for |
|---|---|---|---|
| gamma | $1 \times 10^{-12}$ | an excited nucleus releasing energy in radioactive decay | tracing the flow of fluids and treating cancer |
| X-rays | $1 \times 10^{-10}$ | high-speed electrons being stopped by a target | seeing inside the body |
| ultraviolet | $1 \times 10^{-8}$ | very hot objects and passing electricity through gases | suntan and security marking |

| light | $1 \times 10^{-6}$ | hot objects and passing electricity through gases | vision and photography |
| infra-red | $1 \times 10^{-5}$ | warm and hot objects | heating and cooking |
| microwave | $1 \times 10^{-1}$ | microwave diode or oscillating electrons in an aerial | communications and cooking |
| radio | $1 \times 10^{2}$ | oscillating electrons in an aerial | communications |

All objects give out infra-red radiation, no matter what the temperature. The hotter the object, the more power is emitted and the greater the range of wavelengths.

## Waves or particles

| AQA A | M1 | NICCEA | M2 |
| AQA B | A2 | OCR A | M2 |
| EDEXCEL A | A2 | OCR B | M2 |
| EDEXCEL B | M1, M2 | WJEC | M2 |

The **threshold wavelength**, $\lambda_0$, is the wavelength of the waves that have the threshold frequency.
$$\lambda_0 = c \div f_0$$

The photoelectric effect can be demonstrated using a zinc plate connected to a gold leaf electroscope. An ultraviolet lamp discharges a negatively-charged plate but has no effect on a positively-charged plate.

The word quantum refers to the smallest amount of a quantity that can exist. A quantum of electromagnetic radiation is the smallest amount of energy of that frequency.

The **photoelectric effect** provides evidence that electromagnetic waves have a particle-like behaviour which is more pronounced at the short-wavelength end of the spectrum. In the photoelectric effect electrons are emitted from a metal surface when it absorbs electromagnetic radiation.

The results of photoelectricity experiments show that:

- there is no emission of electrons below a certain frequency, called the **threshold frequency**, $f_0$, which is different for different metals
- above this frequency, electrons are emitted with a range of kinetic energies up to a maximum, $(\frac{1}{2}mv^2)_{max}$
- increasing the frequency of the radiation causes an increase in the maximum kinetic energy of the emitted electrons, but has no effect on the photoelectric current, i.e. the rate of emission of electrons
- increasing the intensity of the radiation has no effect if the frequency is below the threshold frequency; for frequencies above the threshold it causes an increase in the photoelectric current, so the electrons are emitted at a greater rate.

The wave model cannot explain this behaviour; if electromagnetic radiation is a continuous stream of energy then radiation of all frequencies should cause photoelectric emission, it should only be a matter of time for an electron to absorb enough energy to be able to escape from the attractive forces of the positive ions in the metal.

The explanation for the photoelectric effect relies on the concept of a **photon**, a quantum or packet of energy. We picture electromagnetic radiation as short bursts of energy, the energy of a photon depending on its frequency.

A lamp emits random bursts of energy. Each burst is a photon, a quantum of radiation.

> **KEY POINT**
>
> The relationship between the energy, $E$ of a photon, or quantum of electromagnetic radiation, and its frequency, $f$, is:
> $$E = hf$$
> where $h$ is Planck's constant and has the value $6.63 \times 10^{-34}$ J s.

The energy of a photon can be measured in either joules or **electronvolts**. The electronvolt is a much smaller unit than the joule.

The conversion factor for changing energies in eV to energies in joules is $1.60 \times 10^{-19}$ J eV$^{-1}$

> **KEY POINT**
>
> One electron volt (1 eV) is the energy transfer when an electron moves through a potential difference of 1 volt.
> $$1 \text{ eV} = 1.60 \times 10^{-19} \text{ J}$$

Einstein's explanation of photoelectric emission is:

The work function is the **minimum** energy needed to liberate an electron from a metal. Some electrons need more than this amount of energy.

- an electron needs to absorb a minimum amount of energy to escape from a metal. This minimum amount of energy is a property of the metal and is called the work function, $\phi$
- if the photons of the incident radiation have energy $hf$ less than $\phi$ then there is no emission of electrons
- emission becomes just possible when $hf = \phi$
- for photons with energy greater than $\phi$, the electrons emitted have a range of energies, those with the maximum energy being the ones that needed the minimum energy to escape

Radiation below the threshold frequency, $f_0$, no matter how intense, does not cause any emission of electrons.

- increasing the intensity of the radiation increases the number of photons incident each second. This causes a greater emission of electrons, but does not affect their maximum kinetic energy

> **KEY POINT**
>
> Einstein's photoelectric equation relates the maximum kinetic energy of the emitted electrons to the work function and the energy of each photon:
> $$hf = \phi + (\tfrac{1}{2}mv^2)_{max}$$

At the threshold frequency, the minimum frequency that can cause emission from a given metal, $(\tfrac{1}{2}mv^2)_{max}$ is zero and so the equation becomes $hf_0 = \phi$.

## *Particles or waves*

| | | | |
|---|---|---|---|
| AQA A | M1 | NICCEA | M2 |
| AQA B | A2 | OCR A | M2 |
| EDEXCEL A | A2 | OCR B | M2 |
| EDEXCEL B | M1, M2 | WJEC | M2 |

If waves can show particle-like behaviour in photoelectric emission, can particles also behave as waves? Snooker balls bounce off cushions in the same way that light bounces off a mirror, so reflection is not a test for wave-like or particle-like behaviour. Diffraction and interference are properties unique to waves, so particles can be said to have a wave-like behaviour if they show these properties.

All particles have an associated wavelength called the de Broglie wavelength.

Try calculating the de Broglie (pronounced de Broy) wavelength of a moving snooker ball. Is it possible for the ball to show wave-like behaviour?

> **KEY POINT**
>
> The wavelength, $\lambda$, of a particle is related to its momentum, $p$, by the de Broglie equation:
> $$\lambda = h/p = h/mv$$
> where $h$ is the Planck constant.

An electron that has been accelerated through a potential difference of a few hundred volts has a wavelength similar to that of X-rays and gamma rays.

The de Broglie wavelength of such an electron is of the order of $1 \times 10^{-10}$ m.

This wavelength is also similar to the spacing of the atoms in crystalline materials, so these materials provide suitable sized gaps to cause diffraction.

Diffraction patterns formed by a beam of electrons after passing through thin foil or graphite show a set of bright and dark rings on photographic film, similar to those formed by X-ray diffraction.

You may have seen this pattern formed on a fluorescent screen in a vacuum tube.

*a diffraction pattern formed by passing a beam of electrons through graphite*

Electrons can also be made to interfere when two coherent beams overlap. They produce an interference pattern similar to that of light, but on a much smaller scale.

Other particles such as protons and neutrons also show wave-like behaviour.

> Particles and waves are the models that we use to describe and explain physical phenomena. It is not surprising that the real world does not fit neatly into our models.

There are two separate models of how matter behaves. The particle model explains such phenomena as ionisation and photoelectricity, while the wave model explains interference and diffraction. It is not appropriate to classify matter as waves or particles as photons and electrons can fit either model, depending on the circumstances.

## Line spectra

| AQA A | M1 | NICCEA | M2 |
| AQA B | M2 | OCR B | M2 |
| EDEXCEL A | A2 | WJEC | M2 |
| EDEXCEL B | M1, M2 | | |

> Gamma rays are the exception because they are a result of the nucleus losing energy.

With the exception of gamma rays, the emission of electromagnetic radiation is associated with electrons losing energy. A hot solid can radiate the whole range of wavelengths through the infra-red and part of the visible spectrum, the extent of the range depending on its temperature.

When an electric current is passed through an ionised gas, only a small number of wavelengths are emitted. The wavelengths are characteristic of the gas used and are called a line spectrum.

The diagram below shows some of the lines present in the hydrogen spectrum and their wavelengths.

> Not all of these lines lie in the visible spectrum, which extends from 0.4 μm to 0.65 μm.

The existence of line spectra provides evidence that the electrons in orbit around a nucleus can only have certain values of energy, the values being characteristic of an atom. Energy can only be emitted or absorbed in amounts that correspond to the differences between these allowed values.

An energy level diagram shows the amounts of energy that an electron can have. The diagram below is an energy level diagram for hydrogen. Note that on an energy level diagram:

- the energies are measured relative to a zero that represents the energy of an electron at rest outside the atom, i.e. one that is just free
- an orbiting electron has less energy than a free electron, so it has negative energy relative to the zero
- an electron with the minimum possible energy is in the ground state; higher energy levels are called excited states.

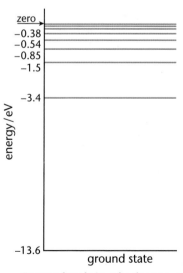

**Energy levels in a hydrogen atom**

Movement of an electron to the ground state results in the emission of a photon with an energy in excess of 10 eV. Photons with this amount of energy give spectral lines in the ultraviolet region of the spectrum.

The lines of the emission spectrum often appear black in the absorption spectrum. The photons of these energies are absorbed as the electrons move to more excited states, and then released when the electrons lose energy. The emitted photons are radiated in all directions, so very little energy is detected in any one direction.

Energy is emitted in the form of a photon when an electron moves from an excited state to a lower energy level. The energy of the photon is equal to the difference in the values of the energy levels. For example, an electron moving from an energy level with –0.38 eV of energy to one with –0.85 eV loses energy equal to (–0.38 – –0.85) eV = 0.47 eV. This corresponds to emitting a photon of frequency $1.13 \times 10^{14}$ Hz, which lies in the infra-red part of the electromagnetic spectrum.

Electrons can gain energy by absorbing photons. As with emission, the only photons that can be absorbed are those that correspond to allowed movements, or transitions, of electrons. An absorption spectrum is produced by shining white light through a sample of a gaseous element. The spectrum that emerges is the full spectrum with the element's emission spectrum missing or of low intensity. This is due to the electrons absorbing photons of just the right energy to allow them to move to a more excited state.

> **KEY POINT**
>
> When an electron moves from an energy level $E_1$ to a lower energy level $E_2$, the energy of the photon emitted is given by
> $$hf = E_1 - E_2$$

## Progress check

1 A photon of green light has a wavelength of $5.0 \times 10^{-7}$ m.
   a Calculate the frequency of the light.
   b Calculate the energy of the photon:
      i   in J
      ii   in eV.

2 Calculate the de Broglie wavelength of an electron, $m_e = 9.1 \times 10^{-31}$ kg, travelling at a speed of $3.0 \times 10^7$ m s$^{-1}$.

3 An electron in a hydrogen atom undergoes a transition from an energy level of –0.54 eV to one of –3.40 eV.
   a Calculate the frequency of the photon emitted.
   b In what part of the electromagnetic spectrum is the emitted radiation?

3 a $6.90 \times 10^{14}$ Hz   b visible (light)
2 $2.43 \times 10^{-11}$ m
1 a $6.0 \times 10^{14}$ Hz   b i $3.98 \times 10^{-19}$ J   ii $2.49$ eV

# Sample question and model answer

In the photoelectric effect, electrons are emitted from a metal surface when it is irradiated with electromagnetic radiation. The graph shows the variation of the maximum photoelectron kinetic energy with the frequency of the radiation incident on the emitting surface.

The equation needs to be written in the form $y = mx + c$ to determine which aspect of the graph determines the Planck constant, $h$.
The graph is a plot of $(\frac{1}{2}mv^2)_{max}$ against $f$, so rearrange the equation with $(\frac{1}{2}mv^2)_{max}$ on the left-hand side and $f$ on the right-hand side.

(a) Use the data from the graph to calculate the Planck constant. [3]

> The photoelectric equation is
> $hf = \theta + (\frac{1}{2}mv^2)_{max}$
> This can be written as
> $(\frac{1}{2}mv^2)_{max} = hf - \theta$  1 mark
> The gradient of a graph of $(\frac{1}{2}mv^2)_{max}$ against $f$ represents the Planck constant.  1 mark
> The gradient of the graph $= 17 \times 10^{-19} \div (34 - 7.0) \times 10^{14} = 6.30 \times 10^{-34}$.
> This gives the value of the Planck constant as $6.30 \times 10^{-34}$ Js.  1 mark

It is important to remember here that a graph is a relationship between two sets of numbers, so the gradient does not have a unit. The gradient of this graph gives the numerical value of the Planck constant; you need to add the correct unit to this numerical value in your final answer.

(b) Determine the minimum energy required to remove an electron from the target metal. [2]

> The minimum energy required to remove an electron is $hf_o$, where $f_o$ is the threshold frequency.  1 mark
> $hf_o = 6.30 \times 10^{-34}$ Js $\times 7.5 \times 10^{14}$ Hz $= 4.73 \times 10^{-19}$ J  1 mark

This value is obtained by using the unrounded value of h from part a. If you use the rounded value, you should obtain an answer of $4.82 \times 10^{-19}$ J. Either answer gains full marks.

(c) Explain how the photoelectric effect produces evidence which illustrates the particulate nature of light. [3]

> There is no emission below the threshold frequency; this shows that energy arrives in packets rather than a continuous stream. 1 mark
> The maximum kinetic energy of the emitted electrons depends on the frequency of the radiation. This shows that the energy of each packet of radiation depends on the frequency.  1 mark
> Increasing the intensity of the radiation does not affect the maximum kinetic energy of the emitted electrons, only the rate of emission. This shows that intensity is related to the number of packets of energy incident each second.  1 mark

Since the question asks how the evidence supports the particulate nature of light, it is important to explain how each piece of evidence quoted supports this theory, rather than just quoting the evidence.

*AEB AS-A/PHYS/7 0635J Jan 1999*

## Practice examination questions

**1** (a) Distinguish between a **transverse** and a **longitudinal** wave and give one example of each. [4]

  (b) Light can be **polarised** when it is reflected at a surface.
  (i) Describe the difference between polarised light and unpolarised light. [2]
  (ii) Explain why sound cannot be polarised. [1]

**2** (a) State the laws that describe the refraction of light. [2]

  (b) Light of frequency $6.00 \times 10^{14}$ Hz is incident on an air–water boundary at an angle of 63°. The speed of light in air is $3.00 \times 10^8$ m s$^{-1}$ and the speed of light in water is $2.25 \times 10^8$ m s$^{-1}$.
  (i) Calculate the wavelength of the light in air. [3]
  (ii) Calculate the refractive index for light passing from air into water. [2]
  (iii) Calculate the angle of refraction in the water. [2]

**3** The diagram shows how the profile of a wave on a rope changes. In a period of $2.50 \times 10^{-4}$ s it moves from position 1 to position 2.

  (a) (i) Write down the wavelength of the wave. [1]
  (ii) Calculate the frequency of the wave. [2]
  (iii) Calculate the speed of the wave along the rope. [3]

  (b) (i) How can you tell from the diagram that the wave is a stationary wave? [1]
  (ii) Suggest how the stationary wave is formed in this example. [2]

  (c) What is the phase difference between:
  (i) the two points marked X on the diagram? [1]
  (ii) the points marked Y and Z on the diagram? [1]

  Give your answers in degrees.

**4** The diagram shows two identical loudspeakers, A and B, placed 0.75 m apart. Each loudspeaker emits sound of frequency 2000 Hz.
Point C is on a line midway between the speakers and 5.0 m away from the line joining the speakers. A listener at C hears a maximum intensity of sound. If the listener then moves from C to E or D, the sound intensity heard decreases to a minimum. Further movement in the same direction results in the repeated increase and decrease in the sound intensity.

speed of sound in air = 330 m s$^{-1}$

(a) Explain why the sound intensity is:
    (i) a maximum at C,
    (ii) a minimum at D or E.            [4]

(b) Calculate:
    (i) the wavelength of the sound,
    (ii) the distance CE.            [4]

*NEAB Particles and Waves (PH02), June 1999, Q 5*

**5** The diagram shows an optical fibre made of a plastic-coated glass.
The refractive index of the glass is 1.58 and that of the plastic is 1.24.

(a) Calculate the refractive index for light striking the glass–plastic interface.   [2]

(b) Calculate the value of the critical angle at this interface.   [2]

(c) The diagram shows two paths of light through the fibre.
Explain how this can affect a pulse of light as it passes along a fibre.   [2]

## *Practice examination questions* (continued)

**6** When electromagnetic radiation is incident on a clean surface of copper, *photoelectric emission* may take place. No emission occurs unless the photon energy is greater than or equal to the *work function*.

(a) Explain the meaning of the terms:
   (i) photoelectric emission [2]
   (ii) photon [2]
   (iii) work function [2]

(b) The work function of copper is 5.05 eV.
   (i) Calculate the minimum frequency of radiation that causes photoelectric emission.
   $e = 1.60 \times 10^{-19}$ C
   $h = 6.63 \times 10^{-34}$ Js [2]
   (ii) Calculate the maximum kinetic energy of the emitted electrons when electromagnetic radiation of frequency $1.80 \times 10^{15}$ Hz is incident on the copper surface. [3]

# Practice examination answers

## Chapter 1  Force and motion 1

1   The vector diagram is shown on the right [1].
    The resultant velocity is 19 m s$^{-1}$ [1]
    at an angle of 18° E of N [1].

2   (a)  450 N × cos 50° [1] = 289 N [1]
    (b)  289 N [1]
    (c)  The light fitting is in equilibrium [1] so the sum of the forces on it is zero [1].
    (d)  The vector diagram is shown on the right [1].
         S = 345 N [1].

3   (a)  Moment  = force × distance to pivot = 60 N × 0.45 m [1]
              = 27 N m [1].
    (b)  F = 27 N m ÷ 0.06 m [1] = 450 N [1].
    (c)  A bigger moment is caused for the same applied force [1]. So the force acting on the
         branch is greater [1].

4   (a)  Horizontal component = 2893 N [1].
         Vertical component = 3447 N [1].
    (b)  pressure = normal force ÷ area [1] = 3447 N ÷ 0.09 m$^2$ [1] = 38300 Pa [1].
    (c)  To reduce the pressure on the ground [1] so that the girder does not penetrate the
         surface [1].

5   (a)  Volume of water = 10.5 m × 4.2 m × 3.5 m = 154 m$^3$ [1]
         Mass = volume × density = 1.54 × 10$^5$ kg [1]
         Weight = mass × g = 1.54 × 10$^6$ N [1].
    (b)  Pressure = normal force ÷ area [1] = 1.54 × 10$^6$ N ÷ 4.41 × 10$^1$ m$^2$ [1] = 3.5 × 10$^4$ Pa [1].
    (c)  Liquid pressure acts equally in all directions [1] so at the bottom of the tank the pressure
         on the sides is equal to that on the base [1].
    (d)  It is the same [1] the pressure depends on the depth only [1]

6   (a)  A ductile material undergoes plastic deformation [1] but a brittle material breaks [1].
    (b)  (i)    340 MPa [1]
         (ii)   up to 300 MPa [1]
         (iii)  E = stress ÷ strain = 3.00 × 10$^8$ Pa ÷ 1.50 × 10$^{-3}$ [1] = 2.00 × 10$^{11}$ Pa [1].
         (iv)   The curve would have the same shape [1]; the values on the axes would be the
                same as the stress-strain graph is the same for all samples of the material [1].

7   (a)  Acceleration = change in velocity ÷ time [1] = 24 m s$^{-1}$ ÷ 80 s [1] = 0.30 m s$^{-2}$ [1].
    (b)  Between 230 and 290 s [1].
    (c)  (i)    20 s [1]
         (ii)   120 s [1].
    (d)  960 m + 1680 m + 720 m + 900 m + 1500 m + 900 m = 6660 m [2] ( for all correct).
    (e)  3360 m – 3300 m = 60 m [1].

8
    (a)  Use s = ut + ½at$^2$ [1] t = √(2 × 0.39 m ÷ 10 m s$^{-2}$) [1] = 0.28 s [1].
    (b)  s = ut [1] = 35 m s$^{-1}$ × 0.28 s = 9.8 m [1].
    (c)  u = s÷t [1] = 15 m ÷ 0.28 s = 54 m s$^{-1}$ [1].

9
    (a)  t = Δv ÷ a [1] = 14 m s$^{-1}$ ÷ 2 m s$^{-2}$ = 7 s [1].
    (b)  distance = average speed × time [1] = 21 m s$^{-1}$ × 7 s = 147 m [1].

10
    (a)  Use v$^2$ = u$^2$ + 2 as [1] u$^2$ = – 2 × – 10 m s$^{-2}$ × 3.5 m = 70 m$^2$s$^{-2}$ [1] u = 8.4 m s$^{-1}$ [1].
    (b)  Time travelling upwards, t = distance ÷ average speed = 3.5 m ÷ 4.2 m s$^{-1}$ = 0.83 s [1]
         Time in air = 1.66 s [1].

## Chapter 2 Force and motion 2

**1** (a) Use $v^2 = u^2 + 2as$ [1] $a = v^2 \div 2s = (60 \text{ m s}^{-1})^2 \div 2 \times 1500 \text{ m}$ [1] $= 1.2 \text{ m s}^{-2}$ [1].
  (b) $F = ma$ [1] $= 70\,000 \text{ kg} \times 1.2 \text{ m s}^{-2}$ [1] $= 84 \text{ kN}$ [1].
  (c) As the aircraft gains speed the size of the resistive forces increases [1]. This causes the resultant force to decrease [1].

**2** (a) (i) The pull of the trailer on the car [1]
      (ii) 150 N backwards [1].
  (b) (i) 130 N forwards [1]
      (ii) $a = F \div m$ [1] $= 130 \text{ N} \div 190 \text{ kg}$ [1] $= 0.68 \text{ m s}^{-2}$ [1].
  (c) (i) 0 [1]
      (ii) 150 N backwards [1].

**3** (a) (i) $40 \text{ kg} \times g = 400 \text{ N}$ [1]
      (ii) The child pulls the Earth [1].
  (b) (i) $400 \text{ N} \times \cos 66°$ [1] $= 163 \text{ N}$ [1]
      (ii) $a = F/m$ [1] $= 73 \text{ N} \div 40 \text{ kg}$ [1] $= 1.8 \text{ m s}^{-2}$ [1].
  (c) Use $v^2 = u^2 + 2as$ [1] $v^2 = 2 \times 1.8 \text{ m s}^{-2} \times 5.5 \text{ m} = 19.8 \text{ m}^2 \text{ s}^{-2}$ [1] $v = 4.4 \text{ m s}^{-1}$ [1].

**4** (a) Momentum before and after collision $= 0.45 \text{ kg} \times 0.60 \text{ m s}^{-1} - 0.30 \text{ kg} \times 0.42 \text{ m s}^{-1}$ [1] $= 0.144 \text{ kg m s}^{-1}$ [1]. Combined speed after collision $= 0.144 \text{ kg m s}^{-1} \div 0.75 \text{ kg} = 0.19 \text{ m s}^{-1}$ [1].
  (b) Momentum of lighter vehicle before collision $= (-) 0.45 \text{ kg} \times 0.60 \text{ m s}^{-1} = (-) 0.27 \text{ kg m s}^{-1}$ [1]
  Speed $= 0.27 \text{ kg m s}^{-1} \div 0.30 \text{ kg} = 0.90 \text{ m s}^{-1}$ [1].

**5** (a) Total momentum before the neutron is absorbed $= 1.7 \times 10^{-27} \text{ kg} \times 1.4 \times 10^7 \text{ m s}^{-1} = 2.38 \times 10^{-20} \text{ kg m s}^{-1}$ [1]. Total momentum after the neutron is absorbed $= 2.38 \times 10^{-20} \text{ kg m s}^{-1}$ [1]. Velocity of nucleus $= p \div m = 2.38 \times 10^{-20} \text{ kg m s}^{-1} \div 4.017 \times 10^{-25} \text{ kg} = 5.9 \times 10^4 \text{ m s}^{-1}$ [1].
  (b) The velocity of the nucleus would be greater [1]. The neutron would rebound and have a greater change in momentum [1] so the change in momentum of the nucleus would also be greater [1].
  (c) $E_K = \frac{1}{2}mv^2$ [1] $= \frac{1}{2} \times 1.7 \times 10^{-27} \text{ kg} \times (1.4 \times 10^7 \text{ m s}^{-1})^2$ [1] $= 1.67 \times 10^{-13} \text{ J}$ [1].

**6** (a) The astronaut gains momentum in the opposite direction to the push on the spacecraft [1]. The spacecraft gains the same amount of momentum, in the direction of the astronaut's push [1].
  (b) The change in momentum of the gas [1] is balanced by an equal and opposite change in momentum of the astronaut [1].

**7** (a) Change in momentum $= 0.080 \text{ kg} \times 32 \text{ m s}^{-1} - 0.080 \text{ kg} \times -22 \text{ m s}^{-1}$ [1] $= 4.32 \text{ kg m s}^{-1}$ [1].
  (b) $F = \Delta p \div t$ [1] $= 4.32 \text{ kg m s}^{-1} \div 0.15 \text{ s}$ [1] $= 28.8 \text{ N}$ [1].
  (c) Momentum is conserved [1]. The wall and the Earth gain equal momentum in the opposite direction [1].

**8** (a) $W = F \times s$ [1] $= 30\,000 \text{ N} \times 18 \text{ m}$ [1] $= 540 \text{ kJ}$ [1].
  (b) 540 kJ [1]
  (c) $P = W \div t = 540 \text{ kJ} \div 24 \text{ s}$ [1] $= 22.5 \text{ kW}$ [1].
  (d) $P_{in} = P_{out} \div$ efficiency [1] $= 22.5 \text{ kW} \div 0.45 = 50 \text{ kW}$ [1].

**9** (a) $\Delta E_p = mg\Delta h$ [1] $= 2.0 \times 10^5 \text{ kg} \times 10 \text{ N kg}^{-1} \times 215 \text{ m}$ [1] $= 4.30 \times 10^8 \text{ J}$ [1].
  (b) $E_k = \frac{1}{2}mv^2 = 4.30 \times 10^8 \text{ J}$ [1] $v = \sqrt{(2 \times 4.30 \times 10^8 \text{ J} \div 2.0 \times 10^5 \text{ kg})}$ [1] $= 65.6 \text{ m s}^{-1}$ [1].
  (c) Kinetic energy of water leaving turbines each second $= \frac{1}{2}mv^2 = \frac{1}{2} \times 2.0 \times 10^5 \text{ kg} \times (8 \text{ m s}^{-1})^2 = 6.4 \times 10^6 \text{ J}$ [1]. Maximum power input $= 4.30 \times 10^8 \text{ W} - 6.4 \times 10^6 \text{ W} = 4.24 \times 10^8 \text{ W}$ [1].
  (d) Efficiency $=$ useful power out $\div$ power in $= 2.5 \times 10^8 \text{ W} \div 4.24 \times 10^8 \text{ W}$ [1] $= 0.59$ [1].

**10** (a) Energy $= \frac{1}{2}Fx = \frac{1}{2} \times 25 \text{ N} \times 0.15 \text{ m}$ [1] $= 1.875 \text{ J}$ [1].
  (b) (i) $\frac{1}{2}mv^2 = 1.875 \text{ J}$ [1] $v^2 = 2 \times 1.875 \text{ J} \div 0.020 \text{ kg} = 187.5 \text{ m}^2 \text{ s}^{-2}$ [1] $v = 13.7 \text{ m s}^{-1}$ [1]
      (ii) $mg\Delta h = 1.875 \text{ J}$ [1] $\Delta h = 1.875 \text{ J} \div 0.20 \text{ N}$ [1] $= 9.4 \text{ m}$ [1].

# Chapter 3  Electricity

**1** (a)  $R = \rho l/A$ [1].
  (b)  Resistivity is a property of a material [1]; the resistance of an object also depends on its dimensions [1].
  (c)  Cross-sectional area, $A = 2.0 \times 10^{-5}$ m² [1] $R = 3.00 \times 10^{-5}$ Ω m $\times 1.0 \times 10^{-2}$ m ÷ 2.0 × $10^{-5}$ m² [1] $= 1.5 \times 10^{-2}$ Ω [1].

**2** (a)  $1/R = 1/10$ Ω $+ 1/10$ Ω [1] $R = 5$ Ω [1].
  (b)  $I = V/R$ [1] $= 6.0$ V ÷ 20 Ω [1] $= 0.30$ A [1].
  (c)  0.30 A ÷ 2 [1] $= 0.15$ A [1].
  (d)  $V = IR = 0.15$ A × 10 Ω [1] $= 1.5$ V [1].

**3** (a)  (i)  The graph should show the resistance decreasing with increasing temperature [1] in a non-linear way [1]
      (ii)  The resistance of a metallic conductor increases with increasing temperature [1] linearly [1]
      (iii)  Ohm's law only applies to metallic conductors [1] at constant temperature [1].
  (b)  (i)  Current in circuit = 9.0 V ÷ 600 Ω = $1.5 \times 10^{-2}$ A [1]. Thermistor p.d. = $1.5 \times 10^{-2}$ A × 500 Ω [1] = 7.5 V [1]
      (ii)  The p.d. across the thermistor decreases [1] as it has a smaller proportion of the circuit resistance [1]
      (iii)  Current in circuit = 9.0 V ÷ 175 Ω = $5.1 \times 10^{-2}$ A [1]. Resistor p.d. = $5.1 \times 10^{-2}$ A × 100 Ω [1] = 5.1 V [1]
      (iv)  Current in circuit = 3.9 V ÷ 300 Ω = 0.013 A [1]. Resistor p.d. = 5.1 V [1]. Resistor value = 5.1 V ÷ 0.013 A = 392 Ω [1].

**4** (a)  The current in the cell and the digital voltmeter is negligible [1] so there is no voltage drop due to internal resistance [1]. With the moving coil voltmeter there is a significant current in the cell; the voltage drop is the p.d. across the internal resistance [1].
  (b)  $R = V/I$ [1] $= 0.05$ V ÷ $1.55 \times 10^{-3}$ A [1] $= 32.3$ Ω [1].

**5** (a)  Current in each lamp, $I = P/V = 6$ W ÷ 12 V = 0.50 A. Current in battery = 4 × 0.50 A = 2.0 A [1].
  (b)  $\Delta q = I\Delta t$ [1] $= 2.0$ A × 60 s [1] $= 120$ C [1].
  (c)  $R = V/I$ [1] $= 12$ V ÷ 2.0 A [1] $= 6.0$ Ω [1].

**6** (a)  (i)  $I^2 = P/R$ [1] $I = \sqrt{(10 \text{ W} ÷ 4.7 \text{ Ω})}$ [1] = 1.46 A [1]
      (ii)  $V = IR = 1.46$ A × 4.7 Ω [1] = 6.9 V [1]
      (iii)  $l = AR/\rho$ [1] $= 2.0 \times 10^{-7}$ m² × 4.7 Ω ÷ $5.0 \times 10^{-7}$ Ω m [1] = 1.88 m [1].
  (b)  A smaller length of material is needed [1] for the same resistance [1].

**7** (a)  Potential divider [1].
  (b)  Apply a varying p.d. to component X [1].
  (c)  Circuit B varies the current and only allows investigation of the characteristics when the diode is conducting [1]. Circuit A allows the diode to be investigated when it is not conducting and when it is conducting [1].
  (d)  (i)  The completed table is:

| resistance /Ω | 10.8 | 3.55 | 1.73 | 1.15 | 0.80 |
|---|---|---|---|---|---|

      [Two marks for all correct.]
      (ii)  The graph is shown on the right. Marks are awarded for:
      Scales and labelling of axes [1]
      Correct plotting [2]
      Drawing a smooth curve [1]
      (iii)  0.785 V [1]
      (iv)  The diode only conducts over a narrow range of voltages [1].

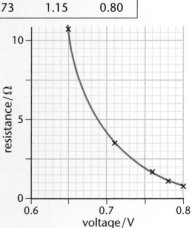

8 (a) $v = I/nAe$ [1] = 5 A ÷ (8.0 × 10²⁸ m⁻³ × 1.25 × 10⁻⁶ m² × 1.60 × 10⁻¹⁹ C) [1] = 3.1 × 10⁻⁴ m s⁻¹ [1].

(b) The cross-sectional area of the tungsten is smaller than that of the copper [1] and the concentration of free electrons is less [1].

(c) Heating is due to collisions between the free electrons and the metal ions [1]. The collisions are more frequent in the tungsten than in the copper [1].

9 (a) Effective resistance of the two parallel resistors = 4 Ω [1].
Circuit current, $I = V/R$ = 3.0 V ÷ 7.5 Ω [1] = 0.40 A [1].

(b) $V = IR$ = 0.40 A × 7.0 Ω [1] = 2.8 V [1].

# Chapter 3  Particle physics

1 (a) 295 K [1].

(b) $n = pV/RT$ [1] = 1.20 × 10⁵ Pa × 1.50 × 10⁻⁴ m³ ÷ (8.3 J mol⁻¹ × 295 K) [1] = 7.4 × 10⁻³ [1].

(c) $T_2 = P_2T_1/P_1$ [1] = 2.00 × 10⁵ Pa × 295 K ÷ 1.20 × 10⁵ Pa [1] = 492 K [1].

(d) (i) ½$m <c^2>$ = 3/2 $kT$ [1] = 1.5 × 1.38 × 10⁻²³ J K⁻¹ × 295 K [1] = 6.11 × 10⁻²¹ J [1]

(ii) 590 K [1] since temperature is proportional to the mean kinetic energy of the particles [1].

2 Energy released when steam condenses to water at 100°C = $ml$ = 1.8 kg × 2.25 × 10⁶ J kg⁻¹ = 4.05 × 10⁶ J [1]. Energy released when water cools from 100°C to 85°C = $mc\Delta\theta$ = 1.8 kg × 4.20 × 10³ J kg⁻¹ K⁻¹ × 15 K = 1.13 × 10⁵ J [1]. Total energy released = 4.16 × 10⁶ J [1].

3 (a) (i) A temperature difference of 10 K can be measured more precisely than one of 1 K [1]

(ii) There is less energy loss to the surroundings [1]

(iii) The energy supplied heats the block and replaces the energy lost to the surroundings [1] so this should give a higher value [1]

(iv) Insulate the block to reduce energy losses [1].

(b) $c = E/m\Delta\theta$ [1] = 24 W × 350 s ÷ (0.80 kg × 14 K) [1] = 750 J kg⁻¹ K⁻¹ [1].

4 (a) $\Delta U$ is the increase in internal energy [1]
$Q$ is the heat supplied to the gas [1]
$W$ is the work done on the gas [1].

(b) The temperature drops [1] as the heat removed from the gas is greater than the work done on it [1].

5 (a) Efficiency = useful power output ÷ power input = 2 kW ÷ 8 kW [1] = 0.25 [1].

(b) $(T_H - T_C) ÷ T_H$ = (1023 K – 423 K) ÷ 1023 K [1] = 0.59 [1].

(c) Some energy is wasted as heat [1].

6 (a) (i) Particles that approach a nucleus are deflected away from it [1]. So the nucleus must have the same sign of charge as the alpha particles [1]

(ii) Many particles pass through undeviated [1].

(b) (i) High energy electrons [1] are used to penetrate nucleons [1]

(ii) Electrons are not repelled by the nucleus [1] and they are not affected by the strong nuclear force [1]

(iii) The proton and neutron are made up of other particles [1]

(iv) The scattering of the electrons shows that there are dense regions of charge within a nucleon [1]

(v) A proton is uud [1]; a neutron is udd [1].

7 (a) 6.02 × 10²³ mol⁻¹ × 1/12 mol [1] = 5.02 × 10²² atoms [1].

(b) (i) No [1] since 99% of carbon is carbon-12 [1]

(ii) 5 × 10¹⁰ [1].

(c) (i) $\lambda$ = 0.25 ÷ N [1] = 0.25 ÷ 5 × 10¹⁰ = 5 × 10⁻¹² s⁻¹ [1]

(ii) $t_{1/2}$ = 0.69 ÷ $\lambda$ [1] = 0.69 ÷ 5 × 10⁻¹² s⁻¹ = 1.38 × 10¹¹ s [1].

8 (a) If the neutron–proton ratio is less than 1 the nucleus undergoes β⁺ decay [1]; if the neutron–proton ratio is greater than 1 the nucleus decays by β⁻ emission [1].

(b) When $^{13}_{6}C$ decays, the neutron–proton ratio changes from 1.17 [1] to 0.86, which is closer to 1 [1]

(c) It becomes greater than 1 [1] and reaches a value of 1.5 for massive nuclei [1].

9 (a) A meson consists of a quark and an antiquark [1]. A baryon is made up of three quarks [1].

(b) $\pi^0$ [1].

# Chapter 4 Electromagnetic radiation

1 (a) In a transverse wave the vibrations are at right angles to the direction of wave travel [1].
Examples include any electromagnetic wave, e.g. radio, light [1].
In a longitudinal wave the vibrations are parallel to the direction of wave travel [1].
Examples include sound or any other compression wave [1].

(b) (i) In an unpolarised wave the vibrations are in all directions perpendicular to the direction of travel [1]. In a polarised wave the vibrations are in one direction only [1].

(ii) Sound is a longitudinal wave; there are no vibrations perpendicular to the direction of travel [1].

2 (a) The incident light, refracted light and normal are all in the same plane [1].
$\sin i \div \sin r$ = constant [1].

(b) (i) $\lambda = v \div f$ [1] = $3.00 \times 10^8$ m s$^{-1}$ $\div$ $6.00 \times 10^{14}$ Hz [1] = $5.00 \times 10^{-7}$ m [1]

(ii) $n = c_{air} \div c_{water} = 3.00 \times 10^8$ m s$^{-1}$ $\div$ $2.25 \times 10^8$ m s$^{-1}$ [1] = 1.33 [1].

(iii) $\sin r = \sin i \div n = \sin 63° \div 1.33$ [1] $r = \sin^{-1} 0.668 = 42°$ [1].

3 (a) (i) 1.2 m [1]

(ii) $f = 1/T = 1 \div 2.50 \times 10^{-4}$ s [1] = 4000 Hz [1]

(iii) $v = f\lambda$ [1] = 4000 Hz $\times$ 1.2 m [1] = 4800 m s$^{-1}$ [1].

(b) (i) The wave profile does not move along the rope [1]

(ii) By superposition [1] of a wave and its reflection [1].

(c) (i) 0° [1]

(ii) 180° [1].

4 (a) (i) The waves are in phase [1] and interfere constructively [1]

(ii) The waves are out of phase [1] and interfere destructively [1].

b) (i) $\lambda = v \div f$ [1] = 330 m s$^{-1}$ $\div$ 2000 Hz = 0.165 m [1]

(ii) Separation of maxima, $x = \lambda D/a$ = 0.165 m $\times$ 5.0 m $\div$ 0.75 m = 1.10 m [1].
CE is half this distance = 0.55 m [1].

5 (a) $_g n_p = {_a}n_p \div {_a}n_g$ [1] = 1.24 $\div$ 1.58 = 0.785 [1].

(b) $\sin C$ = 0.785 [1] $C = \sin^{-1} 0.785 = 52°$ [1].

(c) Light has further to travel along path B than along path A [1]. This causes a pulse to be elongated [1].

6 (a) (i) Emission of electrons [1] when a metal surface is illuminated with electromagnetic radiation [1]

(ii) A quantum of electromagnetic radiation [1] of a particular frequency [1]

(iii) The minimum energy [1] an electron needs to escape from a metal [1].

(b) (i) $f_0 = \phi \div h$ [1] = 5.05 eV $\times$ $1.60 \times 10^{-19}$ C $\div$ $6.63 \times 10^{-34}$ J s = $1.22 \times 10^{15}$ Hz [1]

(ii) $(\frac{1}{2}mv^2)_{max} = hf - \phi$ [1] = $6.63 \times 10^{-34}$ J s $\times$ $1.80 \times 10^{15}$ Hz $-$ (5.05 eV $\times$ $1.60 \times 10^{-19}$ C) [1] = $3.85 \times 10^{-19}$ J [1].

# Index

absolute zero 90–1
acceleration 40, 47–8
aerofoil 65
air resistance 47–8
airbag 50
alpha particle 99–101, 110, 112
alpha particle scattering 99, 109
alternating current 81, 82
amorphous 36
amplitude 117–18
annealing 35
antineutrino 104, 105, 107–8, 112
antinode 128
atomic mass unit 99, 109
atomic model 99
atomic number 99
Avagadro constant 93

background radiation 100
baryon number 104–5, 107
Bernoulli effect 65, 66
beta-minus particle 100–1, 108, 110, 112
beta-plus particle 100–1, 110, 112
binding energy per nucleon 111
Boltzmann constant 93
Boyle's law 92
braking distance 50–1
breaking stress 32, 34
brittle 31, 34, 35, 37
Brownian motion 90

carbon fibre 37
Celsius scale 90–1
centre of gravity 25–6
centre of mass 25–6
charge 74–5, 78–9
chipboard 37
coherent 127
collision 55, 56, 58
composite 36–7
concrete 37
conservation of energy 61, 62
conservation of momentum 54–5
copper 31, 35, 109
couple 29
creep 35
critical angle 122
critical velocity 64
cross-linking 36
crumple zone 50
current 74

de Broglie 132
decay chain 110
decay constant 101–2
decay energy 112
deep inelastic scattering 103
density 30
diffraction 125
digital signal 123
diode 72
direct current 81
disc brake 48

dislocation 34–5
displacement 39
distance 39–40, 59
drag 64
drag coefficient 64
drift velocity 74
drum brake 48
ductile 31, 35

edge dislocation 34
efficiency 62, 63, 97
elastic 31–2
elastic limit 31, 34
electrical power 76
electromagnetic spectrum 130
electromotive force 75, 77–9
energy 60
energy level 133
energy transfer 60–3, 75–6, 79, 94–6, 112
exchange particle 106–7
excited state 133–4
exponential decay 102

fatigue 35
Feynman diagram 106–7
fibre composite 37
fibre optic 122–3
filament lamp 60–1, 72, 75
first law of thermodynamics 96
flavour 104, 108
Fleming's left hand rule 84–5
free electron 74
free fall acceleration 26, 41, 47, 62
frequency 117–19
frequency 82
friction force 25
fundamental particle 103, 105

gain 81
gamma ray 100–1, 112, 130
gas pressure 90–3
glass 31
glass-reinforced plastic 37
gluon 107
gravitational field strength 26
gravitational force 25–6
gravitational potential energy 60, 61–2
graviton 106
ground state 133

hadron 105
half-life 102
heat engine 96–7
heat pump 97
high-carbon steel 34, 35
Hooke's law 31, 34
hysteresis 33

ideal gas 91–3
impulse 57–8
intensity 120
interference 125–7
intermediate vector boson 107
internal energy 94–7
internal resistance 77–8
ionisation 100
isotope 99, 109, 111

kelvin scale 91
kinetic energy 56, 60, 61–2
kinetic theory 92–3
Kirchoff's laws 78–9

laminar 64
laminate 36
lepton 105
lift coefficient 65
light-dependent resistor 79–80
limit of proportionality 31, 34
line spectrum 133
longitudinal wave 117

magnetic field 84–5, 130
malleable 31
mass defect 111
mass number 99
meson 105
mild steel 34, 35
molar gas constant 92
moment 29, 38
momentum 54–8
multipath dispersion 123

necking 34, 35
neutrino 104, 105, 112
neutron 99–101, 105, 107, 109–10
Newton's first law 52
Newton's second law 56–7
Newton's third law 53
node 128
normal contact force 25
nuclear binding energy 111
nucleon 107, 109, 111
nucleus 99

Ohm's law 72
oscilloscope 81–2

pair annihilation 104
pair production 104
parallel 77, 79
particle composite 37
path difference 127
peak voltage 82
peak-to-peak voltage 82
period 117–19
phase 118
phase difference 118
photoelectric effect 131–2
photon 106
pivot 29
plastic 31
plywood 36
polarisation 120
polycrystalline 34
polymer 36
potential difference 75, 76, 78–9
potential divider 71
power 62–3, 76, 120
pressure 30
pre-stressed concrete 37–8
principle of moments 29
projectile 41–2
proton 99–101, 103–5, 107, 109–10

quantum 131
quark 103–5, 107, 108
quenching 35
radioactive series 110
radiocarbon dating 111
recoil 56
reflection of light 120–1
refraction 121
refractive index 121
refrigerator 97
resistance. 71–3, 77–80
resistive forces 47, 52, 64
resistivity 73
resolution of vectors 28
r.m.s voltage 82
rocket propulsion 57
rubber 31, 33, 36

scalar 26
seat belt 49
semi-crystalline 36
sensitivity 81
series 77, 79
slip plane 34
Snell's law 121
specific heat capacity 95
specific latent heat 95
speed 39–40, 47
speed–time graph 40, 47
stationary wave 128
stiffness 31, 32, 33
stopping distance 50–1
strain 32–4
strangeness 104–5
streamline 64
stress 32–4
strong nuclear force 103
superposition 125–6

tempering 35
thermistor 79–80
thermoplastic 36
thermoset 36
thinking distance 50–1
threshold frequency 131–2
time base 81–2
torque 29
towing 50
transverse wave 117
turbulence 64–5
tyre 33, 48–9

ultimate tensile stress 32, 34
uniform acceleration 41–2

vector 26–8
vector addition 26–7
vector subtraction 27
velocity 39–40, 54–6, 63
velocity–time graph 40, 41
wave speed 117–18
wavelength 117–18, 130
weak nuclear force 103
wind tunnel 64
work 59–60, 61, 63, 96
work function 132
work-hardening 35

yield point 34
Young modulus 32